OXFORD MONOGRAPHS ON GEOLOGY AND GEOPHYSICS NO. 10

OXFORD MONOGRAPHS ON GEOLOGY AND GEOPHYSICS

1. De Verle P. Harris: *Mineral resources appraisal: Mineral endowment, resources, and potential supply: concepts, methods, and cases*
2. J. J. Veevers (ed.): *Phanerozoic earth history of Australia*
3. Yang Zunyi, Cheng Yuqi, and Wang Hongzhen: *The geology of China*
4. Lin-gun Liu and William A. Bassett: *Elements, oxides, and silicates: High-pressure phases with implications for the Earth's interior*
5. Antoni Hoffman and Matthew H. Nitecki (eds.): *Problematic fossil taxa*
6. S. Mahmood Naqvi and John J. W. Rogers: *Precambrian geology of India*
7. Chih-Pei Chang and T. N. Krishnamurti (eds.): *Monsoon meteorology*
8. Zvi Ben-Avraham (ed.): *The evolution of the Pacific Ocean margins*
9. Ian McDougall and T. Mark Harrison: *Geochronology and thermochronology by the* $^{40}AR/^{39}AR$ *method*
10. Walter C. Sweet: *The Conodonta: Morphology, taxonomy, paleoecology, and evolutionary history of a long-extinct animal phylum*
11. H. J. Melosh: *Impact cratering: A geological process*
12. J. W. Cowie and M. D. Brasier (eds.): *The Precambrian–Cambrian boundary*
13. C. S. Hutchison: *Geological evolution of south-east Asia*

THE CONODONTA

Morphology, Taxonomy, Paleoecology, and Evolutionary History of a Long-Extinct Animal Phylum

WALTER C. SWEET
The Ohio State University

New York • Oxford
CLARENDON PRESS · OXFORD · 1988

OXFORD UNIVERSITY PRESS

Oxford New York Toronto
Delhi Bombay Calcutta Madras Karachi
Petaling Jaya Singapore Hong Kong Tokyo
Nairobi Dar es Salaam Cape Town
Melbourne Aukland

and associated companies in
Berlin Ibadan

Published by Oxford University Press, Inc.,
200 Madison Avenue, New York, New York 10016

Library of Congress Cataloging-in-Publication Data
Sweet, Walter C.
The Conodonta : morphology, taxonomy, paleoecology, and evolutionary history of a long-extinct animal phylum.

(Oxford monographs on geology and geophysics ; no. 10)
Bibliography: p. Includes index.
1. Conodonts. I. Title. II. Series.
QE899.S94 1988 562′.2 87-34966
ISBN 0-19-504352-9

Printing (last digit): 9 8 7 6 5 4 3 2 1

Printed in the United States of America
on acid-free paper

PREFACE

Conodonts are common fossils. Almost anyone who deals on a regular basis with Paleozoic and Triassic marine rocks has probably seen a few of them. Through the last 30 years conodonts have come to be of exceptional value in biostratigraphy, and they now have pride of place as index fossils in many parts of the geologic column. But conodonts are extinct and are unknown to most neontologists. The sketchiness of information about them in most texts on invertebrate paleontology may result either from the fact that they are microfossils that have only lately come to be important as stratigraphic tools, or from the fact that nothing quite like them exists today, so their zoologic relations are uncertain.

In this monograph, I provide a summary of information about a group with which I have worked all my adult life. Charts that name conodont-based biozones and show stratigraphic range of selected conodont species are included as Appendix B. However, I have purposely avoided a recap of conodont biostratigraphy because it is constantly changing and current views are readily available in various other places. Instead, I focus here on the conodonts as a group of extinct animals, about which it is important to know as much as possible before assessing their distribution biostratigraphically.

Of course, one takes considerable risk in attempting such a summary, particularly of a group of animals known only from its fossil record, because most of what I think I know about conodonts as animals is either conjecture or a highly personal interpretation of a still-imperfect fossil record. So, with the *caveat* that what follows is only one way of viewing an important group, I offer my account of the Conodonta.

For their "witting" or unwitting contributions to what I believe I know about the fascinating group of extinct animals described on the following pages, I am grateful to a long list of my students and faculty colleagues at The Ohio State University, especially Stig M. Bergström, and to members of the Pander Society, an international group of exceptionally good-natured "conodontologists" that has met frequently and informally through the last 20 years to share information about conodonts, argue conclusions, and correct the misapprehensions of its senior members.

Karen Tyler, faculty illustrator at The Ohio State University, drafted nearly all the figures from my very crude copy and assisted with labeling others. Dr. Jerzy Dzik, of the Polish Academy of Sciences, Warszawa, provided about half the stippled drawings of conodonts that grace various figures in Chapter 5. The artistry of these two good friends is plainly evident in their work and is warmly acknowledged.

I am also grateful to Sue Shipley and David Little, of The Ohio State University, for their help in completing various parts of the manuscript and illustrations, and to Mark Kleffner, who graciously compiled information on the ranges of Silurian conodonts and assembled the Silurian chart in Appendix B. The Department of Geology and Mineralology generously assumed much of the expense of drafting and photography.

Columbus, Ohio
February 1988

W.C.S.

CONTENTS

The Conodonta

Natural assemblages of conodont elements, X35. Assemblage at top is holotype of *Scottognathus typicus* (Rhodes), from Pennsylvanian (Missourian) rocks in Illinois. Bottom assemblage is from Lower Carboniferous rocks in Montana and represents *Gnathodus bilineatus* (Roundy), which was recently designated type-species of *Gnathodus*. Black and white prints made from color diapositives kindly furnished by Rodney Norby, Illinois Geological Survey.

1. INTRODUCTION

With just a bit of preparation, almost any marine rock of Paleozoic or Triassic age, from almost anywhere on earth, will yield to the patient investigator an assortment of phosphatic microfossils termed conodonts. Although named and first described in 1856, these tiny fossils were paleontologic curiosities until just a few years ago. Little was known of their geographic or stratigraphic distribution, and they were classified according to a mechanical system based entirely on shape. Furthermore, their relations to other, better-known animal groups were hotly debated.

Nowadays, conodonts are mentioned in most reports on Paleozoic and Triassic biostratigraphy, and their contribution is widely acknowledged. Furthermore, enough has been learned about their anatomy, associations, and patterns of development to make them the honest subjects of a wide variety of truly paleobiologic studies. Thirty years ago only a handful of paleontologists claimed more than a passing acquaintance with conodonts. Today, more than 200 persons devote a major part of their waking hours to the study of these tiny fossils, and only a few of the world's major oil companies or geologic surveys are without at least one conodont specialist. The reasons for this paleontologic success story are numerous, but they have to do primarily with the fact that the conodonts have proved to be as successful at solving biostratigraphic problems in the Paleozoic and Triassic as the foraminifers have been in the interval from the Jurassic to the present.

In this monograph I will introduce the conodonts; show how at least some of them may be reconstructed from the jumbled bags of bones by which they are commonly represented in the fossil record; and look at major features of their long geologic history, with an eye to suggesting something useful about their relations to other animals and about their patterns of deployment and evolutionary development. The literature on conodonts is now rather large (and growing at the rate of about 300 new papers each year), so it will not be possible to explore every aspect of conodont paleobiology in great depth. But this book is not intended as a comprehensive review for specialists. It is an attempt to summarize current knowledge of conodonts for the nonspecialist, who may well have been wondering why this group of tiny fossils has come to be so important in recent years.

1.1 History of Discovery and Study

Sometime between 1833 and 1844, the Russian paleontologist Christian Heinrich Pander discovered tiny, lustrous, toothlike fossils in washed residues of Lower Ordovician and Silurian clastic rocks from Estonia, and on the surface of at least one slab of Carboniferous rock collected within the present city limits of Moscow. However, these microscopic fossils were not illustrated or discussed in print until 1856, when they were described as the remains of an otherwise unknown group of Paleozoic fishes that Pander named *Conodonten* (conodonts in English).

In the years between 1833 and 1856, Pander evidently gave a lot of thought to the tiny fossils he named conodonts. During part of that time, he had great trouble with his eyes and was not able to use his microscope. However, what he saw when he was able to study his specimens microscopically convinced him that he was looking at the teeth and jaws of a previously unknown group of fishes, a group with no modern analogue. Thus, in descriptions of specimens in his collection he used terminology appropriate to the teeth and jaws of fishes, and some of that terminology survives today.

Pander's 1856 monograph not only provided the first descriptions of the hard parts of a previously unknown group of animals and a name for the group itself, but it also began debates that were to run on for more than a century about the zoologic affinities of conodonts and about the appropriate taxonomic base. That is, although Pander more or less confidently concluded that conodonts were an extinct group of fishes, he also candidly pointed out that, in the absence of a living analogue of that group, he could not determine whether the body of a single individual contained teeth that were all of the same morphology or whether that body might have included teeth of many different shapes. He chose to regard the shape of individual specimens as the basis for the classification he proposed, which included 56 species and 14 genera, but it is clear that he was not entirely convinced that the alternate was not also a possibility.

For most of the time between 1856 and 1926 conodonts were regarded as paleontologic curiosities, and little was added to what Pander wrote about them in his famous monograph. However, lively but inconclusive debate continued as to their zoologic affinities. Many authors sided with Pander in regarding them as the hard parts of an extinct group of fishes, but others concluded that they were crustacean remains, the radular teeth of mollusks, or the jaws of a group of annelid worms. George Jennings Hinde, who wrote one of his first papers in 1879 about Ordovician and Devonian conodonts he collected as a young man in Canada and the northeastern United States, agreed with Pander that conodonts were fish remains, but he took a rather different point of view about the taxonomic base. That is, one of Hinde's specimens, a slab of black shale from the Devonian of New York, exposed a number of differently shaped specimens on its surface. Hinde regarded all these as components of the skeletal apparatus of a single conodont species, which he named *Polygnathus dubius* (literally, "doubtful many jaws"). Thus, by 1926 there were not only many different opinions as to the zoologic affinities of conodonts, but there were also two different opinions (Pander's and Hinde's) as to the appropriate taxonomic base. Pander and most other authors favored the idea that individual conodonts formed hard parts that were all the same shape; Hinde favored the idea that at least some conodonts had skeletal components of several different shapes. Ironically, Hinde turns out to have been right, even though everyone today agrees that the specimen on which he based his concept of *Polygnathus dubius* represents bits and pieces of the skeletons of several different species, not just one! Hinde was right, but for the wrong reasons.

In 1926, the American paleontologists E. O. Ulrich and R. S. Bassler described many new genera and species of conodonts, presented a new suprageneric classification, and suggested that this group of microfossils might be of considerable utility in helping decipher the stratigraphy of widespread rock units such as the Devonian-Mississippian Chatanooga Shale. Ulrich and Bassler stated their belief that conodonts are the teeth and dermal plates of primitive fishes, not necessarily all of the same group, and also concluded ". . . that the prevalent opinion that many kinds of these teeth may occur in the same mouth, or plates in the same dermal armor, is not warranted by the facts." In brief, they came down firmly on the side of Pander in both the matter of zoologic affinity and taxonomic base. Having Ulrich on your side in 1926 was of substantial importance.

Although neither Ulrich nor Bassler wrote anything further about conodonts, it is clear that their 1926 paper encouraged other American paleontologists to take up the study of these fossils. Of these, the most influential were certainly a pair of vertebrate paleontologists at the University of Missouri, Edwin B. Branson and Maurice G. Mehl. Their *Conodont Studies*, a volume of systematic essays published by the University of Missouri Press in 1933 and 1934, initiated two decades of intensive study of conodonts at Missouri. This resulted in making Ordovician through Pennsylvanian representatives of the group much better known than they had been previously and also produced a cadre of students trained to recognize and deal with conodonts as an important group of microfossils.

In the introduction to their *Conodont Studies*, Branson and Mehl wrote that ". . . nearly all recent students of conodonts have classed them as teeth of primitive fishes, and we concur in that classification." They regarded the evidence for this statement, and for their further conviction that conodonts are polyphyletic, as conclusive, but they put off describing that evidence until a later time. Unfortunately, that time never came. However, they did allude to the fact that their evidence of the fish nature of conodonts is the fact that some specimens are attached to a substance ". . . that appears bony but does not have the structure of ordinary bone."

Branson and Mehl also joined Pander, Ulrich and Bassler, and all previous students of conodonts except Hinde in using the shape of individual "teeth" as the guide to their classification. That is, like their predecessors, they adopted *form taxonomy*. However, like Ulrich and Bassler, they thought it likely that specimens of different shape functioned in different ways—some as teeth and jaws, others as body scales, and additional ones as denticles on spines.

Between 1933 and 1950, the Missouri school flourished under the leadership of Branson and Mehl, and the prediction of Ulrich and Bassler that conodonts might one day be of great use stratigraphically encouraged other American students to collect and describe these previously enigmatic fossils. As a consequence, the literature on conodonts—a mere 200 articles in the late 1920s—more than tripled between 1930 and 1950. It is not easy to single out any of these contributions as more important than the others, but it is not difficult to identify those that introduced significant new trends.

In 1934, for example, Hermann Schmidt in Germany and Harold Scott in the United States independently reported the discovery of clusters of morphologically different conodont elements on the surfaces of Carboniferous black shale slabs. Like Hinde before them, Schmidt and Scott regarded these *natural assemblages* as the more or less complete apparatuses of individual conodonts—an opinion that was roundly criticized by Branson and evidently accepted with great reservations by other students of conodonts. In more recent years natural assemblages have played an important role in the development of conodont taxonomy.

In 1941, Samuel Ellison, a student of the Missouri school, and Roy Graves, one of his students at the Missouri School of Mines, jointly reported on a group of Pennsylvanian conodont elements they had collected from samples of the Dimple Limestone in Texas, which they had dissolved in dilute acetic acid. Ellison and Graves do not describe the "acetic acid method" they used, which suggests that they did not regard it as novel. However, they were evidently the first to use a laboratory procedure that, a decade later, had become routine in research institutions around the world. The significance of this is that prior to 1941, and for about a decade after, most collections of conodont elements had been assembled from shales and other readily disaggregated rocks, and carbonate rocks were generally ignored. However, as Ellison and Graves demonstrated in 1941, as Branson and Mehl reported in 1944, and as nearly everyone knows today, much larger collections of well-preserved specimens can be isolated readily from carbonates by dissolving them slowly in 10 to 15 percent acetic or formic acid. This, in turn, means that, unlike many other types of microfossils, conodonts may be collected easily and relatively inexpensively from both carbonate and siliciclastic rocks. Thus their distribution in stratigraphic sections of mixed lithologic type may be determined with considerable precision.

The increased size of collections made possible by more widespread use of organic acids encouraged micropaleontologists interested in conodonts to broaden their studies through consideration of sections in which there were no shales or mechanically reducible clastic rocks. By 1959, conodonts had been collected in some variety from rocks that range in age from Late Cambrian to Late Triassic, and there was one report of distinctive elements from Upper Cretaceous rocks in west Africa. Furthermore, conodonts were known from marine rocks in this time interval from six of the seven continents. The stratigraphic range of conodonts had been considerably broadened through improvements in laboratory techniques, and geographic distribution had been extended and collections greatly increased in size. No longer were investigators satisfied with the generally meager numbers of specimens liberated by laborious washing of shales. Most insisted on much larger samples taken at closely spaced intervals from numerous stratigraphic sections.

As the number of large, stratigraphically well-controlled collections grew through the 1950s and early 1960s, so did concern about suitability of the taxonomic base. This concern led in the middle 1960s to proposals that shook the foundations of conodont research and still provoke strident debate in some quarters. This interval in the history of conodont research might be characterized as the *era of multielement taxonomy.*

In multielement taxonomy the taxonomic base is not morphology of discrete specimens but composition of recurrent groups of elements such as those that form the natural assemblages identified in 1934 by Schmidt and Scott. As already noted, Pander and Hinde both considered such groups as potential bases for their taxonomy and Hinde actually diagnosed and named a species (*Polygnathus dubius*) on the basis of a presumed cluster from the Devonian of New York. Although Hinde's work seems to have been ignored or to have provoked largely negative responses from subsequent students of conodonts, I suspect that the papers by Schmidt and Scott, coupled with an elegant study by Rhodes in 1952, were so convincing that all serious students understood the biological inadequacy of Panderian form taxonomy by the early 1950s. However, as long as collections were small and widely spaced stratigraphically and as long as interest in conodonts was primarily biostratigraphic, it was possible to rationalize continued use of form taxonomy by expressing doubt that we would ever find natural assemblages to support appropriate diagnosis of more than a handful of conodont species. In short, a taxonomy based on clusters of elements, or natural assemblages, although regarded as biologically more suitable, was seen as a practically unattainable goal.

As collections of discrete elements grew in size and as stratigraphic gaps between them diminished, investigators in several laboratories noted that those collections could be divided internally into groups of elements of diverse morphologic types that formed recurrent associations through reappearance in successive samples in the same stratigraphic intervals. In 1958, for example, Huckriede identified several such recurrent groups in collections of discrete elements from Triassic rocks. In 1964 Walliser described nine such associations in his collections from the Silurian. In addition, Walliser listed the criteria by which one might determine which specimens in collections of discrete elements form recurrent groups. In the same year, Lindström described additional evidence of architectural constancy that might be useful in establishing and recognizing recurrent groups. It is of interest, however, that none of these authors took the next step and actually diagnosed a species on the basis of any of the recurrent groups they recognized. All of them reverted to form taxonomy in the systematic sections of their respective studies.

In reports that appeared in 1966, Webers and Bergström and Sweet plunged headlong into the jungle so neatly avoided by Huckriede, Walliser, and Lindström. As Walliser had done, they identified within their Ordovician collections a number of recurrent groups regarded as the likely skeletal apparatuses of single conodont species. Walliser reverted to form taxonomy at this point, but Webers and Bergström and Sweet presented revised diagnoses of species based on these recurrent groups. In addition, nomenclature for the newly diagnosed *multielement species* was revised in strict accordance with the Rules of Zoological Nomenclature, which resulted in long, complex synonymies and some confusing descriptive terminology. In 1969 the Ordovician recurrent groups were subjected to a thorough statistical treatment by Kohut, whose results strengthened arguments supporting their original interpretation as the skeletal apparatuses of single species; and, in 1970 and 1972, respectively, Sweet and Von Bitter demonstrated that other types of clustering techniques might be used effectively to identify recurrent groups and evaluate them taxonomically.

In 1969 Stig Bergström and I predicted publicly and confidently that multielement taxonomy would replace form taxonomy in conodonts in 5 years or so. In 1971 an international conference of conodont workers in Marburg, Germany, concluded after much soul-searching that multielement taxonomy was more desirable than form taxonomy for conodonts and was also attainable even in the absence of natural assemblages to support every diagnosed species. A complete multielement taxonomy is still not in place, however, and there are a number of workers who still prefer its much simpler predecessor. However, a revised version of Volume W of the now-venerable *Treatise on Invertebrate Paleontology* (Clark et al., 1981) issued in 1981 (but written largely before 1976) is couched mostly in terms of multielement taxonomy, and a majority of current reports on conodonts at least give lip service to this more sophisticated mode of classification.

In 1967 students of conodonts attending an international symposium on the Devonian System in Calgary, Alberta, were impressed with the fact that data provided by conodonts contributed to a very large number of the reports presented and with the difficulty they experienced in keeping abreast of the burgeoning literature and research interests of a rapidly increasing group of conodont students. To solve these problems, the group founded an informal organization, *The Pander Society*, open to anyone interested in conodonts and with the single purpose of sharing information on current conodont research. Subsequently, The Pander Society has grown in membership to more than 250; has become the official working group on conodonts of the International Palaeontological Association; and has distributed an annual newsletter that includes reports on the research activities of members, addresses of conodont workers and, in recent years, current bibliographies of the conodont literature. The Pander Society meets annually in North America and at 3- to 5-year intervals at various sites in Europe.

In sorting out significant events of the post-1950 era in the history of conodont research, I should mention that students of conodonts have been blessed more or less regularly with bibliographers, who have kept track of the literature about conodonts and have periodically assembled and published lists of it. Grace Holmes began this service in 1928; Robert Fay continued it through 1948, with a useful catalog published in 1952; and Sidney Ash and Sam Ellison have continued with the published bibliographies of the conodont literature that are listed among the references at the end of this chapter. In more recent years, the bibliography of conodont literature has been updated annually in a supplement to The Pander Society's newsletter. At the end of 1987, my file included the titles of more than 7000 books, articles, and papers on conodonts.

1.2 Achievements

The study of conodonts is very different in the 1980s from what it was in 1950, when I began. For example, in the last 30 years interest in conodonts has again become international. For much of the time between 1926 and 1950, study of conodonts was largely an American endeavor; now there is active and increasingly well-informed interest in nearly every part of the globe. At a meeting of The Pander Society in 1985, the approximately 150 participants represented 31 different countries in Africa, Asia, Australia, Europe, and North and South America. Study has become international again, and there is also close coordination and an unusual degree of cooperation among students of conodonts. These achievements, perhaps more than any others, have enabled the rapid growth of knowledge about conodonts in recent decades.

As noted earlier, assessment of large, stratigraphically comprehensive collections on an international scale has also caused a shift in taxonomic base from the morphology of single skeletal elements to the composition and relationships within recurrent groups of skeletal elements. This has resulted in a taxonomy for conodonts that, while admittedly more complex than its predecessor, is probably closer to biologic "truth" than the form taxonomy of Pander, Ulrich and Bassler, and Branson and Mehl. Such a taxonomy is obviously a collective effort and, like any other, will always be ripe for modification and the subject for debate. Nevertheless, students of conodonts now have the framework within which to make biologically meaningful statements about conodont paleoecology, biogeography, and evolution, and that must be regarded as an important achievement.

Distribution of conodonts has enabled biostratigraphers to divide stratified marine rocks of Late Cambrian through Triassic age into about 150 biozones, which form a framework within which it is now possible to make exceptionally refined analyses of the various aspects of Paleozoic and Triassic history. This is the achievement for which students of conodonts have attained the greatest notoriety, of course, and the one that has gained them grudging access to the company of foraminiferologists and others who have for so long reserved to themselves the title of "micropaleontologist."

Pander made some remarkable observations about the internal structure of conodonts in 1856, but much of our present knowledge is based on work done since 1950. That knowledge influences part of the classification of conodonts discussed in Chapter 4 and a good deal of the discussion of skeletal anatomy in Chapter 2. A particularly important recent discovery (Epstein, Epstein, and Harris, 1977) is the close relationship between the color of conodont elements and the temperature to which enclosing rocks have been raised by burial at different depths for varying lengths of time. As ambient temperature rises and is maintained, carbon in the organic matter of conodont elements becomes fixed, and the color darkens from pale yellow to brown to black and, ultimately, with the loss of carbon, to crystal clear. Experimental calibration of this change has produced a *Color Alteration Index* (CAI), which is very useful in petroleum exploration because it aids in identifying areas in which rocks have been heated too high to retain oil or gas and because the method is much less expensive than those previously used by industry and is operable over a broader temperature range.

Among the achievements of the past 30 years I should certainly list the recent discovery (Briggs, Clarkson, and Aldridge, 1983; Aldridge, Briggs, Clarkson, and Smith, 1986) in Dinantian rocks in Scotland of several specimens that surely seem to represent more or less complete conodonts. These fossils show the Dinantian species to have been an elongate, wormlike creature with a bulbous head that contains a natural assemblage of conodont elements; a trunk that seems to have included chevron-shaped features that are reminiscent of the muscle blocks (or myotomes) of the primitive chordate *Branchiostoma* (=*Amphioxus*); and a prominent caudal fin. The Scottish specimens will have starring roles in Chapter 3; here it is sufficient to note that they are thus far the most convincing candidates for complete conodonts to have come down the pike, and their discovery and description rank high among recent achievements.

1.3 Pending Problems

Despite the rapid advance in our knowledge about conodonts in the last 30 years, more information is clearly needed on a number of subjects. There will be occasion to address several of these problems in later chapters; here I merely list them.

First, students of conodonts have assembled an objective data base that has enabled development of a biologically significant taxonomy and a closely controlled biostratigraphy for the Paleozoic and Triassic. Thus we should be in a better position than nearly anyone else to evaluate the various schemes of evolutionary mode currently being debated. However, there has been little interest so far in considering our data base in this way or in using it even to synthesize and evaluate phylogenies for most major conodont groups. Although this may not properly be a problem, it is certainly a challenge.

It should also be pointed out that we still have no definitive answers to the question, "How and where did conodonts live?" It seems certain that the little creatures were exclusively marine, but it is by no means established that they were swimmers, floaters, or bottom dwellers (or if some may represent each of these modes of life), and no one has as yet proposed a certain means of determining mode of life from features of the distributional record. I continue to regard the common occurrence of conodonts in the black-shale facies as an indication that the dominant mode of life was pelagic. Also, the prominent tail fin of the more or less complete Scottish specimens is a further suggestion that the animals were mobile and thus probably mostly nektonic. However, this may not apply to all conodonts, and the elongate, prominently "tailed" hagfish (or slime eel), which many hold to be the nearest living relative of conodonts, is a largely sedentary creature that spends most of its life in burrows in the muddy substrate of relatively deep marine water. Clearly, we need to address the mode-of-life question intensively before we delve further into heady subjects such as conodont ecology or biogeography.

On a more mundane level, but nevertheless of signal importance, is the business of filling in a number of gaps in the current stratigraphic record of the Conodonta. Lower Ordovician species are well known in the high-latitude faunas characteristic of European localities, but in North America there have been only a few comprehensive reports, and these leave a number of important taxonomic and phylogenetic questions unanswered. The Silurian is also a problem, particularly the part of it above the *Pterospathodus amorphognathoides* Zone. This part of the Silurian was evidently a time of major marine regression, with widespread shallow-water environments, evaporites, and conditions hostile to development of rocks from which conodonts might be extracted easily and in abundance. Nevertheless, a number of very important events in the evolutionary history of the Conodonta took place during the late Silurian, and we badly need well-documented collections as the basis for finding out what happened. Taxonomy of Carboniferous and Permian conodonts is in less robust health than is that of earlier forms or that of most Triassic lineages. Only a few studies have addressed multielement taxonomy of Mississippian conodonts with conviction, and a few excellent recent studies suggest that there are major surprises in store for anyone who undertakes detailed studies of Permian conodonts. I recommend these stratigraphic studies strongly.

Finally, we have probably reached the stage at which monographic studies of the taxonomy and phylogeny of major lineages are in order. To be sure, a number of such studies have been completed, and my debt to them is plainly in evidence in Chapter 5. We need still more of them, and existing collections will probably be adequate as the basis for many.

References

Aldridge, R. J., Briggs, D. E. G., Clarkson, E. N. K., and Smith, M. P. (1986). The affinities of conodonts—new evidence from the Carboniferous of Edinburgh, Scotland. *Lethaia* **19**(4), 279–291.

Ash, S. R. (1961). Bibliography and index of conodonts, 1949-1958. *Micropaleontology* **7**, 213–244.

Bergström, S. M., and Sweet, W. C. (1966). Conodonts from the Lexington Limestone (Middle Ordovician) of Kentucky and its lateral equivalents in Ohio and Indiana. *Bull. Am. Paleont.* **50**(229), 271–441.

Branson, E. B., and Mehl, M. G. (1933–1934). Conodont Studies. *Univ. Missouri Studies* **8**, 1–300.

Briggs, D. E. G., Clarkson, E. N. K., and Aldridge, R. J. (1983). The conodont animal. *Lethaia* **16**, 1–14.

Clark, D. L., Sweet, W. C., Bergström, S. M., Klapper, G., Austin, R. L., Rhodes, F. H. T., Müller, K. J., Ziegler, W., Lindström, M., Miller, J. F., and Harris, A. G. (1981). Conodonta. In *Treatise on Invertebrate Paleontology* (ed. R. A. Robison), Pt. W, Suppl. 2, W1–W202. Geol. Soc. America and Univ. Kansas, 202 pp.

Ellison, S. P., Jr. (1962). Annotated bibliography, and index, of conodonts. *Texas Univ. Publ.* **6210**, 128 pp.

——— (1963). Supplement to annotated bibliography, and index, of conodonts. *Texas J. Sci.* **15**, 50–67.

———, and Graves, R. W., Jr. (1941). Lower Pennsylvanian (Dimple Limestone) conodonts of the Marathon region, Texas. *Univ. Missouri School of Mines and Metallurgy Bull.*, Tech. ser. **14**(3), 1–21.

Epstein, A. G., Epstein, J. B., and Harris, L. D. (1977). Conodont color alteration—An index to organic metamorphism. *U. S. Geol. Surv. Prof. Paper* **995**, 27 pp.

Fay, R. O. (1952). Catalogue of conodonts. *Univ. Kansas Paleont. Contr. (Vertebrata)* Art. **3**, 206 pp.

Hinde, G. J. (1879). On conodonts from the Chazy and Cincinnati group of the Cambro-Silurian, and from the Hamilton and Genesee-shale divisions of the Devonian in Canada and the United States. *Quart. J. Geol. Soc. London* **35**, 351–369.

Holmes, G. B. (1928). A bibliography of the conodonts with descriptions of early Mississippian species. *Proc. U. S. Nat. Mus.* **72**, Art. 5, 38 pp.

Huckriede, R. (1958). Die Conodonten der Mediterranen Trias und ihr stratigraphischer Wert. *Paläont. Z.* **32**, 141–175.

Kohut, J. J. (1969). Determination, statistical

analysis, and interpretation of recurrent conodont groups in Middle and Upper Ordovician strata of the Cincinnati Region (Ohio, Kentucky, and Indiana). *J. Paleont.* **43**(2), 392–412.

Lindström, M. (1964). *Conodonts.* Elsevier, Amsterdam, London, New York, 196 pp.

Pander, C. H. (1856). Monographie der fossilen Fische des silurischen Systems der russisch-baltischen Gouvernements. *Konigl. Akad. Wiss., St. Petersburg,* 91 pp.

Rhodes, F. H. T. (1952). A classification of Pennsylvanian conodont assemblages. *J. Paleont.* **26**, 886–901.

Schmidt, H. (1934). Conodonten-Funde in ursprünglichem Zusammenhang. *Paläont. Z.* **16**, 76–85.

Scott, H. W. (1934). The zoological relationships of the conodonts. *J. Paleont.* **8**, 448–455.

Sweet, W. C. (1970). Uppermost Permian and Lower Triassic conodonts of the Salt Range and Trans-Indus Ranges, West Pakistan. In *Stratigraphic Boundary Problems, Permian and Triassic of West Pakistan* (ed. Bernhard Kummel and Curt Teichert). *Univ. Kansas, Dept. Geology, Spec. Publ.* **4**, 207–275.

———, and Bergström, S. M. (1969). The generic concept in conodont taxonomy. *Proc. N. Am. Paleont. Conv.* **1**, 29–42.

Ulrich, E. O., and Bassler, R. S. (1926). A classification of the toothlike fossils, conodonts, with descriptions of American Devonian and Mississippian species. *U.S. Natl. Mus. Proc.* **68, Art. 12**, 63 pp.

Von Bitter, P. H. (1972). Environmental control of conodont distribution in the Shawnee Group (Upper Pennsylvanian) of eastern Kansas. *Univ. Kansas Paleont. Contr.* **59**, 105 pp.

Walliser, O. H. (1964). Conodonten des Silurs. *Abh. Hess. Landesamtes Bodenforsch.* **41**, 1–106.

Webers, G. F. (1966). The Middle and Upper Ordovician conodont faunas of Minnesota. *Minn. Geol. Surv. Spec. Publ.* **4**, 123 pp.

2. SKELETAL ANATOMY

For reasons listed toward the end of Chapter 1, it is now clear that conodonts were primarily soft-bodied animals. Their only mineralized parts formed a cephalic apparatus, at least partly of epidermal origin, that probably functioned to grasp prey and aid in the intake of food. In most, if not all, conodonts, components of the exoskeletal apparatus were discrete objects that became dissociated from others on death of the animal. Although collections at various places in the world house several hundred more or less completely preserved apparatuses, discrete specimens are the ones on which most of our information about the Conodonta is based.

Pander diagnosed the *Conodonten* in terms of the physical characters of discrete specimens. However, everywhere in his monograph except on the page bearing the diagnosis, he referred to the "teeth," "jaws," or "remains" of conodonts and thus used that word both for the animals as a whole and for their hard parts. I am not aware that this dual usage of conodont has ever puzzled anyone, but the potential for confusion exists. In the rest of this book I will use the word conodont only for an entire individual of the Conodonta. The term "conodont element" (or "skeletal element," or just plain "element") will be used for discrete components of the cephalic apparatus.

2.1 Composition of Conodont Elements

From what appear to have been rather primitive wet-chemical analyses, Pander concluded that the skeletal elements of conodonts are composed entirely of calcium carbonate (*reinem kohlensaurem Kalk*). A few years later, however, Harley (1861) wrote that "...they are composed of phosphate and carbonate of lime, the former, contrary to my expectations, the more abundant constituent." Hinde (1879), like most others, felt that most of Harley's specimens were not conodonts and continued to maintain that conodont elements are composed of carbonate of lime. He cited this evidence to counter conclusions that the little fossils might represent a group related to annelids or "naked mollusks," whose jaws or radular elements are of purely organic composition.

In 1926 P. V. Roundy reported that conodont elements "... appear to be a phosphatic carbonate of lime." Six years later Stauffer and Plummer asserted, without citing further evidence, that "conodont teeth are probably chiefly calcium phosphate." Branson and Mehl seem not to have worried much about the chemical composition of conodont elements but, in 1944, one of their doctoral students, Samuel Ellison, demonstrated from X-ray analyses that the substance involved is a member of the apatite isomorphous group. Because of this, conodont elements are relatively heavy (2.84 to 3.10) and are less soluble in acetic, formic, and citric acids than the matrix of carbonate rocks enclosing them. This is important, of course, because it means that conodont elements may be extracted from carbonate rocks by prolonged soaking in such acids—a capacity that sets them aside from many other fossils—and they may also be separated from large acid-insoluble residues by use of heavy liquids or other types of gravity separation techniques.

The most definitive, and also the most recent, contribution to the subject of conodont-element composition is a monograph by Pietzner et al. (1968), whose studies led them to the following formula for the mineral matter of conodont elements.

$$Ca_5\, Na_{0.14}\, (PO_4)_{3.01}\, (CO_3)_{0.16}\, F_{0.73}\, (H_2O)_{0.85}$$

They interpreted the mineral to be francolite, a carbonate apatite in which the OH and CO_3 ions substitute for phosphate and do not occupy lattice positions as they do in hydroxyapatites. Traces of at least 39 other chemical elements have been identified in various places

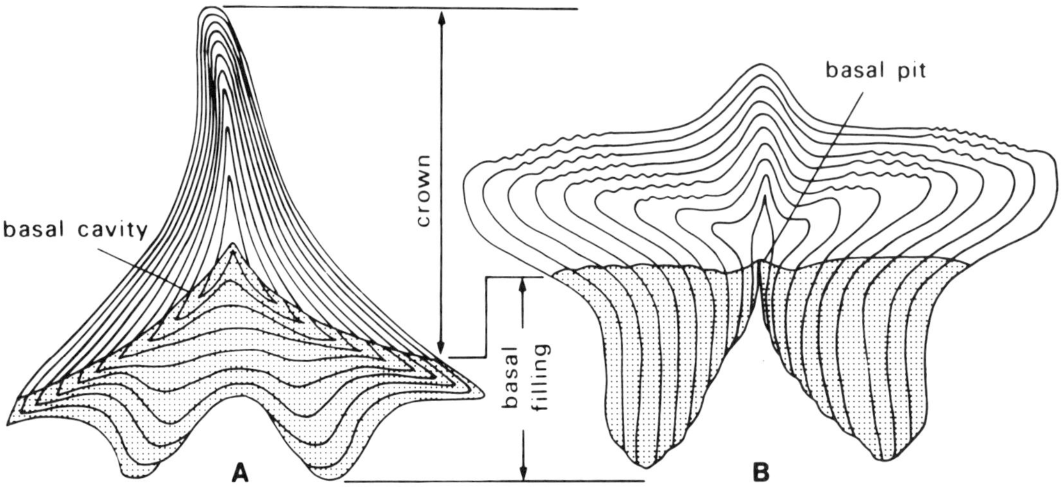

Fig. 2.1. Complete conodont elements in cross section. The crown of the element in (A) has a conical basal cavity, whereas that of the one in (B) has a tiny basal pit, which is bordered by wide zones of recessive basal margin.

in the hard parts of conodonts, and some of these (e.g., neodymium) have lately come to be of considerable significance in interpreting conditions in the water bodies in which conodonts lived.

If a conodont element is dissolved slowly in a weak solution of hydrochloric acid, a fragile, demineralized ghost remains. This delicate structure is organic in composition and indicates that mineralization took place within an organic framework, which is retained within the element. This retained organic material is responsible for the fact that conodont elements darken gradually as they are heated and more and more of their organic carbon becomes fixed. It has been discovered that the color of conodont elements is a remarkably accurate index to the temperature to which enclosing rocks have been raised.

2.2 Structure of Skeletal Elements

Although the skeletal elements of conodonts vary greatly in external morphology, a number of studies indicate that they are remarkably similar in internal structure. Major structural features are shown diagrammatically in Fig. 2.1.

Well-preserved conodont elements consist of two distinct parts for which the terms *crown* and *basal filling* are commonly used. Internally, both are laminated—a fact noted for crowns by Pander, who was also the first to comment on features revealed in thin sections of these tiny objects. The two components differ primarily in that laminae of the basal filling tend to be thicker and less densely mineralized than those of the crown. The surface between crown and basal filling is a surface of weakness along which the two components of the element readily separate, and this explains the fact that most discrete specimens lack the less resistant basal filling.

In all of the complete skeletal elements studied thus far in thin section, laminae of crown and basal filling are continuous. This is shown in two dimensions in Fig. 2.1 and can be visualized in three dimensions with just a bit of imagination. The significance of this fact, which is substantial, was brought out about 30 years ago in a series of papers by Gross (1957, 1960). Gross pointed out that in order for lamellae to be continuous from crown to basal filling, the element as a whole must have grown through successive external apposition of laminae, not (as in teeth) by addition of laminae internally. In short, elements became progressively larger through ontogeny and, most significantly, must have been surrounded by tissue at times of growth. To Gross and others, this has meant that conodont elements grew in a manner quite unlike that of fish teeth and thus that the Conodonta were probably not fish, as Pander and many others have thought them to be.

Elements of the cephalic apparatus of most conodonts were dissociated upon death by decay of the tissues that held them together in life or by digestion of those tissues in the gut of a predator. In only a few cases have complete apparatuses been preserved. After dissociation most conodont elements were moved about with other sedimentary particles on the seafloor and that agitation was apparently sufficient in most cases to separate crowns from basal fillings. In any event, most collections of conodont elements consist entirely (or almost entirely) of crowns, and we have little or no information on the internal structure or external morphology of the basal fillings for most species. I suspect that the basal fillings of a great many species, perhaps the majority, were only slightly mineralized—or not mineralized at all. They are mostly unknown as isolated objects in rocks that contain lots of discrete crowns, and they do not occur in any of the complete apparatuses that have been described, which seem to have undergone little, if any, physical distortion subsequent to death of the animals. Whatever the explanation, nearly all we know about conodonts is derived from a study of isolated crowns.

Crowns of conodont elements, like basal fillings, are internally laminated. In morphologically simple forms, like the one in Fig. 2.1A, individual laminae are conical in shape, and the entire crown has the structure of a nested stack of paper drinking cups. In the conical specimen of Fig. 2.1A, the basal filling is still attached. However, if the two components of the element were separated, the surface over which they were joined would be that of a conical indentation in the abapical part of the crown. This indentation is termed a *basal cavity*, and it occupies a part of the crown known as its *base*. Note in Fig. 2.1A that basal edges of successively younger crown lamellae extend somewhat farther toward the basal margin than their predecessors. Thus, the surface along which crown and basal filling were originally attached is completely enclosed within the base of the crown. In the element depicted in Fig. 2.1B, however, only the first few lamellae overlapped basally to form a very tiny basal cavity, whereas later lamellae became progressively shorter than their predecessors as the crown grew larger. Basal cavities of crowns with features like this are termed *basal pits*, and the remainder of the attachment surface is described as a *zone of recessive basal margin*.

In most parts of the crowns of typical conodont elements, lamellae are 0.2 to 1.2 μm thick. However, at places of accelerated growth in elements that grew laterally to form platelike structures, or toward the extremities of those that built elongate comblike shapes, lamellae may be as much as 5 μm thick. As noted previously, each lamella of a complete element surrounds all its predecessors. We can see edges of lamellae along the attachment surfaces of crowns, however, and outlines of some of them are commonly visible in transmitted light within the thinner parts of many crowns. Each lamella is built on a framework of organic material, and the tiny apatite crystallites that mineralize this framework are oriented with their prism surfaces parallel to the direction of growth.

In reflected light crowns of conodont elements are shiny and translucent. They are pale amber in color if they are from rocks that have not been heated above 80°C for any length of time; otherwise they may range in color from reddish yellow to brown to black if they have been held for any appreciable time at higher temperatures. Elements from some metamorphic rocks are gray, white, or even crystal clear. Evidently elements such as this have been deeply buried and kept at high temperatures for such long periods of time that they have lost even the fixed carbon responsible for the dark colors of specimens from lower-temperature rocks.

Thermally unaltered crowns may be uniformly translucent and of the same pale yellowish color throughout, or one may distinguish within them more or less irregular, fuzzy-edged masses that are opaque and of whitish appearance. Crowns of the first type are said to be *hyaline*; those of the latter are termed *albid*. Many crowns have both hyaline and albid areas, and internal distribution of the *white matter* that distinguishes the albid parts is commonly the same or nearly the same in elements of the same conodont species.

The nature of the white matter in albid crowns has interested students of conodonts since the time of Pander. From recent studies of conodont ultrastructure, it seems that white matter is composed of very finely crystalline apatite that is permeated by tiny cavities of spherical or irregular shape. Masses of white matter lack the lamination that is obvious in hyaline parts of the same crowns, but they grade laterally through the fuzzy-margined areas into parts of the crown that are hyaline and distinctly laminated. These relationships indicate that the white matter of albid areas formed by recrystallization of formerly hyaline parts of the crown. Interestingly, studies by Lindström and Ziegler (1971) of albid areas in the crowns of a few panderodontide conodonts showed that despite recrystallization to form white matter and blurring or obliteration of original lamellae, orientation of the 0001 crystal surfaces in the original hyaline matter is retained in the resulting white matter. Further, such surfaces apparently grew together during episodes of recrystallization so that the albid areas are crossed by many almost continuous transverse surfaces along which breakage occurs quite readily. Finally, it should be noted that masses of white matter are commonly concentrated in the axial regions of crown cusps and denticles and are always covered externally by at least a few unaltered hyaline lamellae.

Although some authors have regarded the white matter of albid crowns as a result of secondary alteration and have minimized its significance in conodont biology or taxonomy, it is nevertheless true that elements judged to represent comparable ontogenetic stages of the same species commonly show closely similar patterns in the distribution of their white matter. Of course, other groups of species formed elements that lack white matter entirely or have thin and distinctive albid areas confined to axial regions of their cusps and denticles. Since it also seems that conversion of originally hyaline laminae to white matter took place gradually during the lifetime of a conodont, it is reasonable to suppose that recrystallization of the apatite and resorption of some of the original phosphate or carbonate to leave the white-matter voids was under control of the tissues that pervaded the elements. In short, the distinctive shapes of white-matter masses in albid parts of conodont crowns and the sequence in which these masses form ontogenetically are almost surely physical expressions of genetic control and thus are significant in taxonomy and in interpreting features of the life histories of conodonts.

Branson and Mehl called attention to the fact that the almost completely hyaline crowns of a group of conodonts known best from Middle Ordovician rocks tend to break to form fractures with frayed instead of clean margins. They intepreted this to mean that internal structure consisted of elongate bundles of fibers instead of lamellae. Elements with such distinctive "fibrous" stucture were taken to represent a taxonomically separate group of conodonts, the order Neurodontiformes. Subsequent studies have shown, however, that the hyaline crowns of neurodontiform elements are built of rather thick lamellae, which are made up of long, rodlike crystallites that are oriented parallel to the axis of growth. Broken surfaces parallel to the axis of growth reveal that the rodlike components of lamellae are closely packed prisms that break at different levels. Thus, although Branson and Mehl may have erred in describing the crowns of neurodontiform conodonts as lacking internal lamellae, they were apparently correct in identifying a group distinguished by rather different internal structure. Many of Branson and Mehl's fibrous, or neurodontiform, conodonts are now included in the order Prioniodinida.

2.3 Shapes of Element Crowns

Although their internal features may be much the same, conodont elements occur in a wide and confusing variety of shapes. Because those shapes are the best clues we have as to the function of elements and are also the primary basis for conodont taxonomy, a good deal of attention must be paid to sorting out and recognizing major shape categories and to the descriptive terminology for those categories.

A classification of shape categories is given in Table 2.1, which also includes the names ap-

TABLE 2.1. *Shape categories of conodont elements*

Major shape category	Primary divisions	Secondary divisions
Coniform (simple cones)	Geniculate Nongeniculate	
Ramiform (bars)	Alate Tertiopedate Digyrate	 Breviform Extensiform
	Bipennate Dolabrate Quadriramate Multiramate	
Rastrate		
Pectinform (blades, plates, platforms)	Stellate	Stelliplanate Stelliscaphate
	Pastinate	Pastiniplanate Pastiniscaphate
	Carminate	Carminiplanate Carminiscaphate
	Angulate	Anguliplanate Anguliscaphate
	Segminate	Segminiplanate Segminiscaphate Bisegminiscaphate Trisegminiscaphate

plied to those categories in Volume W (Supplement 2) of the *Treatise on Invertebrate Paleontology* (Clark et al., 1981).

Pander, the first to describe conodont elements, divided his specimens into two major categories, *einfache Zahne* and *zusammengesetzte Zahne* (i. e., "simple or single teeth" and "compound or composite teeth"). That division is still useful, although it has been customary for many years to divide the "compound or composite" category into two or three divisions termed *bars, blades,* and *plates* (or *platforms*).

In the shape classification adopted by *Treatise* authors, Pander's simple or single teeth are *coniform* (cone-shaped) *elements*; barlike compound or composite teeth are *ramiform* (ray-shaped) *elements*; and blade- and plate- (or platform-) like elements are grouped together as *pectiniform* (comb-shaped) *elements.* Elements described in Table 2.1 as *rastrate* have been regarded either as peculiarly modified coniform elements by some authors or as compound elements by others. They were not specifically identified in the *Treatise* classification of shape categories but are here regarded as a separate category for descriptive purposes.

Coniform elements are further divided into two categories (*geniculate* and *nongeniculate*) on the basis of the type of curvature. Seven major types of ramiform elements are recognized on the basis of the number of processes and their relationship to the cusp. Pectiniform elements are divided into five major groups on the basis of the number and arrangement of primary processes. If pectiniform elements develop platformlike lateral extensions, they may be further described in terms of the resulting shape of the attachment surface (planate or scaphate). Typical representatives of each of these categories are shown in Figs. 2.2, 2.3, 2.4, 2.5, and 2.6, which also include information on terms used to describe features of each of these element types.

2.3.1 Coniform crowns

Coniform elements, which are also known in the literature as "simple cones" or just as "cones," are basically of conical shape. Pander divided elements like this into two parts, a more or less expanded *base*, which encloses a subconical *basal cavity* (the *Cavitas Pulpae* of

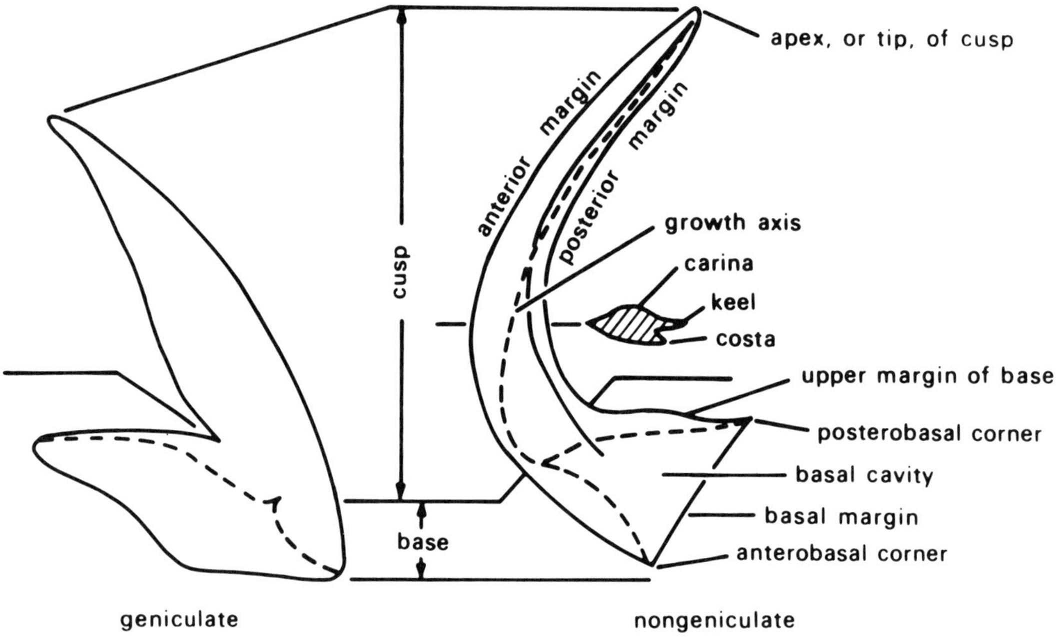

Fig. 2.2. Orientation and terminology of the two major types of coniform crowns. Diagram on right redrawn from Treatise on Invertebrate Paleontology, courtesy of the Geological Society of America and University of Kansas.

Pander); and a *cusp* (Pander's *Spitze*), which is solid, tapers to the apex of the unit, and has a long axis that is either straight or is curved to various degrees. For descriptive purposes, coniform elements are oriented with their tips up and with the upper edge of the base horizontal. To coniform crowns oriented in this way, one may then apply the descriptive terminology indicated in Fig. 2.2.

In the *Treatise* only two major types of coniform crowns are distinguished. In *geniculate* coniform elements, the upper edge of the base joins the posterior margin of the cusp in an acute angle, whereas in *nongeniculate* coniform elements there is a smooth transition from the upper margin of the base into the posterior edge of the cusp. However, as anyone who has worked with coniform elements will attest, the morphologic situation within this group is more complex than a twofold shape classification would indicate. I do not believe it should be necessary to propose additional specialized terms here. Instead, I suggest that it may be possible to utilize for coniform crowns some of the descriptive terms used now only for ramiform elements. For example, bilaterally symmetrical coniform elements with conspicuous lateral costae might well be described as *alate geniculate* (or *alate nongeniculate*) elements, and asymmetric forms with lateral costae that twist in opposite directions basally might be termed *digyrate coniform elements.*

2.3.2 *Ramiform crowns*

In ramiform elements, known in much of the conodont literature before 1982 as "bars," one or more of the sides or edges of the base are extended away from the cusp to form *processes*, whose upper edges are commonly (but not invariably) serrate. Individual serrations on a process are *denticles*, but some processes may be *adenticulate.*

Ramiform elements are oriented in the same manner as coniform elements, with the cusp apex up and the upper margin of that portion of the base posterior to the cusp horizontal. A process that is continuous with the concave, or posterior, side of the cusp is a *posterior process*; one that is a basal projection of the anterior side of the cusp is an *anterior process*; and extensions of the base to the sides of a ramiform element are termed *lateral processes.*

Seven major types of ramiform elements

have been recognized. Names for these shape categories are listed in Table 2.1. Typical forms are shown diagrammatically in Fig. 2.3, which also includes some information on terminology. *Alate ramiform elements* are bilaterally symmetrical, lack an anterior process, but have a posterior process and a lateral process on each side of the cusp. In some descriptive reports, alate ramiform elements have been termed "trichonodelliform" or "hibbardelliform" because the genera *Trichonodella* and *Hibbardella* (and several others) were originally form-taxonomic concepts based on alate ramiform-type specimens.

Tertiopedate ramiform elements also have a posterior process and a lateral process on each side of the cusp, but the lateral processes are not symmetrically disposed with respect to the cusp and the posterior process is commonly long and conspicuously denticulated.

Digyrate ramiform elements are like alate elements, but they are individually asymmetrical; only rarely is the posterior process well developed and denticulate; and lateral pro-

Fig. 2.3. Ramiform crowns and their orientation. Terms used to describe them include the following: ac = anticusp; ap = anterior process; bc = basal cavity; bp = basal pit; c = cusp; d = denticle; ilp = inner lateral process; lp = lateral process; olp = outer lateral process; pp = posterior process; and zrm = zone of recessive basal margin. Diagrams redrawn from Treatise on Invertebrate Paleontology, courtesy of the Geological Society of America and University of Kansas.

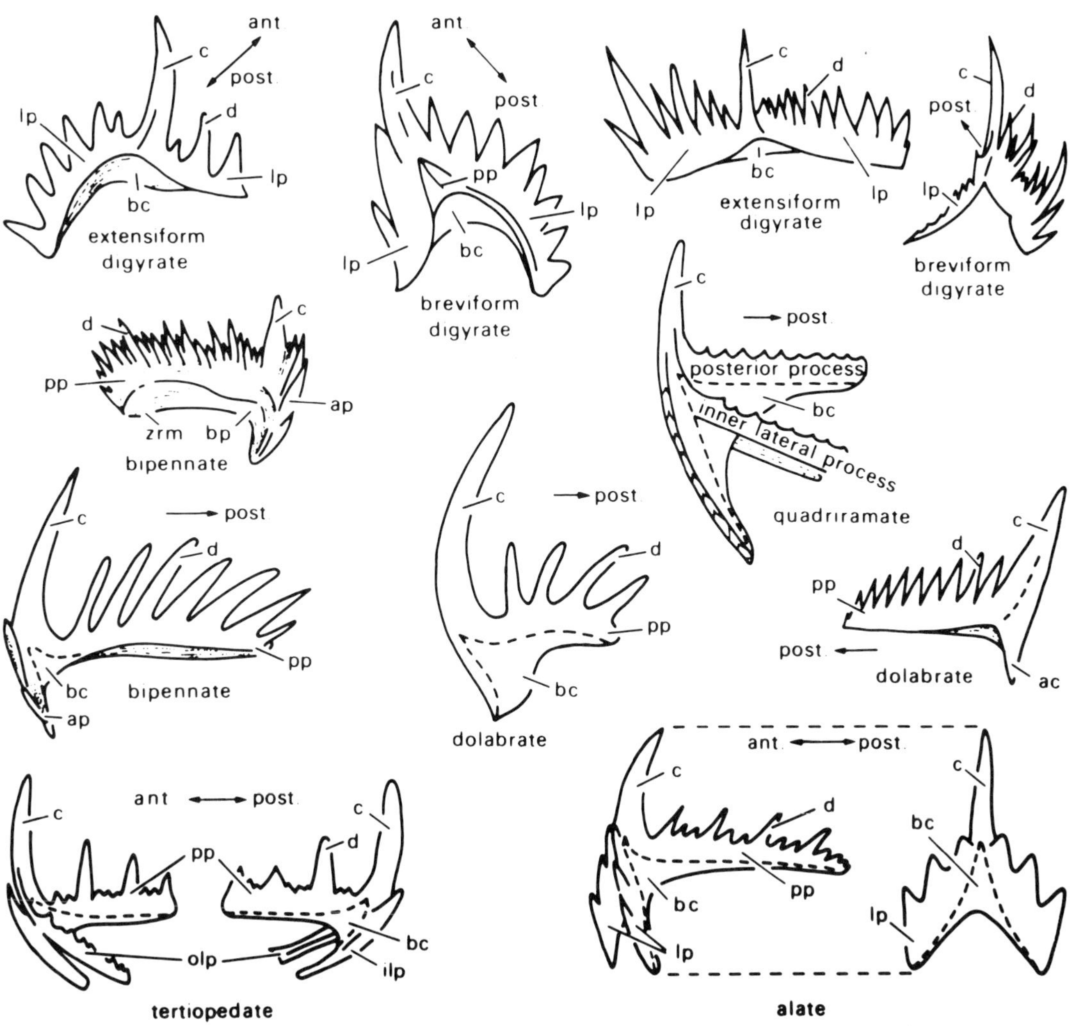

cesses are unequally developed, with distal extremities that commonly curve in opposite directions. Because the genera *Oulodus*, *Plectospathodus*, *Zygognathus*, and *Enantiognathus* were originally form-taxonomic concepts based on digyrate ramiform type-specimens, some of the descriptive literature includes terms such as "oulodontiform," "plectospathodontiform," "zygognathiform," and "enantiognathiform" for the digyrate components of various multielement apparatuses. In discussions that follow in Chapter 5, I find it quite useful to distinguish between *extensiform* and *breviform* digyrate elements. The former have long lateral processes that commonly extend conspicuously outward from the cusp before curving anteriorly or posteriorly at their distal ends. The latter tend to have much shorter processes, which generally begin their anterior or posterior curvature at the point of juncture with the cusp. *Oulodus* and *Plectospathodus* were created in form taxonomy for extensiform digyrate elements, whereas *Zygognathus* and *Enantiognathus* were form-taxonomic designations for breviform digyrate elements.

Bipennate ramiform elements have two processes, which are anterior and posterior in position with respect to the cusp. In most bipennate ramiform elements, the anterior process is shorter than the posterior process and commonly curves, or is deflected, laterally. Not uncommonly, the anterior process lacks denticles and is represented only by a flangelike extension of the cusp margin. In a few cases (e.g., in the skeletal apparatuses of species of *Periodon* and *Erismodus*) the anterior process is denticulate and the posterior process adenticulate. Bipennate ramiform elements have been termed "hindeodelliform" or "cordylodontiform" in some systematic papers, although it is hoped that this practice will cease with availability of the nomenclaturally neutral term bipennate.

Dolabrate ramiform elements have only a posterior process and are commonly pick shaped in lateral aspect. In this respect, they differ from geniculate coniform elements only in having a denticulate posterior process. In fact, it might be appropriate to use the expression *dolabrate coniform element* for the forms now described as geniculate, if for no other reason than to emphasize the probable homology between elements of this type in the skeletal apparatuses of various conodont species. Elements of this sort have also been termed "cyrtoniodontiform" and "neoprioniodontiform" in descriptive works that appeared in the interval of transition between form taxonomy and multielement taxonomy.

Ramiform elements with more than three primary processes are rare. Those few with four processes (anterior, posterior, and two lateral) are appropriately described as *quadriramate ramiform elements*, whereas those with more than four processes (should any ever be discovered) might be termed *multiramate ramiform elements*.

2.3.3 Rastrate crowns

Elements of the panderodontide genera *Belodina*, *Pseudobelodina*, *Culumbodina*, *Parabelodina*, and *Plegagnathus* are basically coniform units that lack structures comparable to the processes of typical ramiform elements but commonly develop one to several denticles on the concave posterior margin of the cusp. No special descriptive term was proposed in the *Treatise* for such elements, but in a systematic report I published in 1979 the need for one was recognized and the word *rastrate* was coined. Rastrate elements (Fig. 2.4) have compressed, individually asymmetric crowns, with a rela-

Fig. 2.4. Rastrate crowns and their orientation. Terms used to describe them include the following: c = cusp; d = denticle; f = furrow; and h = heel.

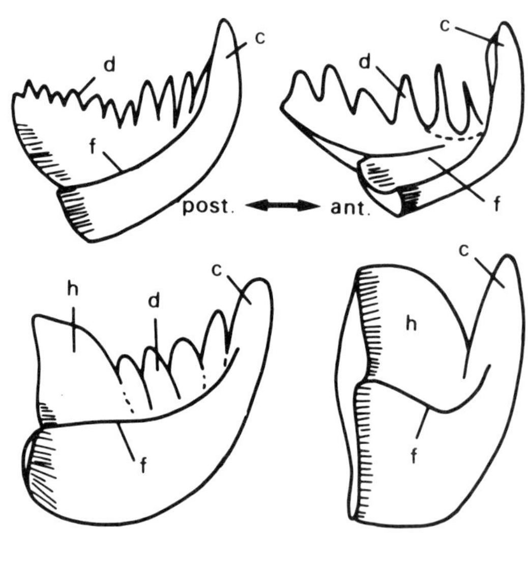

tively deep basal cavity and a posterior margin characterized by a prominent flangelike "heel" at the posterior end and, typically, a series of several denticles between the heel and the tip of the cusp. A few rastrate elements lack denticles between heel and cusp apex and are thus superficially like geniculate coniform elements. A few others (e.g., the upper pair in Fig. 2.4) lack a "heel."

2.3.4 *Pectiniform crowns*

In 1879 Hinde used the term "pectinate teeth" for comb-shaped conodont elements of the sort

Fig. 2.5. Pastinate and stellate pectiniform elements, their scaphate and planate derivatives, and terms used to describe them. (See Fig. 2.3 for explanation of abbreviations.) Diagrams redrawn from Treatise on Invertebrate Paleontology, courtesy of the Geological Society of America and University of Kansas.

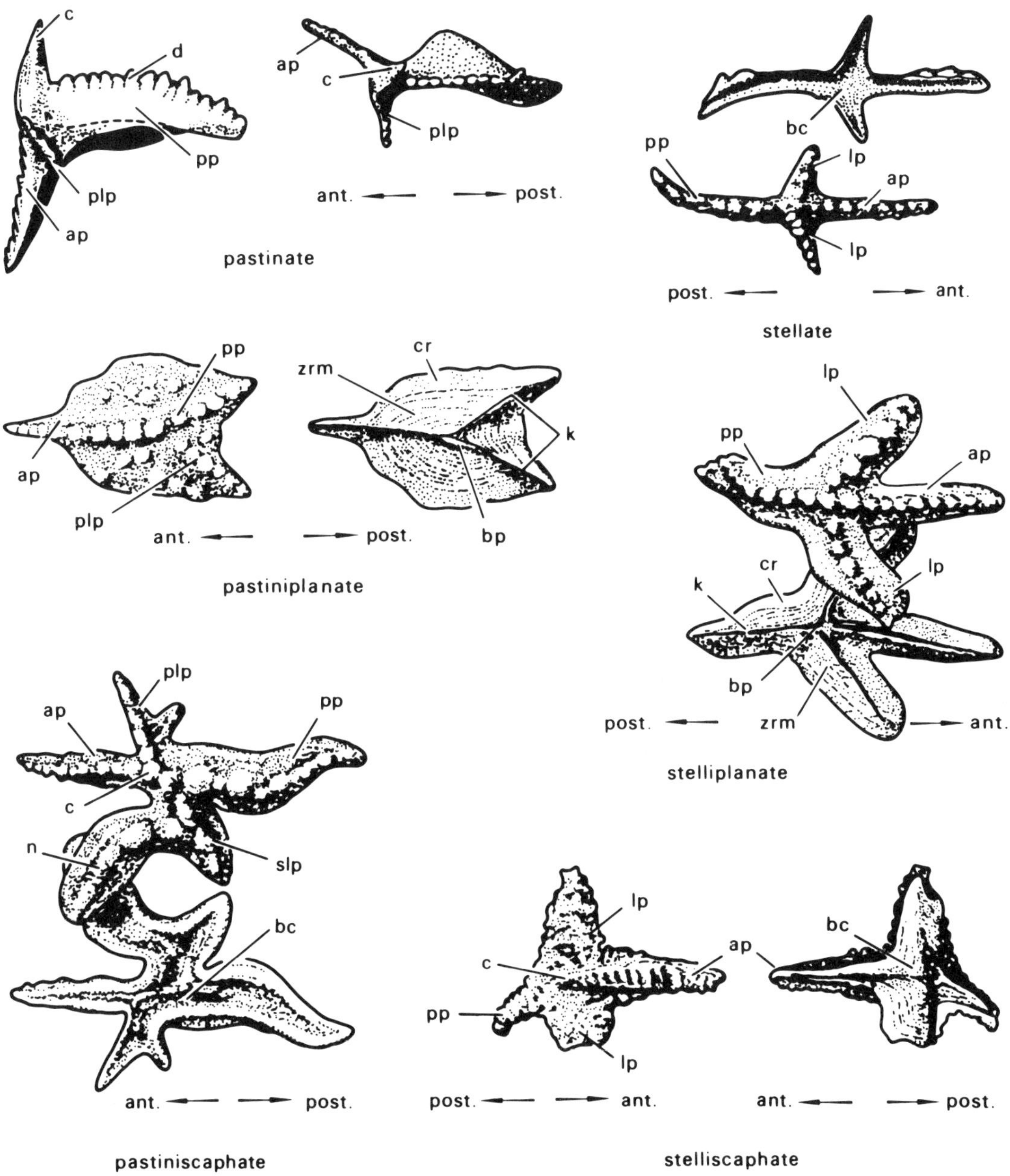

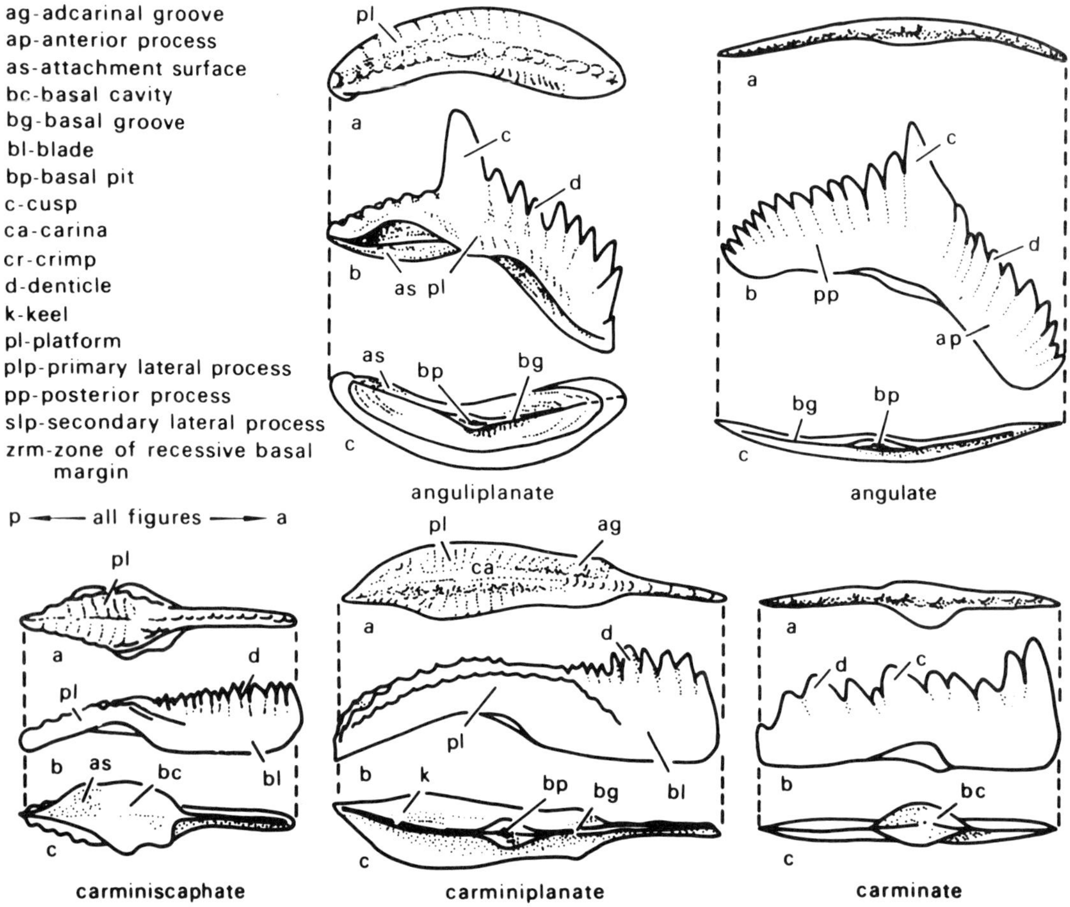

Fig. 2.6. Angulate and carminate pectiniform elements, their scaphate and planate derivatives, and terms used to describe them. From Treatise on Invertebrate Paleontology, courtesy of the Geological Society of America and University of Kansas.

termed "blades" by many subsequent authors. In the 1981 *Treatise* volume on conodonts, however, I broadened definition of the term (in the form "pectiniform") to include elements dubbed "plates" or "platforms" by previous authors because it was by then quite clear that blades, plates, and platforms all occupied similar positions in various skeletal apparatuses and thus should be described in a way that makes that homology obvious. In addition, by regarding all these types of crowns as variants of a single major shape category, it was possible to devise a descriptive terminology based on just a few new words that could be used, as appropriate, in different combinations. Terms proposed in the *Treatise* for pectiniform elements are listed in Table 2.1. Typical pectiniform elements are shown in Figs. 2.5, 2.6, and 2.7, which also include some of the terms used to describe elements of these types.

Comb-shaped crowns that lack plate- or platformlike lateral expansions are all similar in having a cusp that is situated above the apex of a basal cavity or pit and one to as many as four processes that extend away from the cusp. In most platformless pectiniform elements, processes are laterally compressed and higher than processes of ramiform elements. On the lower parts of those processes extensions of the attachment surface may be enclosed in groove- or troughlike extensions of the basal cavity, which expands significantly beneath the cusp; or the cavity may be reduced to a basal pit beneath the cusp and the attachment surface beneath the processes may have a narrow medial groove that is flanked laterally by zones of inverted basal margin of varying width.

Platformless pectiniform elements are divided in Table 2.1 into five major categories, depending primarily on the number and arrangement of processes. *Stellate pectiniform elements* (Fig. 2.5) have at least four primary processes, of which two are anterior and posterior. One or more of these processes may branch distally to form *secondary processes.*

Pastinate pectiniform elements (Fig. 2.5) have three primary processes, which are anterior, posterior, and lateral in position. In pastinate components of the skeletal apparatuses of some Ordovician conodonts (e.g., *Neomultioistodus, Multioistodus, Phragmodus, Plectodina*) the anterior process is commonly adenticulate and is represented only by a conspicuous flangelike rim on the anterior margin of the cusp. Pastinate elements of this type have been termed "dichognathiform" in some of the descriptive literature because the genus *Dichognathus* was based on such elements in the heyday of form taxonomy.

Carminate and *angulate pectiniform elements* (Fig. 2.6) have two primary processes, which are regarded as anterior and posterior in position. The categories differ only in that the longitudinal axis of carminate elements is straight, or essentially so, in lateral view, whereas that of angulate elements is arched beneath the cusp. Carminate elements were commonly termed "spathognathodontiform" and

Fig. 2.7. Segminate pectiniform elements, their scaphate and planate derivatives, and terms used to describe them. Upper sets of diagrams from Treatise on Invertebrate Paleontology, courtesy of the Geological Society of America and University of Kansas.

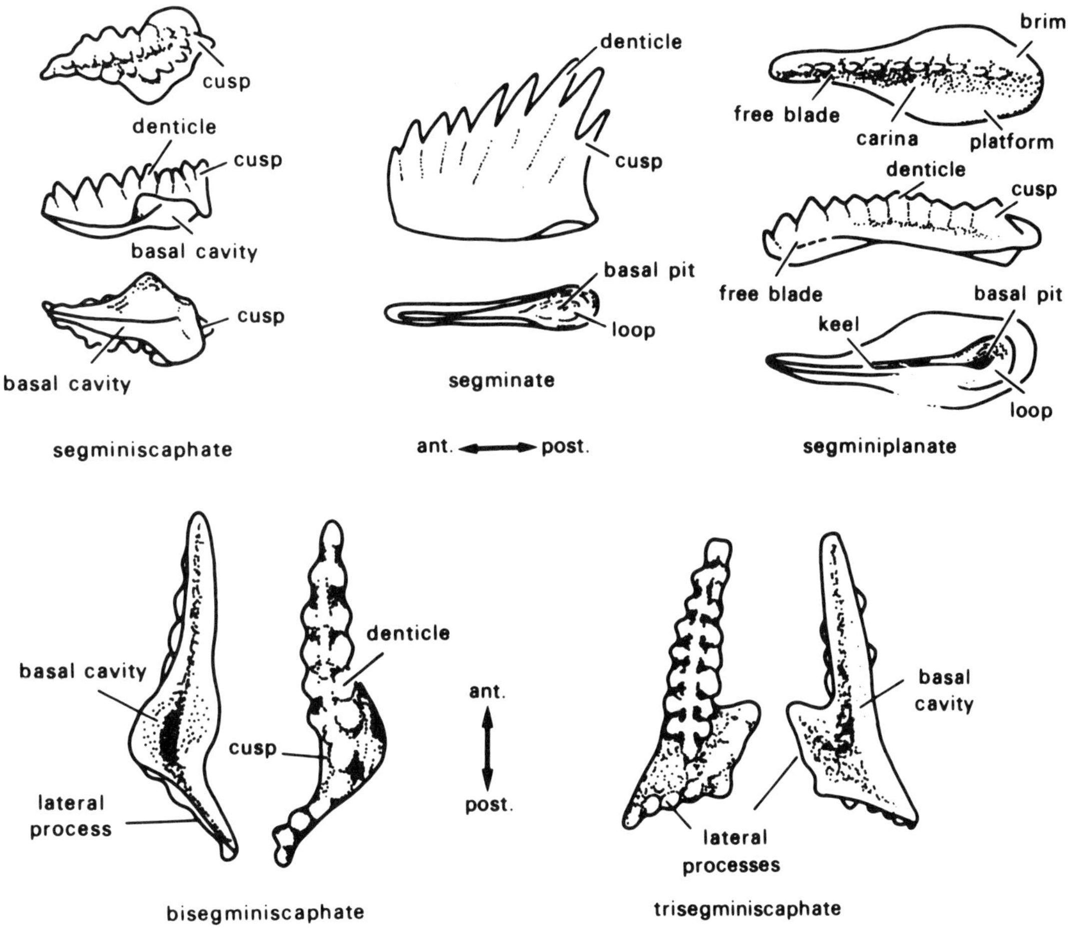

angulate ones "ozarkodiniform" in the descriptive literature of the 1970s, when the language of form taxonomy was being adapted for the purposes of multielement taxonomy.

Segminate pectiniform elements (Fig. 2.7) have only an anterior process, with the cusp at its posterior end. Some types may also have a lateral process or two, and in Table 2.1 and Fig. 2.7 I introduce the terms *bisegminiscaphate* and *trisegminiscaphate* for such forms, which are platformed elements with a broad, deep basal cavity, one or two lateral processes, and no posterior process. I am not aware that there are platformless or planate versions of these pectiniform elements but, if they should turn up, there are now terms for them!

Platformed analogues of stellate, planate, carminate, angulate, and segminate pectiniform elements are named according to the shape of the attachment surface. If the element has a capacious basal cavity that occupies much or all of its underside, the attachment surface is said to be *scaphate*. However, if the attachment surface consists of a basal pit beneath the cusp and a broad adjacent area of recessive basal margin, the surface is said to be *planate*. Thus a platformed pectiniform element with three primary processes and a basal cavity that occupies its entire underside would be termed *pastiniscaphate*, whereas the term *carminiplanate* would identify an unarched platformed element with only two processes and an attachment surface consisting of a basal pit and a broad surrounding area of recessive basal margin. Combinations of terms describing number and arrangement of primary processes (stellate, pastinate, etc.) and nature of attachment surface (scaphate or planate) lead to the 12 terms for platformed pectiniform elements listed in Table 2.1.

2.4 Symmetry- and Curvature-Transition Series

As collections of ramiform elements are sorted into shape categories, one often discovers that there is an almost continuous transition in form between elements of one type and those of another. Characteristically, bilaterally symmetrical alate elements grade by way of slightly asymmetric alate forms into closely similar digyrate or tertiopedate elements. These, in turn, may be traced into bipennate or even dolabrate forms by way of numerous morphologic intermediates. Lindström (1964) was the first to recognize such *symmetry-transition series*, and he described and illustrated several types. Subsequently other series have been worked out, and their recognition has come to be of great practical importance in reconstructing the skeletal apparatuses of conodont species that are represented only by discrete elements. I illustrate several symmetry-transition series in Fig. 2.8, and numerous others are shown elsewhere in this book.

As a rule, the symmetry-transition series involves just three or four of the six or seven types of elements that make up the complete apparatus of most conodonts. Shapes of elements in the symmetry-transition series are conservative, and it is commonly difficult to distinguish the apparatuses of congeneric species through study of those elements alone. However, the conservative nature of elements

Fig. 2.8. The symmetry-transition series in the apparatuses of three conodont species. Alate elements on the right grade by way of the digyrate or tertiopedate elements in the center row into the bipennate or dolabrate elements on the left.

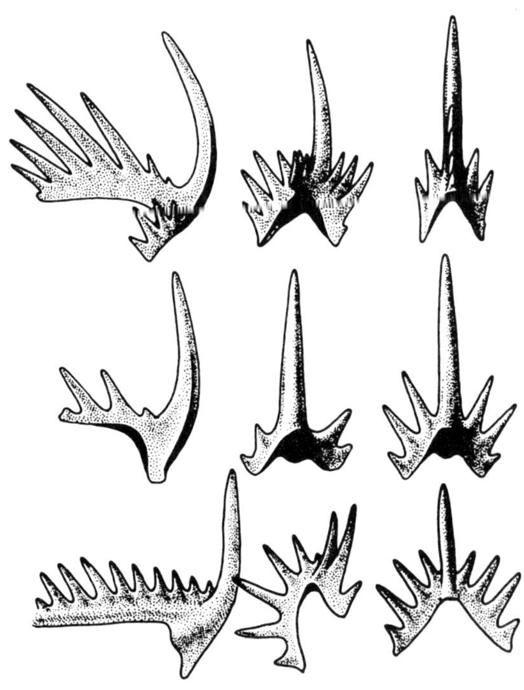

in the symmetry-transition series may be of considerable help in distinguishing between conodonts that belong in different suprageneric stocks. Thus it is important to pay careful attention to the correct assignment of transition-series components during the systematic study of any collection of discrete elements.

In a few conodont species with apparatuses composed solely of coniform elements, it has been noted that positions in a symmetry-transition series may be occupied by elements that are variable with respect to longitudinal curvature. Fåhraeus and Hunter (1986) have recently termed these variable groups *curvature-transition series* and have illustrated those they regard as characteristic of the apparatuses of *Panderodus gracilis*, *Protopanderodus varicostatus*, and *Drepanoistodus* sp. aff. *D. suberectus*.

In the skeletal apparatus of *Panderodus gracilis*, as reconstructed by Fåhraeus and Hunter, four of the five components form a symmetry-transition series (Fig. 2.9). Specimens in row A of Fig. 2.9 are all bilaterally symmetrical, alate coniform elements, but they differ from one another by slight variations in curvature; specimens in row B, which are slightly asymmetric versions of the alate specimens in row A, also exhibit some variation in curvature, as do those in row C, which are otherwise even more asymmetric versions of elements in row B. Fåhraeus and Hunter thus recognize five curvature-transition series within a skeletal apparatus that also exhibits a three- or four-member symmetry-transition series.

Fåhraeus and Hunter evidently conclude that each individual of *Panderodus gracilis* had a skeletal apparatus in which there were five morphologically distinct components, each of which was represented by a six-member curvature-transition series. Thus there must have been at least 30 elements in the skeleton of a complete individual. Fåhraeus and Hunter regard the nature and number of such curvature-transition series as features of potential suprageneric value in conodont taxonomy. However, they do not seem to have given serious consideration to the possibility that they might merely be charting the breadth of morphologic variation within the particular populations they sampled.

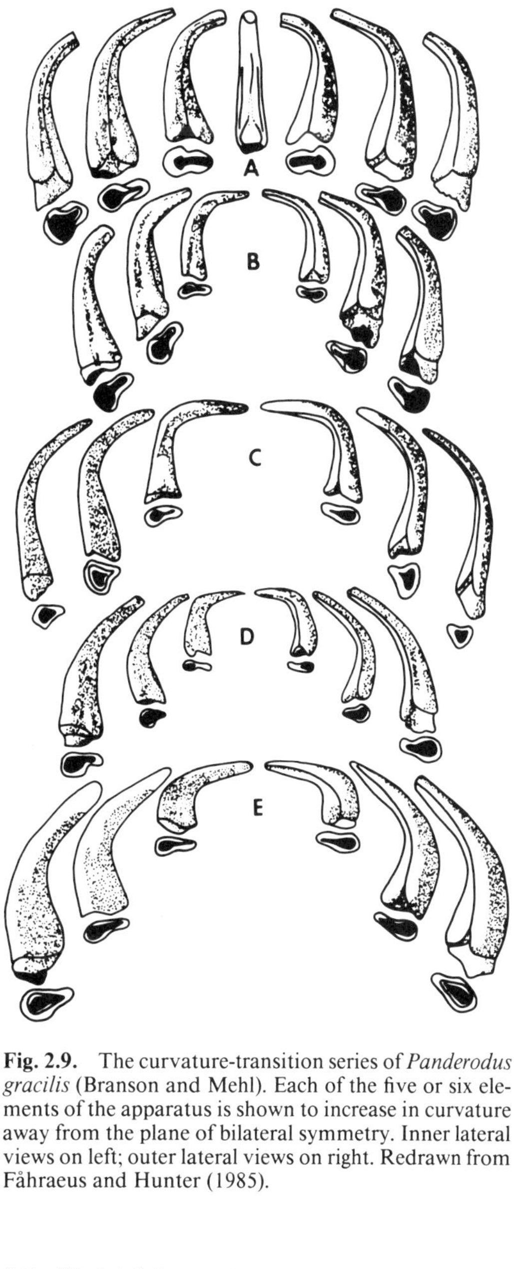

Fig. 2.9. The curvature-transition series of *Panderodus gracilis* (Branson and Mehl). Each of the five or six elements of the apparatus is shown to increase in curvature away from the plane of bilateral symmetry. Inner lateral views on left; outer lateral views on right. Redrawn from Fåhraeus and Hunter (1985).

2.5 Skeletal Apparatuses

We have thus far considered the composition, classification, and descriptive terminology for discrete components of the skeletal apparatuses of conodonts. However, it has been known for at least the last 50 years that in many conodonts, perhaps in all of them, the skeletal apparatus included elements of several different shape categories. Thus, in order to compare one conodont with another and to build a biologically reasonable taxonomy for the group, we also must consider the patterns common to skeletal apparatuses. And, of course, there will need to be some terminology for them, as well.

More than 500 more or less complete apparatuses have now been discovered, and many other collections contain clusters of elements held together by phosphatic cement. The latter also seem to represent at least partial original assemblies. From these *natural assemblages* and *fused clusters* we have learned that the number of major skeletal patterns was probably small, and various schemes have been invented to classify the apparatuses represented. For the most part, those schemes involve no more than three or four categories.

In the *Treatise*, I suggested that the skeletal apparatuses of conodonts be described simply as *unimembrate* or *multimembrate* and that the latter category be subdivided into *bimembrate, trimembrate, quadrimembrate, quinquimembrate, seximembrate*, and *septimembrate* categories, depending on the number of morphologically distinct element types of which they are composed. By and large this purely descriptive scheme has been successful, and the terminology just mentioned is beginning to be widely used. Discovery or determination that two different conodonts had septimembrate apparatuses does not imply, however, that the components are the same—only that each has an apparatus of seven morphologically different types of elements.

It is probably a good idea to note here that a septimembrate skeletal apparatus may have many more than seven discrete-element components. That is, most of the components of the skeletal apparatus of conodonts occur in mirror-image pairs and, in a given apparatus, there may be several pairs of any one morphologic type. Thus it is likely that a complete septimembrate apparatus would have included at least 14 discrete elements (one pair of each of the seven morphologically different types of elements), although the number might have been as low as 13 if one of the types was alate and thus not demonstrably paired, and the total number of elements might have been much larger than 13 or 14 if those at any site were represented in the apparatus by more than a single pair.

As noted in the *Treatise*, it is important to

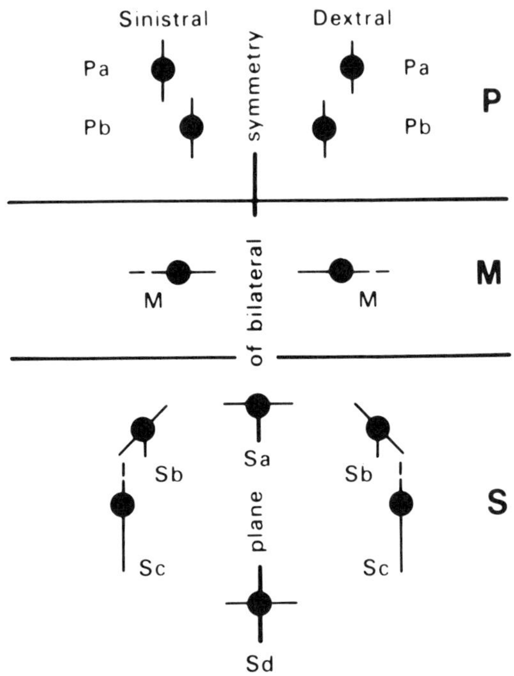

Fig. 2.10. Scheme of locational notation for elements in a septimembrate apparatus. The map is not intended to show actual positions of various types of elements; it merely provides a system for designating major positions. Redrawn and modified from Treatise on Invertebrate Paleontology, courtesy of the Geological Society of America and University of Kansas.

have a scheme for identifying and naming homologous positions within the apparatus. In the *Treatise* I discussed several of the systems that have been devised to do this and concluded by recommending the scheme illustrated diagrammatically in Fig. 2.10. That scheme is based primarily on the fact that the skeletal apparatuses of most species represented by complete natural assemblages can be separated readily into three principal categories, which are lettered P, M, and S in Fig. 2.10.

In natural assemblages P positions are occupied by pectiniform elements or by specialized ramiform elements (e.g., extensiform digyrate elements) and, as a rule, there are two types of these in each apparatus. These may be designated Pa and Pb elements (or the positions they occupy may be termed the Pa and Pb positions). The M position is characteristically occupied by a single pair of pick-shaped (or "makelliform") elements, which may be arched dolabrate, bipennate, or breviform digyrate ramiform elements or geniculate coniform types. Elements in the three major positions in the S series form a symmetry-transition series of the sort first recognized by Lindström (1964). The Sa position is occupied by an alate coniform or ramiform element; the Sb position by a digyrate or tertiopedate element; and the Sc position by a bipennate or dolabrate ramiform element. However, the nomenclature just mentioned (and illustrated in Fig. 2.10) provides formal identification for just the midpoint and the two extremes in the S series. In the apparatuses of many conodonts, however, there may be more than three morphologically distinct components of the S series and, to describe and locate them, it may be necessary to invent intermediate categories such as Sa-b or Sb-c. In Fig. 2.10 there is also a position in the S series labeled Sd, which is occupied in some reconstructed septimembrate apparatuses by quadriramate ramiform elements. It is not appropriate to use Sd as the designation for intermediate positions in the symmetry-transition series, as some authors have recently done. In most skeletal apparatuses the Sd position is not filled at all.

2.6 Symmetry of Elements, Element Pairs, and Apparatuses

It takes very little experience with collections of discrete conodont elements to discover that, like components of the human skeleton, most of them come in mirror-image pairs. The only ones that depart very obviously from this pattern are alate coniform and ramiform elements, which are themselves bilaterally symmetrical, and a few straight pectiniform elements that are developed in the same way on both sides. Indeed, it was not until almost 1960 that it became obvious that the right and left (or dextral and sinistral) members of some element pairs are not exact mirror-image replicas of one another; and the increasingly widespread practice of multielement taxonomy has led to the discovery that bilateral asymmetry was rather common in the skeletons of at least certain groups of conodonts.

Lane (1968) suggested that four major types, or classes, of symmetry could be recognized among various element pairs. These are illustrated diagrammatically in Fig. 2.11. In Lane's scheme, *Class I symmetry* is exhibited by elements that are themselves bilaterally symmetrical. For such elements (e.g., alate coniform or ramiform elements) no pairs can be established, and it is possible that they occurred singly in skeletal apparatuses or were paired with other individually symmetrical Class I elements. *Class II symmetry* is a feature of individually asymmetric elements that are represented by approximately equal numbers of "rights" and "lefts" in large collections of discrete specimens. As Lane (1968) noted, the bulk of known types of conodont elements apparently belong in this class.

Class III symmetry is exhibited by conodont element pairs in which the "right" and "left"

Fig. 2.11. Various symmetry classes exhibited by elements typical of (I) *Rhipidognathus*; (II) *Xaniognathus*; (IIIa) *Eognathodus*; (IIIb) *Amorphognathus*.

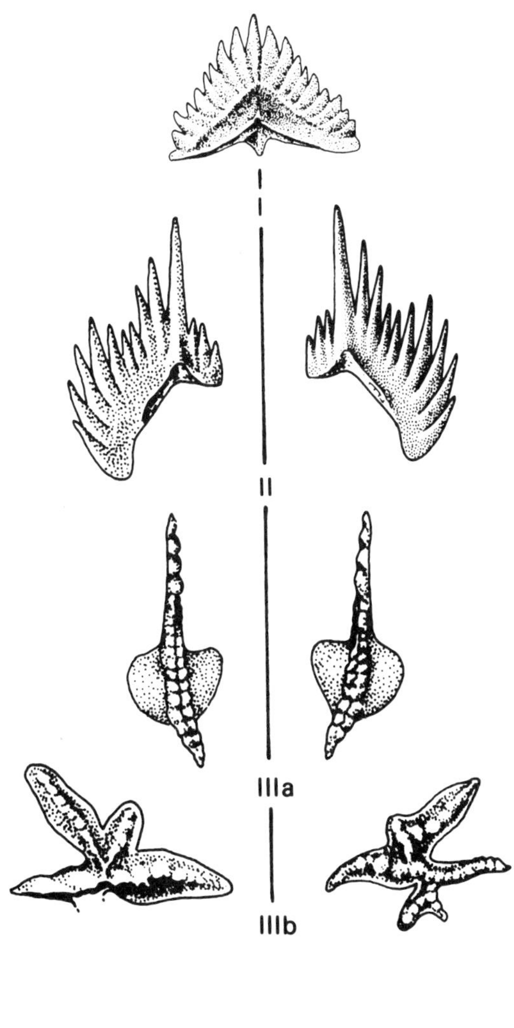

members differ morphologically. The difference may be slight or considerable and of two different sorts. As an example of one sort of Class III symmetry, Lane cited the Pa elements of *Eognathodus sulcatus*, which are laterally bowed carminiscaphate pectiniform elements that develop a double row of denticles on their upper surfaces. Right- and left-handed members of a pair are identified by differences in lateral bowing (Fig. 2.11), but the bladelike anterior process of both members of a pair joins the left-hand denticle row. Lane (1968) termed this type of asymmetry *Class IIIa symmetry*.

A second type of Class III symmetry involves right- and left-handed members that are morphologically quite distinct and can be established as parts of an original pair only by similar abundance and geographic and stratigraphic co-occurrence. Examples of this variety of Class III symmetry (dubbed *Class IIIb* by Lane) are considerably more numerous now than they were in 1968. The complex, stelliscaphate Pa elements of the Late Ordovician *Amorphognathus ordovicicus* serve as an excellent example (Fig. 2.11). Note that right and left (or dextral and sinistral) members of the pair are similar (but surely not identical) in outline and in number and arrangement of primary and secondary processes. However, the anterior process of the dextral member of the pair is developed as a conspicuous blade, whereas that of the sinistral member is not. The two versions occur in approximately equal numbers in representative collections and, with respect to arrangement of the processes, the "bladed" version is always dextral and the "nonbladed" version always sinistral.

Lane (1968) also recognized a *Class IV symmetry*, which includes individually asymmetric elements that lack features, such as lateral bowing, by which dextral and sinistral versions can be recognized. As Lane recognized, this symmetry class is only theoretical, and the need for it has largely disappeared as multielement taxonomic analysis of large collections has come more and more into vogue. That is, tabulated frequencies are commonly sufficient to indicate that elements with Class IV symmetry paired with other such elements in the skeletal apparatus; hence such elements may really have Class II or IIIa symmetry, but right and left members of a pair just cannot be identified.

The commonly observed mirror-image (or Class II) symmetry of conodont-element pairs is primarily responsible for the generally held opinion that conodonts were themselves bilaterally symmetrical. Opposite members of element pairs, wherever those pairs may have been situated in a conodont's skeleton or body, must then have been on opposite sides of the plane of bilateral symmetry. The now-acknowledged existence of element pairs with Class III symmetry might appear to throw a monkey wrench into such an interpretation—or to suggest that conodonts with such pairs in their skeletal apparatuses might not have been bilaterally symmetrical. However, bilateral asymmetry of certain components of the skeleton need not mean that other components were asymmetrically arranged or that the animal as a whole departed greatly from bilateral symmetry. Asymmetry is merely another feature to be dealt with in any interpretation of the mode of life or development of those conodonts in which it has been established.

References

Clark, D. L., Sweet, W. C., Bergström, S. M., Klapper, G., Austin, R. L., Rhodes, F. H. T., Müller, K. J., Ziegler, W., Lindström, M., Miller, J. F., and Harris, A. G. (1981). Conodonta. In *Treatise on Invertebrate Paleontology* (ed. R. A. Robison), Pt. W (Miscellanea), Suppl. 2, 202 pp. Geol. Soc. America and Univ. Kansas.

Ellison, S. P., Jr. (1944). The composition of conodonts. *J. Paleont.* **18**, 133–140.

Fåhraeus, L. E., and Hunter, D. R. (1986). The curvature-transition series: Integral part of some simple-cone conodont apparatuses (Panderodontacea, Distacodontacea, Conodontata). *Acta Palaeont. Polonica* **30(3/4)**, 177–189.

Gross, W. (1957). Über die Basis der Conodonten. *Paläont. Z.* **31**, 8–91.

——— (1960). Über die Basis bei den Gattungen *Palmatolepis* und *Polygnathus (Conodontida). Paläont. Z.* **34**, 40–58.

Harley, J. (1861). On the Ludlow bone-bed and its crustacean remains. *Quart. J. Geol. Soc. London* **17**, 542–552.

Hinde, G. J. (1879). On conodonts from the Chazy and Cincinnati Group of the Cambro-Silurian, and from the Hamilton and Genesee Shale divisions of the Devonian, in Canada and the United States. *Quart. J. Geol. Soc. London* **35**, 351–368.

Lane, H. R. (1868). Symmetry in conodont element-pairs. *J. Paleont.* **42**, 1258–1263.

Lindström, M. (1964). *Conodonts.* Elsevier, Amsterdam, London, New York, 196 pp.

———, and Ziegler, W. (1971) Feinstrukturelle Untersuchungen an Conodonten. 1. Die Überfamilie Panderodontacea. *Geol. Palaeont.* **5**, 9–33.

Pander, C. H. (1856). Monographie der fossilen Fische des silurische Systems der russisch-baltischen Gouvernements. *Akad. Wiss., St. Petersburg*, 91 pp.

Pietzner, H., Vahl, J., Werner, H., and Ziegler, W. (1968). Zur chemischen Zusammensetzung und Mikromorphologie der Conodonten. *Paleontographica*, **128**, 115–152.

Roundy, P. V. (1926). The micro-fauna. In, P. V. Roundy, G. H. Girty, and M. I. Goldman, Mississippian formations of San Saba County, Texas. 2. *U. S. Geol. Surv. Prof. Paper* **146**, 5–17.

Sweet, W. C. (1979). Late Ordovician conodonts and biostratigraphy of thae western Midcontinent Province. *Brigham Young Univ. Geol. Studies* **26**, 45–85.

——— (1981). Macromorphology of elements and apparatuses. Pp. W5-W20. In *Treatise on Invertebrate Paleontology* (ed. R. A. Robison), Pt. W, Suppl. 2, Conodonta. Geol. Soc. America and Univ. Kansas, 202 pp.

3. WHOLE-ANIMAL ANATOMY

It has been clear since 1856, when Pander diagnosed them for the first time, that conodonts were largely soft-bodied animals. Their fossil record, however, is made up almost entirely of tiny phosphatic elements like those described in Chapter 2 and, until quite recently, there has been little direct evidence on which to base concepts of whole-animal anatomy.

Even now, objective evidence of the soft anatomy of conodonts consists of just five specimens, four from a thin bed in the Lower Carboniferous (Mississippian) near Edinburgh, Scotland, and one from a Silurian dolomite in Wisconsin. Several additional specimens from Mississippian rocks in Montana were thought by Melton and Scott (1973) to be conodonts and formed the basis for their phylum Conodontochordata. It is the opinion of virtually everyone else, however, that those specimens are not conodonts, but a previously unrecognized (and possibly distinct) group of conodont eaters. They are not discussed here. Also, in 1976 Conway Morris suggested that poorly preserved impressions in a previously undescribed lophophorate from the famous Burgess Shale of British Columbia might originally have included conodontlike elements. So, for a time, this Middle Cambrian specimen was seriously considered as a general anatomic model for conodonts. However, if *Odontogriphus* (as the specimen was named) was a conodont, it would surely have been a paraconodont, and that group is omitted from consideration in this book. *Odontogriphus* was evidently also quite different in anatomy from the chaetognaths, with which paraconodonts seem to be most closely associated in morphology.

3.1 The Scottish Carboniferous Specimens

The four Scottish Dinantian specimens, which evidently represent at least two different genera, also preserve different portions of the body apparently in somewhat different aspects. However, both the animals reconstructed from these fragments were elongate, laterally compressed, and wormlike in form.

The most completely preserved specimen (Briggs et al., 1983; Fig. 3.1) is 40.5 mm (1.6 in.) long and between 1.8 and 1.95 mm (0.07 to 0.08 in.) wide. A bulbous, bilobed impression at one end encloses a complete conodont apparatus and is regarded as an anterior cephalic lobe. On one margin of the opposite end, judged to be posterior, are traces of what is interpreted to be a ray-supported fin.

The anterior third of the trunk in the best-preserved specimen (Fig. 3.1) lacks diagnostic features except for a subparallel pair of indistinct linear impressions near its anterior end. In the posterior two thirds, however, the axis of the trunk is marked by a distinct linear impression, on one side of which is a series of some 33 irregular structures that slope toward the posterior. These structures are more distinct in the three less well-preserved specimens (Fig. 3.2) described by Aldridge et al. (1986), in which they may be seen to have a V-like shape, with apices of the V's pointing to the anterior. In one of the fragmentary Scottish specimens, the V-shaped structures extend almost to the anterior end of the trunk. However, in that specimen apices of the anteriormost V's appear to be directed posteriorly, perhaps as a result of tortion of the specimen just before burial and compression.

In the description of the first (and still the most instructive) of the Scottish specimens (Briggs et al., 1983; Fig. 3.1), the V-shaped impressions were rather tentatively taken to suggest "segmentation" and were compared provisionally with the myotomes (or muscle blocks) of amphioxus and fish. But these interpretations were also identified as "equivocal." In a later paper (Aldridge et al., 1986) describing additional specimens from the Scottish

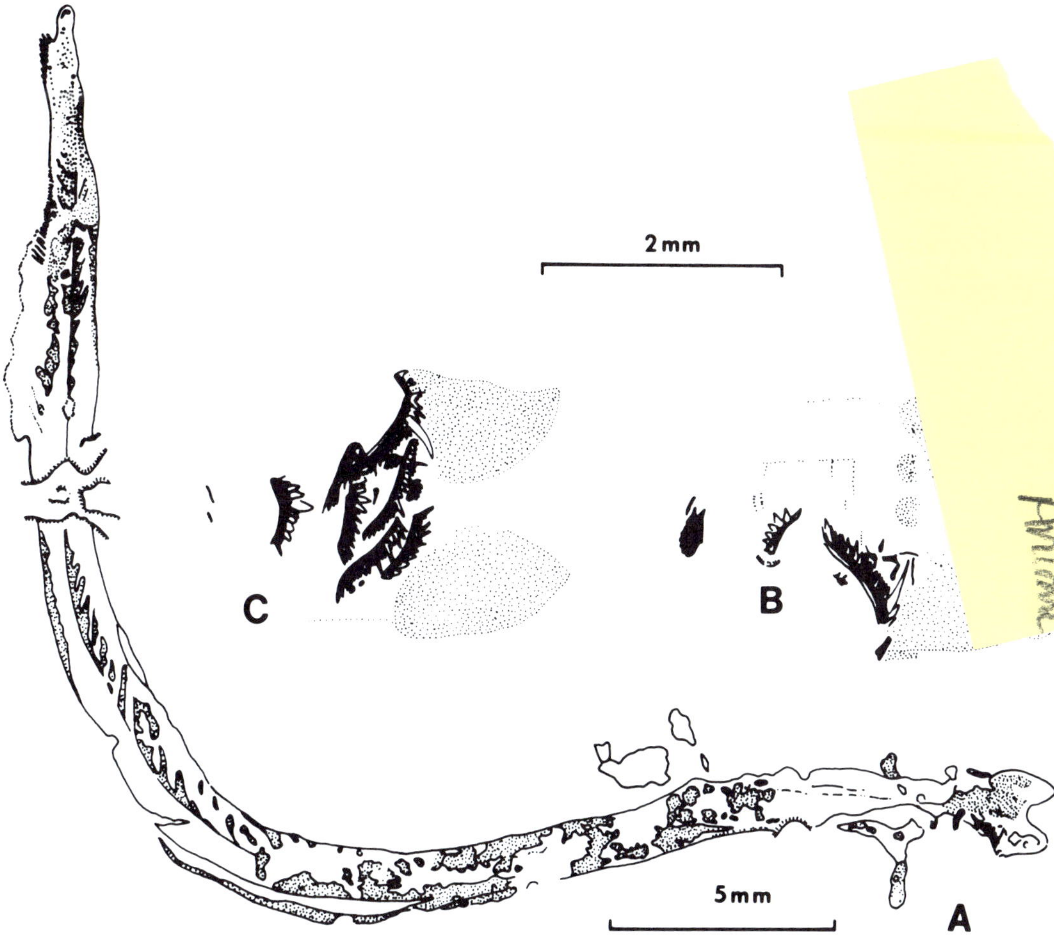

Fig. 3.1. *Clydagnathus*? sp. cf. *C. cavusformis.* (A) Camera-lucida drawing of part; (B) and (C) enlarged views of head showing conodont assemblage in part (B) and counterpart(C). Specimen from "shrimp-band" in Granton Sandstones (Dinantian), Edinburgh, Scotland. Redrawn from Briggs, Clarkson, and Aldridge (1983).

Dinantian locality, these V-shaped structures are identified, with no further discussion, as "somites." This interpretation was obviously quite influential in the much more subjective discussion of anatomy in the 1986 paper, even though it is in no way required by the evidence.

Outlines of the anterior end are preserved in only one of the Scottish Carboniferous specimens (Fig. 3.1). Although this end is termed the "head" by Briggs et al. (1983) and Aldridge et al. (1986), there is no evidence in it of a brain. The term *cephalic lobe* may thus be more appropriate. As noted previously, the cephalic lobe of the best-preserved Scottish specimen encloses a complete and apparently little distorted conodont apparatus, which permits the specimen to be identified as *Clydagnathus*? sp. cf. *C. cavusformis.*

In the only Scottish specimen to preserve the cephalic lobe, right and left components of the skeletal apparatus are superimposed instead of arrayed on opposite sides of a line of bilateral symmetry. It is thus reasonable to suppose that the cephalic lobe of this specimen is preserved in lateral aspect, that is, that the plane of the bedding approximates the plane of bilateral symmetry. The deep cleft at the anterior end of the cephalic lobe may thus mark the position of the mouth, and the position of the apparatus within the cephalic lobe may indicate the general position of a buccal cavity. If the entire specimen shown in Fig. 3.1 has been flattened

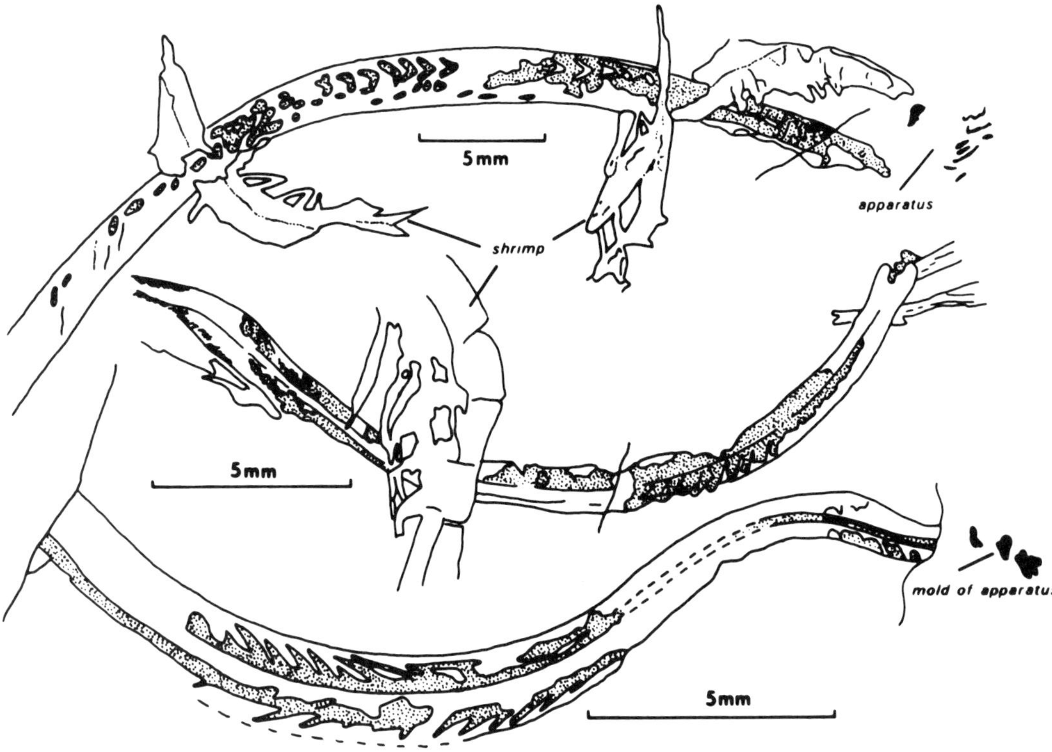

Fig. 3.2. Three additional specimens from the "shrimp-band" in the Dinantian Granton Sandstone of Scotland. Redrawn from Aldridge, Briggs, Clarkson, and Smith (1986).

parallel to its plane of bilateral symmetry, and if the ray-supported structure along one margin of the tail is a dorsal fin, as Briggs et al. (1983) suggested it might be, then it seems likely that the M and S series of elements in the cephalic apparatus were oriented with their denticulate margins directed anteroventrally. The right and left components of the Pa and Pb series were evidently some distance posterior to those of the S series and situated on opposite sides of the plane of bilateral symmetry. Their denticulated margins or surfaces were apparently opposed, but their longitudinal axes may well have been normal to and not parallel with the long axis of the animal. This interpretation of element arrangement agrees in large measure with one made by Jeppsson in 1971 and also with the one assumed by Nicoll in several recent papers (e.g., 1987).

Although the best-preserved Scottish Carboniferous specimen convincingly relates a complete skeletal apparatus to the remainder of a vermiform body and may also indicate the situation and orientation of the apparatus within the cephalic lobe, it is unfortunately mute with respect to several other anatomic matters. For example, none of the elements in the apparatus appears to retain a basal plate, and there is no indication of musculature associated with the apparatus. Even more puzzling is the absence of anything that might be interpreted as the remnant of a structure comparable to the basal plate of the hagfishes, which serves as a cartilaginous substrate for the lingual teeth and as the site of attachment of the muscles that actuate that complex apparatus. In short, the Scottish specimens provide no information as to whether the conodont apparatus was enclosed within epithelium or exposed, and no insight into whether elements of the apparatus were movable and might thus have served as graspers and teeth or whether they were immovable and thus may have served only as more or less passive supports for tentacles or ciliated epithelium in a buccal filter.

The trunk of all four of the Scottish specimens is 1.2 to 1.8 mm wide and preserves evidence of two distinctive anatomic features. The most striking consists of a series of V-shaped structures, which were apparently situated on the sides of the animal and whose apices point anteriorly. In the most completely preserved specimen (Briggs et al., 1983; Fig. 3.1) these structures are rather indifferently preserved and are confined to the posterior half of the trunk. In the three more fragmentary specimens (Fig. 3.2) described by Aldridge et al. (1986), the V's are better preserved and may be seen to extend nearly to the anterior end of the trunk. The shape, serial repetition, and lateral situation of these V-shaped structures are reminiscent of the myotomes (or muscle blocks) of *Branchiostoma* (or "amphioxus") (Fig. 8.1) and of other chordates such as the hagfish *Myxine* (Fig. 8.2).

But chordates are not the only organisms that might yield such V-shaped impressions on compression. For example, nemertines (phylum Rhynchocoela) such as *Nectonemertes* (Fig. 8.1) are characterized by a long, straight intestine, which bears numerous paired lateral diverticula that might, on flattening and compression, produce patterns similar to those exhibited by various parts of the several Scottish Dinantian conodonts. Nemertines, of course, are pseudometameric but acoelomate invertebrates, probably derived from flatworms and not known to have a fossil record.

The second feature recorded in various ways and with varying fidelity in the trunk of all four Scottish specimens consists of a line, or a pair of subparallel lines, more or less coincident with the longitudinal axis of the body and extending from a point just behind the cephalic lobe on the anterior to the posterior end of the body. Aldridge et al. (1986) note that these lines might represent a number of different structures. For example, they might be traces of a notochord, a nerve chord, the gut, a major longitudinal blood vessel, or a mesentery or septum dividing the body cavity into compartments. I agree that a gut is the anatomic feature most likely to be represented by most of the length of the axial lines in most specimens. However, evidence available thus far does not require that all segments of the axial structure in every specimen represent the same organ or organ system. For most of its length in all of the Scottish specimens that show it clearly, the axial structure joins the apices of the V-shaped structures of the trunk. No such relationship is shown in the anteriormost part of the trunk of the best-preserved specimen (Briggs et al., 1983; Fig. 3.1), however, and the axial structure in that specimen may not be continuous with, or represent, the same anatomic feature, as the discontinuous pair of axial lines in the posterior half of the trunk. Aldridge et al. (1986), who allowed their interpretation of the Scottish specimens as primitive chordates to restrict their survey of anatomic possibilities, suggest that the double lines in the anterior parts of several specimens might represent some parts or all of a nerve chord, notochord, or dorsal aorta. I suggest that one might also consider interpreting this pair of lines as the incomplete trace of a nemertine rhynchocoel, the tubular cavity within which living representatives of the curious group of invertebrates store their eversible proboscis. Such an interpretation, of course, would not rule out interpreting axial lines in more posterior parts of the trunk as traces of a gut with numerous lateral diverticula.

If the axial structure in the posterior part of the trunk of the Scottish specimens is interpreted as a gut, it appears to have extended to the posterior extremity of the tail in the two specimens that retain it (Figs. 3.1 and 3.2). That would mean that these animals lacked a postanal segment, or true tail, features that are characteristic not only of chordates but of individuals in virtually all the deuterostome phyla. On the other hand, if one assumes a nemertine model, even with tongue in cheek, the problem vanishes, for in that group—as in most invertebrates with a complete digestive tract—the anal end of the digestive tube is situated at the posterior extremity of the body. I agree with Aldridge et al. (1986) that anatomic interpretation of the axial structures of the Scottish specimens is inconclusive; however, I suggest that, in combination with other features, those structures might well turn out to be as significant in reconstructing details of conodont soft anatomy as the V-shaped structures

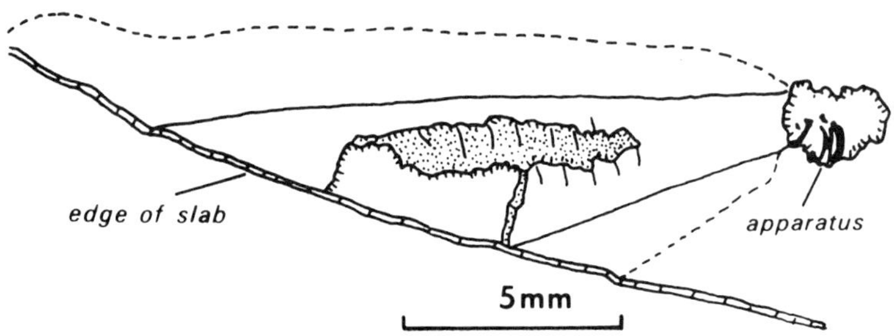

Fig. 3.3. A partially preserved specimen of *Panderodus unicostatus* from the Silurian Brandon Bridge Formation at Waukesha, Wisconsin. Reproduced with permission from *Palaeobiology of Conodonts* (Aldridge, ed.), published by Ellis Horwood Ltd., Chichester, England, 1987).

to which they seem to have paid paramount attention.

Two of the Scottish specimens include the posterior end of the body. In one of these (Fig. 3.1) the margin interpreted by Briggs et al. (1983) as dorsal preserves two separate strips that are reasonably interpeted to be either now discontinuous segments of an originally single, ray-supported fin, or two fins, the posteriormost possibly a caudal one. In the only other specimen in which the posterior end of the body is preserved, short, posteriorly inclined fin rays occur on both margins of the body, although they extend farther anteriorly on one side than the other. No caudal fin is clearly set off, which may suggest that the posterior margin of the first-described, and still most instructive, specimen (Fig. 3.1) is incomplete or was distorted somewhat prior to burial and compaction.

3.2 The Waukesha Specimen

In two papers that appeared in 1985, Mikulic, Briggs, and Kluessendorf described components of an exceptionally preserved biota of dominantly soft-bodied animals from the Silurian Brandon Bridge Formation of Waukesha, Wisconsin. Among the components of that biota is a single specimen (Fig. 3.3) that includes an almost complete apparatus of coniform elements in association with indifferently and incompletely preserved traces of the soft parts of the body. The skeletal apparatus, which has apparently not been deformed, is composed of elements that indicate the animal is referable to *Panderodus unicostatus* (cf. Smith, Briggs, and Aldridge, 1987). The Waukesha specimen is about 17 mm long, is preserved as a thin, apatitic film at the broken edge of a slab of argillaceous dolomite and represents the cephalic lobe and the anterior part of the trunk of a specimen of unknown original length.

The 11 coniform elements of the cephalic apparatus are oriented with their long axes transverse to that of the specimen's trunk and with their tips pointing to the posterior. Smith, Briggs, and Aldridge (1987) suggest that this orientation indicates that the apparatus was originally composed of two opposed halves, one on each side of the animal's longitudinal axis. If this was so, the cephalic lobe and the contiguous trunk of the Waukesha specimen are compressed in a plane normal to the plane of bilateral symmetry. Furthermore, consideration of the orientation of elements in the preserved apparatus suggests that the body was originally dorsoventrally compressed. Had it been laterally compressed (as appears to have been the case with the Scottish Carboniferous specimens) the apparatus would have been deformed in a very different way on compaction of the enclosing sediments.

The Waukesha specimen provides very little other information on the soft-body anatomy of *Panderodus unicostatus* other than vague, transverse lineation that suggests original segmentation. As noted by Smith et al. (1987), the faint lines interpreted as possible segment boundaries are straight and directly transverse to the long axis of the trunk. There is no indication of the V-shaped pattern characteristic of the Scottish specimens.

3.3 Histology of Demineralized Tissue

Fåhraeus and Fåhraeus-van Ree (1987) have identified cells, fibrous material (thought to be at least partly collagen), and luminae of various shapes in scraps of demineralized tissue derived from the pectiniform elements of Silurian specimens of *Ozarkodina confluens.* Cells of various types have been identified; some are relatively large, with nuclei 3 to 5 μm across and long dimensions as much as 20 μm; others are smaller, contain numerous nucleoli, and have a strongly eosinophilic cytoplasm.

Luminae, which have the form of irregular spherical spaces, and single and branching canals, are related by Fåhraeus and Fåhraeus-van Ree to the irregular frothy spaces within conodont elements that have been termed *white matter.* Fibrous matter is sparsely distributed among cells, occurs in bundles and, for the most part, lacks nuclei. Some is dense, basophilic, and apparently associated with elongated sinuous cells. The fibrous matter without nuclei is considered to be collagen. Fåhraeus and Fåhraeus-van Ree emphasize the fact that only 20 to 25 percent of the tissue fragments they studied is collagen, whereas collagen accounts for the entire matrix of most other types of mineralized tissue. They also point out that the large number of cells in demineralized conodont tissue indicates that it "... is not bone or dental tissue as we know it," because bone "... is either totally non-cellular with an extracellular matrix largely consisting of collagen ... or, in cellular bone, it is dominated by bundles of collagen with the occasional 'trapped' osteocyte." They go on to state that the demineralized conodont tissue of their study is also unlike that found in living inarticulate brachiopods or mollusks. They do not exclude the possibility that primitive vertebrates might have formed such tissues, nor do they rule out the possibility that the tissues they studied represent a "... unique approach to biomineralization."

3.4 Summary

Existing evidence on whole-animal anatomy is sparse but may be taken to confirm the conventional wisdom that conodonts were primarily soft-bodied animals with a slender, vermiform body a couple of inches long that may have been either laterally compressed or dorsoventrally depressed. The anterior end of the slender, elongate body is developed as a distinct cephalic lobe in known Silurian and Dinantian specimens and includes the apparatus of mineralized elements by which conodonts are best known. The slender trunk is divided into serially repeated transverse structures that occur on either side of or surround an axial structure that may represent the gut, a notochord, a doral aorta, a longitudinal septum or mesentery, or some combination of these structures. The transverse structures, which are V shaped in Carboniferous specimens (but not in the Silurian one) have been compared with the myotomes (or muscle blocks) of amphioxus and fish and thus taken as evidence of body segmentation. Those interpretations are not required, however, and the transverse structures might just as well represent the paired digestive diverticula that characterize many acoelomate nemertine worms and, as such, indicate only pseudometamerism. The posterior end of two of the five known whole-bodied fossil conodonts is bluntly aciculate, and its margin bears traces of ray-supported finlike structures on one or both sides. Linear axial structures extend to the posterior margin in both specimens. If these represent traces of the gut, the Carboniferous conodonts may not have had a postanal segment, or true tail.

The five whole-bodied fossil conodonts now known raise almost as many questions as they provide answers. It is a comfort to know that the skelctal apparatus is apparently associated with the cephalic lobe, for most of us had concluded from seat-of-the-pants functional analysis that this would turn out to be the case many years before the famous Scottish specimens were discovered. It is also no surprise that the specimens indicate an elongate, vermiform shape, since that is a shape common to mobile aquatic animals. However, even though the skeletal assemblages included in the whole-bodied fossils seem to have been only slightly distorted by compression, they all apparently lack basal bodies, and it is not possible to determine how they may have been related to muscular systems or whether they were completely enclosed in buccal epithelium or situated in epithelial pockets within the buccal cavity. All these matters impinge directly on our interpretations of the function of the cephalic apparatus and indirectly on conclusions with respect to the systematic position of the Conodonta.

References

Aldridge, R J., Briggs, D.E.G., Clarkson, E. N.K., and Smith, M. P. (1986). The affinities of conodonts—new evidence from the Carboniferous of Edinburgh, Scotland. *Lethaia* **19**(4), 279–291.

Briggs, D.E.G., Clarkson, E.N.K. and Aldridge, R. J. (1983). The conodont animal. *Lethaia* **16**, 1–14.

Conway Morris, S. (1976). A new Cambrian lophophorate from the Burgess Shale of British Columbia. *Palaeontology* **19**, 199–222.

Fåhraeus, L. E., and Fåhraeus-van Ree, G. E. (1987). Soft tissue matrix of decalcified pectiniform elements of *Hindeodella confluens* (Conodonta, Silurian). 105–110 in *Palaeobiology of Conodonts* (ed. R. J. Aldridge). Ellis Horwood, Chichester, 180 p.

Jeppsson, L. (1971). Element arrangement in conodont apparatuses of *Hindeodella* type and in similar forms. *Lethaia* **4**, 101–123.

Melton, W. G., and Scott, H. W. (1973). Conodont-bearing animals from the Bear Gulch Limestone, Montana. 31–65 in *Conodont Paleozoology* (ed. F. H. T. Rhodes). *Geol. Soc. Am. Spec. Paper* **141**, 296 pp.

Mikulic, D. G., Briggs, D. E. G., and Kluessendorf, J. (1985a). A Silurian soft-bodied biota. *Science* **228**, 715–717.

———, ———, and ———. (1985b). A new exceptionally preserved biota from the Lower Silurian of Wisconsin, USA. *Phil. Trans. Roy. Soc. London*, Ser. B, **311**, 78–85.

Nicoll, R. S. (1987). Form and function of the Pa element in the conodont animal. 77–90 in *Palaeobiology of Conodonts* (ed. R. J. Aldridge). Ellis Horwood, Chichester, 180 pp.

Pander, C. H. (1856). Monographie der fossilen Fische des silurischen Systems der russisch-baltischen Gouvernements. *Konigl. Akad. Wiss., St. Petersburg*, 91 pp.

Smith, M. P., Briggs, D. E. G., and Aldridge, R. J. (1987). A conodont animal from the lower Silurian of Wisconsin, USA, and the apparatus architecture of panderodontid conodonts. 91–104 in *Palaeobiology of Conodonts* (ed. R. J. Aldridge). Ellis Horwood, Chichester, 180 pp.

4. TAXONOMY

4.1 Form Taxonomy

In Chapter 1, I pointed out that Christian Pander, who discovered and named conodonts, wrestled in his 1856 monograph with the choice of taxonomic base. He wrote (my rough translation):

> We don't know what sort of teeth we have before us; whether they belong to the jaws, the palate, the lips or the tongue; whether each particular shape constitutes sufficient basis for establishment of its own genus or species; or whether different forms could have come from one and the same body.

In the absence of a living analogue, and unable to answer the questions he raised about function or anatomic position or about morphologic monotony or variability of elements in the same body, Pander chose an admittedly utilitarian classification based on shape of individual elements. This approach to the classification of biotic groups is sometimes termed, rather pejoratively, *form taxonomy*.

To begin with, Pander assigned coniform elements in his collections to an informal category termed "simple teeth" (*Einfache Zahne*), and ramiform and pectiniform elements to another, termed "composite teeth" (*Zusammengesetzte Zahne*). The simple teeth (Fig. 4.1) were then further divided on the basis of cross section, into 7 categories, which Pander described and named as genera. Finally, each of these was subdivided on the basis of cusp curvature into groups regarded as species. Composite teeth in Pander's collections were represented by 7 obviously different mophologic categories, among which he recognized 18 species on the basis of features such as mode of denticulation, presence or absence of carinae, angle between processes, character of platform extensions, and the like. Altogether, then, Pander recognized and named 7 genera in each major category (simple and composite teeth) and, collectively, 56 species, all based on an analysis of shape, or form.

Pander's classification of conodonts was proposed largely as a means of giving order to the systematic description of specimens in his collections. He made no attempt to give it profound biologic meaning, nor did he use it as the basis for discussing evolutionary development or addressing other biologic questions. Evidently the classification was to serve only until discovery of more complete specimens made it possible to determine if all elements were morphologically alike in the body of a single individual, or if the body of the same individual might have had an array of morphologically different elements within it.

Although Harley, Smith, Hinde, and Hadding contributed to conodont taxonomy between 1856 and 1926 by describing discrete elements of distinctive shapes as genera, no further consideration of suprageneric categories seems to have been made until 1926, when Ulrich and Bassler published their influential taxonomic scheme. As you will recall from earlier discussions in this book, Ulrich and Bassler stated their firm conclusion that conodonts represented not only the teeth, but also the dermal plates of extinct groups of fishes. In addition, and probably more significantly, they also concluded from their studies of recent fishes that, except for the fact that they may occur as right and left members of a pair, all teeth in the mouth of the supposed living relatives of the conodonts are essentially alike. Thus they wrote ". . . each kind is characteristic of some particular genus and species." In short, they professed complete confidence that a classification based on the shape of discrete elements, a form taxonomy, was a true expression of biologic relationships.

In their 1926 paper Ulrich and Bassler ignored the one genus (*Periodon*) proposed by Hadding in 1913, but distributed all the others known at that time, and 15 new ones, among 4 suprageneric categories described and named as families. Pander's "simple teeth" category

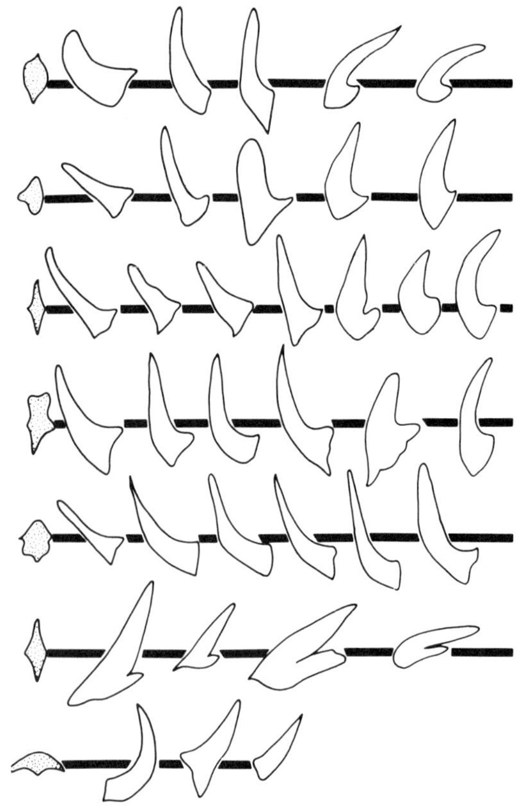

Fig. 4.1. Pander's simple-cone genera and species. The cross section at the end of each horizontal line indicates the major generic characters. Species were based on curvature of the cusp and are shown in lateral outlines along the heavy lines to the right of the cross sections.

was formally identified as the family Distacodidae (which should have been Distacodontidae), and his "composite teeth" category was divided into the families Prioniodidae (rcctc: Prioniodontidae), Prioniodinidae, and Polygnathidae. The first two of these were erected for ramiform elements and platformless pectiniform elements of various shapes; the latter included all genera and species based on platformed pectiniform elements, which were interpreted as dermal plates, not teeth.

Between 1933 and 1944 Branson and Mehl devoted their attentions largely to describing previously unknown types of conodont elements and diagnosing and naming the genera and species they were thought to represent. In 1938 Branson included conodonts in the subclass Conodonti (of the class Pisces, phylum Vertebrata), but in 1944 Branson and Mehl defined the conodonts as an extinct order of primitive fishes (the Conodontophoridia) and described 73 genera, which were assembled into 8 families and 2 suborders. The two suborders (Neurodontiformes and Conodontiformes) were established for conodonts whose elements differ in internal structure. The suborder Neurodontiformes included 13 Ordovician genera whose elements have (or were thought to have) a fibrous internal structure, whereas the suborder Conodontiformes was established for the remaining genera, which are represented by elements that have a lamellar internal structure.Because conodont elements were thought to be fish teeth, the many differently shaped ones known in 1944 were grouped into categories defined as families on the basis of the supposed relationship between the variously shaped teeth and the jaw rami on which they were thought to have been borne. The three genera assigned to the Coleodontidae, for example, are represented in the fossil record by internally fibrous elements with relatively deep, troughlike undersides, which were thought to have clasped the jaw of the fish that formed them.

The mechanical classification of conodonts on the basis of the shapes of individual, discrete elements surely reached its acme in Volume W of the *Treatise on Invertebrate Paleontology*, which appeared in 1962. In that highly personal volume, written almost entirely by Wilbert Hass, 147 genera of conodonts were assigned to a single order (the Conodontophorida), which was divided into 7 families. The features of internal structure that Branson and Mehl used to recognize the suborders Neurodontiformes and Conodontiformes were regarded as insignificant or nonexistent by Hass, who consequently did not recognize those suborders. On the other hand, three families (Coleodontidae, Prioniodontidae, and Idiognathodontidae) were divided into subfamilies so that the number of suprageneric categories recognized by Hass totaled 17 (compared with 8 in Branson and Mehl's scheme or 4 in that of Ulrich and Bassler).

4.2 Multielement Taxonomy

Multielement taxonomy, an alternative to form taxonomy, was also recognized as a possiblity by Pander but was not put into practice until 1879, when Hinde employed it for a species from the Devonian of New York. In the "multielement" way of looking at things, the skeletal apparatus of a conodont is made up of numerous parts, not necessarily all alike in morphology. The base for taxonomy is thus a group of elements that can be shown, or can reasonably be inferred, to have occurred together in the body of an individual conodont. An apt analogy, of course, is with the vertebrate skeleton, which consists of numerous morphologically different parts.

Hinde (1879), the first to practice multielement taxonomy, based his concept of *Polygnathus dubius* on an assortment of elements partly exposed on the surface of a slab of black shale. He took physical proximity on the shale slab to indicate original association, although the name he chose for the first species to be diagnosed in a multielement sense suggests that he was not completely convinced that all elements came from the same individual.

Other, more convincing "natural assemblages" of elements were described independently in 1934 by Hermann Schmidt in Germany and by Harold Scott in the United States. The assemblages available to those paleontologists are all from Carboniferous rocks of about the same age and are made up of about the same types of conodont elements. Because the composition of Schmidt's and Scott's assemblages is closely similar, and because each had several specimens of similar or identical composition, chance association was easily ruled out, and the assemblages were described and interpreted as the skeletal remains of individual conodonts. E. B. Branson and Mehl (1936) and C. C. Branson (1957) attempted to dismiss Scott's specimens as coprolites, but Scott replied (to Branson and Mehl) that ". . . it would be strange indeed to find a group of animals with such a balanced diet that the excretal material would consist time after time of one pair of prioniodids, one pair of spathognaths, one pair of prioniodells, and approximately four pairs of hindeodells."

There are now more than 500 "natural assemblages" of conodonts known from Cambrian, Devonian, and Carboniferous rocks, although not all have been described and illustrated. In addition, fused clusters of elements now and then appear in acid residues, and an essentially complete "natural assemblage" is preserved in the head of the Scottish "conodont animal" discussed in detail in Chapter 3. So, since at least 1934, there has been no good reason to defend a form taxonomy for conodonts as a natural one. The practice persisted through 1962, I suspect, because most students of conodonts were more concerned with biostratigraphic applications than with biologic taxonomy.

4.2.1 *Multielement methodology*

The major difficulty with putting a general scheme of multielement taxonomy into practice for conodonts has been the fact that "natural assemblages" and fused clusters are relatively rare and are not uniformly distributed stratigraphically, whereas the collections available to most systematicists are jumbled assortments of elements that may have been components of several different multielement apparatuses. Thus, except for species represented by natural assemblages, a means must be devised for determining which elements in collections of discrete specimens were originally associated in the same skeletons.

Perhaps the most instructive way of reconstructing the skeletal apparatuses of conodonts from large collections of discrete elements is a basically empirical method. This procedure was applied as early as 1958 by Huckriede, who noted that in his collections of Triassic conodonts certain forms occurred more or less regularly with certain others. From his observation that certain types of ramiform elements were regularly associated with one or the other of the two most common platformed pectiniform elements, Huckriede proposed that the "*Conodonten-Satz*" (literally, conodont-set) of *Gondolella navicula* (the platformed pectiniform element) includes a pair of angulate pectiniform elements (referred then to *Ozarkodina tortilis*), and ramiform elements that were included in 1958 in 9 form species assigned to 6 form genera. Huckriede compared the association he recognized empirically with natural assemblages described by Schmidt (1934, 1950) and Rhodes (1952) and evidently noted agreement in at least major features. He applied similar reasoning to a reconstruction of the "*Satz*" of *Polygnathus tethydis*, with results that are closely similar to those stated many years later by other authors.

Huckriede, in short, made empirical use of his familiarity with large collections of discrete elements to reconstruct recurrent groups whose members could be compared directly with components of the well-known natural assemblages that served as anatomic models. This procedure is the one employed by most taxonomists today.

Walliser (1964) also used the empirical method to recognize recurrent groups of elements in his large collections of Silurian conodonts. He termed those groups "apparatuses" and clearly recognized them as the probable skeletons of 'natural' conodont species. However, the Silurian conodonts he considered in his monograph were ultimately described in more conventional form-taxonomic terms.

Webers (1966) and Bergström and Sweet (1966) applied essentially the same empirical methodology used by Huckriede and Walliser in taxonomic analysis of their large collections of Middle and Upper Ordovician conodonts. However, the recurrent groups they recognized (the *Sätze* of Huckriede, the "apparatuses" of Walliser) were explicitly treated as the representatives of species, and the *International Rules of Zoological Nomenclature* were followed in determining which of the many available names to use for the species represented by these recurrent groups of elements. By adopting this practice, Webers and Bergström and Sweet soon found that their own names were epithets in conodont circles! That is, prior to 1966 (and to a large extent subsequently) most paleontologists who studied conodonts did so to solve stratigraphic problems—and a major revision of the taxonomic base and of the names applied to genera and species in the revised taxonomy was not greeted with enthusiasm, to say the least!

Kohut (1969), Sweet (1970), and Von Bitter (1972) have applied various clustering methods, based on different binary similarity coefficients, to large collections of discrete elements in attempts to isolate recurrent groups that might be evalated as the skeletal apparatuses of conodont species. In effect, they substituted mechanical procedures for the empirical methods of Huckriede, Walliser, Webers, and Bergström and Sweet, with the result that it became possible in at least some cases to subdivide what might be regarded as provincial or facies faunas into their constituent multielement species and, furthermore, to reach some conclusions as to the numerical ratios between elements and pairs of elements in the apparatuses of those species. General features of the clustering techniques are worth a brief survey, although it should be kept in mind that these procedures work only if the investigator is dealing with a reasonably large number of samples that include many of the same forms and are from rocks whose features indicate that reworking was minimum.

Table 4.1 summarizes the occurrence of 15 different types of conodont elements in 10 samples. The multielement taxonomist would be interested in developing from data in that table a concept of the extent of co-occurrence of groups, or clusters, of elements that might be

TABLE 4.1. *Distribution of 15 element types in 10 samples*

	Element types														
Samples	1	2	3	4	5	6	7	8	9	10	11	12	13	14	15
1	X	X	X	X	X	X	X	X	X	X	X	X	X	X	X
2	X		X	X	X	X		X	X		X		X	X	
3			X	X	X	X	X	X	X		X			X	
4	X	X			X	X				X	X	X			X
5	X		X	X	X	X	X	X	X		X			X	
6			X	X	X		X	X	X		X			X	
7	X	X	X	X	X	X	X	X	X	X	X	X	X	X	X
8					X	X			X						
9	X		X	X	X	X		X		X	X	X		X	X
10	X	X		X	X	X			X	X	X	X	X		X

Note: X = represented in sample by at least one specimen.

TABLE 4.2. *Jaccard coefficients for element-type pairs from Table 4.1*

Type	1	2	3	4	5	6	7	8	9	10	11	12	13	14	15
1	—	57	56	67	70	78	33	56	50	71	78	71	57	56	71
2		—	22	33	40	44	29	22	33	80	44	80	60	22	80
3			—	88	70	60	71	100	67	33	78	33	38	100	33
4				—	80	70	63	88	78	44	89	38	50	88	44
5					—	90	50	70	80	50	90	50	40	70	50
6						—	40	60	70	56	80	56	44	60	56
7							—	71	63	29	56	33	29	71	29
8								—	67	33	78	33	38	100	33
9									—	30	70	30	50	67	30
10										—	56	100	50	33	100
11											—	56	44	78	56
12												—	50	33	100
13													—	38	50
14														—	33
15															—

Note: Decimal points omitted: "100" = 1.00; "50" = 0.50, etc.

considered to represent the skeletal apparatuses of various conodont species. Several different coefficients have been developed for use in stating the degree of association between two entities, and each student has a preference for one or another. Kohut (1969), for example, used an index of affinity developed by Fager and McGowan (1963), whereas Von Bitter (1972) employed the Jaccard coefficient. Clusters derived from the data of Table 4.1 are essentially the same whichever similarity coefficient is used; so, because they are easier to calculate than Fager values, I have derived Jaccard coefficients for all possible pairs of the 15 element types in that table. The Jaccard coefficient, as used by Von Bitter, may be given as

$$J = \frac{a}{a + b + c}$$

in which a is the number of samples in which both the element types considered occur, b is the number in which one of them is present alone, and c is the number of samples in which only the second element type occurs. Obviously, if the two types of elements do not occur together in any of the samples available $J = 0$; but if each of the two is represented in every sample that contains either one of them, $J = 1$. Of course, if $J = 1$ for a pair of element types, we must consider the possibility that those element types were originally parts of the same skeletal apparatus, even though they may differ greatly in morphology.

In Table 4.2 I give the Jaccard coefficients for all possible pairs of the 15 element types whose distribution is given in Table 4.1. Note that values of those coefficients vary from a low of 0.22 to a high of 1.00. What we need to do now is rearrange the data of Table 4.2 to reveal clusters of element types more closely related to each other than to other clusters. This may be done in several ways. Kohut (1969) chose, in effect, to rearrange the similarity-coefficient matrix itself, as I have done in Table 4.3, by moving the highest values in each row and column to the diagonal median of the matrix. Von Bitter (1972), on the other hand, used the unweighted pair-group method to develop a dendrogram that graphically displays the relationship between pairs and clusters of pairs of element types. I have derived the dendrogram of Fig. 4.2 from the similarity-coefficient data of Table 4.2.

In both the rearranged matrix of Table 4.3 and the dendrogram of Fig. 4.2, note that there are three well-defined clusters of element types. Two of these clusters include six morphologically different types of elements, whereas one has only three. Furthermore, both the rearranged matrix and the dendrogram reveal the same three clusters, which may appropriately be identified as *recurrent groups*.

In a multielement view of the conodont world recurrent groups are the basis for taxonomy if their components also exhibit several other important features. For example, if a recurrent group is to be interpreted as the skeletal apparatus of a conodont species, one might rea-

TABLE 4.3. *Rearranged matrix of Jaccard coefficients from Table 4.2*

Type	14	3	8	4	11	7	5	6	9	1	13	10	12	15	2
14	—	100	100	88	78	71	70	60	67	56	38	33	33	33	22
3		—	100	88	78	71	70	60	67	56	38	33	33	33	22
8			—	88	78	71	70	60	67	56	38	33	33	33	22
4				—	89	63	80	70	78	67	50	44	38	44	33
11					—	56	90	80	70	78	44	56	56	56	44
7						—	50	40	63	33	29	29	33	29	29
5							—	90	80	70	40	50	50	50	40
6								—	70	78	44	56	56	56	44
9									—	50	50	30	30	30	33
1										—	57	71	71	71	60
13											—	50	50	50	57
10												—	100	100	80
12													—	100	80
15														—	80
2															—

Note: Decimal points have been omitted: "100" = 1.00; "50" = 0.50, etc.

sonably expect its component elements to have a similar or identical mode of denticulation and microstructure; a similar or identical microsculpture and distribution of "white matter"; a similar or identical mode of development of attachment surface and basal cavity; and a similar range in size. It is also helpful if components of a recurrent group can be shown to include an array of elements comparable in number and general development to those in the natural assemblages that have served for many years as the only objective anatomic models for conodonts. Slavish insistence on identity in apparatus composition for all conodonts, however, would merely deny the possibility that other patterns might have existed in the past.

Kohut (1969) contended, with some support from the natural assemblages known at that time, that the components of a recurrent group should also show some regular relationship to one another in rank or ratio in order to be considered parts of the skeletal apparatus of a conodont species. And he was able to demonstrate that statistically significant constancy in rank was a feature of all the best-represented clusters derived from the Ordovician materials he studied. However, it would be difficult to support rank or ratio constancy as a general criterion for interpreting a recurrent group as the apparatus of a conodont species, since it has become increasingly clear that events operated against such constancy in the ontogeny of some conodont species or in the sedimentary environments to which they eventually contributed their skeletal remains. In short, this yardstick must also be applied with caution in many situations.

Fig. 4.2. A dendrogram depicting relations between groups of element types. Jaccard values from Table 4.2. Note that the same clusters appear here and in the rearranged matrix of Table 4.3.

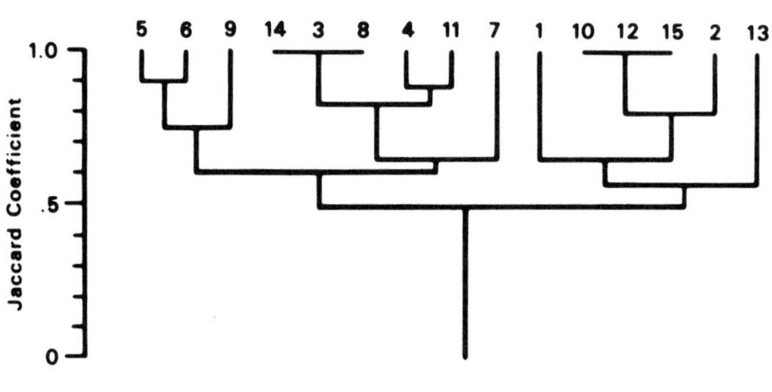

4.3 Multielement Classifications

Lindström (1970) compiled the first classification of conodonts that recognized recurrent groups of morphologically different element types as the basic taxonomic units. Conodonts were grouped into 2 orders (Westergaardodinida Lindström and Conodontophorida Eichenberg), with the latter (the "conodonts proper") further divided into 8 superfamilies (Distacodontacea, Panderodontacea, Chirognathacea, Prioniodinacea, Prioniodontacea, Bryantodontacea, Gondolellacea, and Polygnathacea). These major groups were defined, insofar as possible, in multielement terms, as were the families into which they were divided. Altogether, Lindström was able to assign a total of 73 genera more or less confidently to one or another of the 21 families he recognized.

The taxonomic scheme outlined by Lindström (1970) obviously influenced the one developed by authors of revised Volume W of the *Treatise on Invertebrate Paleontology* (Clark et al., 1981), which was written between 1971 and 1976 but was not published until late 1981. *Treatise* authors included all conodonts in a single class of a phylum Conodonta and followed Lindström in assigning the animals I regard as "true conodonts" to one order (the Conodontophorida), which included 44 families distributed among 9 superfamilies. The 179 genera of "true conodonts" recognized were diagnosed, insofar as possible, in multielement terms and described in the terminological scheme described in Chapter 2 of this book, which was developed specially for the *Treatise*.

4.4 A Revised Multielement Classification

At the time of this writing, skeletal apparatuses of species that represent more than 100 genera have been recognized from natural assemblages or fused clusters or have been reconstructed from recurrent groups identified in large collections of discrete specimens. For the most part, the classification adopted in the *Treatise* provides a satisfactory framework within which to assess relationships of the species so reconstructed or recognized. However, discoveries since 1976, when the *Treatise* manuscript was essentially complete, dictate some rearrangement or redefinition of certain categories established in that volume. Personal prejudice dictates others. Thus the classification adopted in this book and discussed in some detail in Chapter 5 will appear, at least superficially, to be quite different from that in the *Treatise*. Some brief comment is thus in order.

First, I have become increasingly skeptical as to the affinities and relationships of the tiny, weakly phosphatized fossils termed protoconodonts and paraconodonts and lumped together in the *Treatise* as the order Paraconodontida. These superficially conodontlike forms are quite different in internal structure from the ones described in Chapter 2, for which Bengtson (1976) has provided the general term "euconodont." Although Bengtson has shown how paraconodonts might have evolved from protoconodonts, neither he nor anyone else has yet documented the steps by which euconodonts might have developed from paraconodonts. Szaniawski (1982), on the other hand, has demonstrated that conical elements of one of the protoconodonts (*Phakelodus tenuis*) are essentially identical in internal structure to the grasping spines of modern arrow worms such as *Sagitta*. Thus, *Phakelodus tenuis*, a protoconodont, may well have been an early representative of the phylum Chaetognatha or of some now-extinct group closely related to the chaetognaths, and not a conodont at all. Szaniawski (1987), whose careful histologic studies are certainly important in this regard, continues to hold that the proto-, para-, and euconodonts (i.e., the Paraconodontida and Conodontophorida of the *Treatise* classification) form parts of a single evolutionary lineage. He does not regard the Conodonta (in which he would include the Paraconodontida) as chactognaths, despite considerable histologic similarities, but as representatives of a group that developed from a common ancestor. It is my conclusion, which is further developed in Chapter 8, that the Conodonta, Paraconodonta, and Chaetognatha are similar but divergent groups, all possibly of phylum rank, that developed in the general radiation of major structural plans that characterized the Cambrian history of the biosphere.

It would be an unjustified extrapolation from *Phakelodus* to assign all 15 genera included by

Treatise authors in the Paraconodontida to the Chaetognatha, but it is equally unjustified to include these primitive forms in the Conodonta merely on the basis of a broad similarity in external form and minor and partly conjectural homologies in histologic details. Because there is as yet no definite connection between the paraconodonts and forms with euconodont elements, I have omitted them from discussion in this book. Should a connection be definitely established through future studies, such as those of Szaniawski (1987), the paraconodonts will probably be best regarded as a separate class.

Stoutly phosphatized coniform elements with the *euconodont*like structure described in Chapter 2 appear first in Upper Cambrian rocks and characterize marine strata until the very end of the Triassic. Genera and species based on such elements appear to represent two rather different major stocks, but they were grouped in the *Treatise* as the order Conodontophorida. Here, however, the Conodonta is divided into 2 classes, Conodonti [1] and Cavidonti, which are further divided into 7 orders, for only 2 of which (Prioniodontida, Ozarkodinida) are order-group names already available [2].

The orders Proconodontida and Belodellida, of the class Cavidonti, include 18 genera of primitive conodonts assigned in the *Treatise* classification to the superfamilies Proconodontacea and Furnishinacea and to the families Pygodontidae, Ansellidae, and Belodellidae, which were included in suprageneric categories here identified as the orders Prioniodontida and Panderodontida. Cavidont species have uni- to multimembrate skeletal apparatuses composed of surficially smooth (or only weakly striated) elements with deep basal cavities. Processes, if developed, rarely extend very far beyond the bases of elements, and denticulation customarily consists only of ragged and mostly inconspicuous serration. The class includes conodonts of Late Cambrian age, the oldest thus far discovered, but was apparently extinct by the end of the Devonian.

The class Conodonti is divided into five orders. The Protopanderodontida is established for 20 genera of conodonts that built uni- or multimembrate apparatuses of surficially striated, but laterally unfurrowed, coniform elements. For the most part, these are the Distacodontacea of the *Treatise* classification, augmented by transfer of a number of genera assigned by *Treatise* authors to other superfamilies and reduced in scope by removal of a few others. This major group of the Conodonta probably includes the ancestors of all other orders of the Conodonti.

Species of the 8 conodont genera included in the Panderodontida built distinctive bi- to quinquimembrate skeletal apparatuses of deeply excavated coniform and rastrate elements, characterized by conspicuous longitudinal furrows or fissures. Panderodontides, which surely developed from an ancestral stock in the Protopanderodontida, appear first in Lower Ordovician rocks and range upward well into the Devonian.

The Prioniodontida of the classification proposed here includes conodonts with skeletal apparatuses that are basically sexi- or septimembrate and include pastinate elements or their platformed analogs in one or both of the P positions. As thus conceived, this major conodont group is similar in scope and content to the superfamily Prioniodontacea of the *Treatise* classification. However, I also include in it an important group (the Oistodontidae) assigned by *Treatise* authors to the Distacodontacea, and I exclude two other groups (Pygodontidae and Pterospathodontidae), which are assigned to the Proconodontida and Ozarkodinida, respectively. Altogether, I have found a place for 43 genera in various families of the Prioniodontida.

The Prioniodinida combines the superfamilies Hibbardellacea and Gondolellacea of the *Treatise* taxonomy and also includes some of the genera included by *Treatise* authors in the superfamily Chirognathacea. Prioniodinide conodonts built sexi- to septimembrate apparatuses with extensiform or breviform digyrate elements in one or both of the P positions. Elements tend to be robust and set with peglike denticles; only late in the history of the order did elements in the P positions develop platformlike extensions or brims. I recognize 34 genera in the 5 families I include in this order.

The Ozarkodinida of the scheme followed here is essentially the superfamily Polygnathacea of *Treatise* authors. This large order includes conodont species characterized by sexi-

or septimembrate apparatuses with carminate and angulate pectiniform elements (or their platformed analogues) in P positions. The ozarkodinides, which appear to have been the most diverse of all the conodonts, did not appear until late in the Ordovician and persisted into only the earliest part of the Triassic. At present, this is the largest conodont order, with 55 genera distributed among 12 families.

Chapter 5 is devoted to a description of the principal components of each of these conodont orders and to a discussion of the distribution and geologic history of these groups. A synopsis of the classification followed here is also included as Appendix A. Apparatuses of typical species of many conodont genera are illustrated at appropriate places in Chapter 5, and illustrated discussions of additional species are to be found in the *Catalogue of Conodonts* (Ziegler, 1973, 1975, 1977, 1981), of which four volumes have been issued to date.

References

Bergström, S. M., and Sweet, W. C. (1966). Conodonts from the Lexington Limestone (Middle Ordovician) of Kentucky and its lateral equivalents in Ohio and Indiana. *Bull. Am. Paleont.* **50**(229), 271–441.

Branson, C. C. (1957). Comment on the Moore/Sylvester-Bradley "Parataxa Plan." *Bull. Zool. Nomencl.* **15**, 169.

Branson, E. B. (1938). Stratigraphy and paleontology of the Lower Mississippian of Missouri, Part I. *Univ. Missouri Studies* **13**(3), 1–208.

———, and Mehl, M. G. (1933). Conodont studies number 1. *Univ. Missouri Studies* **8**(1), 5–72.

———, (1936). Geological affinities and taxonomy of conodonts. *Geol. Soc. Am. Proc. Abstr.* **1935**, 436.

———, (1944). Conodonts, pp. 235–246, pls. 93, 94. In *Index Fossils of North America* (ed. H. W. Shimer and R. R. Shrock). Wiley, New York.

Clark, D. L., Sweet, W. C., Bergström, S. M., Klapper, G., Austin, R. L., Rhodes, F. H. T., Müller, K. J., Ziegler, W., Lindström, M., Miller, J. F., and Harris, A. G. (1981). Conodonta. In *Treatise on Invertebrate Paleontology* (ed. R. A. Robison), Pt. W, Suppl. 2. Geol. Soc. America and Univ. Kansas, 202 pp.

Eichenberg, W. (1930). Conodonten aus dem Culm des Harzes. *Paläont. Z.* **12**, 177–182.

Fager, E. W., and McGowan, J. A. (1963). Zooplankton species groups in the North Pacific. *Science* **140**(3566), 453–460.

Fåhraeus, L. E. (1983). Phylum Conodonta Pander, 1856 and nomenclatural priority. *Syst. Zool.* **32**(4), 455–459.

Hadding, A. (1913). Undre dicellograptusskiffern i Skåne jämte några därmed ekvivalenta bildningar. *Lunds Univ. Årsskr., Avd. 2* **9**(15), 1–90.

Harley, J. (1861). On the Ludlow bone-bed and its crustacean remains. *Q. J. Geol. Soc. London* **17**, 542–552.

Hass, W. H. (1962). Conodonts. Pp. W3–W69. In *Treatise on Invertebrate Paleontology* (ed. R. C. Moore), Pt. W. Geol. Soc. America and Univ. Kansas.

Hinde, G. J. (1879). On conodonts from the Chazy and Cincinnati group of the Cambro-Silurian, and from the Hamilton and Geneseeshale divisions of the Devonian in Canada and the United States. *Q. J. Geol. Soc. London* **35**, 351–369.

Huckriede, R. (1958). Die Conodonten der mediterranen Trias und ihr stratigraphischer Wert. *Paläont. Z.* **32**, 141–175.

Kohut, J. J. (1969). Determination, statistical analysis, and interpretation of recurrent conodont groups in Middle and Upper Ordovician strata of the Cincinnati Region (Ohio, Kentucky, and Indiana). *J. Paleont.* **43**(2), 392–412.

Lindström, M. (1970). A suprageneric taxonomy of the conodonts. *Lethaia* **3**, 427–445.

Pander, C. H. (1856). Monographie der fossilen Fische des silurischen Systems der russisch-baltischen Gouvernements. *Akad. Wiss. St. Petersburg*, 91 pp.

Rhodes, F.H.T. (1951). A classification of Pennsylvanian conodont assemblages. *J. Paleont.* **26**, 886–901.

Schmidt, H. (1934). Conodonten-Funde in ursprunglichem Zusammenhang. *Paläont. Z.* **16**, 76–85.

——— (1950). Nachtrage zur Deutung der Conodonten. *Decheniana* **104**, 11–19.

Scott, H. W. (1934). The zoological relationships of the conodonts. *J. Paleont.* **8**, 448–455.

Smith, J. (1907). On the occurrence of conodonts in the Arenig-Llandeilo formations of the Southern Uplands of Scotland. *Trans. Glasgow Nat. Hist. Soc.*, n. s. **7**(3), 235–252.

Stoll, N. R., Chm., Dollfus, R. Ph., Forest, J., Riley, N. D., Sabrosky, C. W., Wright, C. W., and Melville, R. V., Secretary (1964). International code of zoological nomenclature adopted by the XV International Congress of Zoology. *Int. Trust Zool. Nomencl.*, London, 176 pp.

Sweet, W. C. (1970). Uppermost Permian and Lower Triassic conodonts of the Salt Range and Trans-Indus ranges, West Pakistan, pp. 207–275. In *Stratigraphic boundary problems, Permian and Triassic of West Pakistan* (ed. B. Kummel and C. Teichert). Univ. Kansas, Dept. Geology, Spec. Publ. 4.

Ulrich, E. O., and Bassler, R. S. (1926). A classification of the toothlike fossils, conodonts, with descriptions of American Devonian and Mississippian species. *U. S. Nat. Mus. Proc.* **68**(12), 1–63.

von Bitter, P. H. (1972). Environmental control of conodont distribution in the Shawnee Group (Upper Pennsylvanian) of eastern Kansas. *Univ. Kans. Paleont. Contrib.* **59**, 1–105.

Walliser, O. H. (1964). Conodonten des Silurs. *Abh. hess. Landesamtes Bodenforsch.* **41**, 1–106.

Webers, G. F. (1966). The Middle and Upper Ordovician conodont faunas of Minnesota. *Minn. Geol. Surv. Spec. Publ.* **4**, 1-123.

Ziegler, W., ed. (1973). Catalogue of conodonts, Vol. 1. *E. Schweizerbart'sche Verlagsbuchhandlung*, Stuttgart, 504 pp.

——— (1975). Catalogue of conodonts, Vol. 2. *E. Schweizerbart'sche Verlagsbuchhandlung*, Stuttgart, 404 pp.

——— (1977). Catalogue of conodonts, Vol. 3. *E. Schweizerbart'sche Verlagsbuchhandlung*, Stuttgart, 574 pp.

——— (1981). Catalogue of conodonts, Vol. 4. *E. Schweizerbart'sche Verlagsbuchhandlung*, Stuttgart, 445 pp.

Notes

1. The name Conodonti was introduced by Branson (1938 for a class-group category (subclass Conodonti) that was regarded as a division of the chordate class Pisces. To the best of my knowledge, Branson's Conodonti is the first name used formally for a class-group division, so it is adopted here despite the fact that priority is not extended to names of this group by the *International Rules of Zoological Nomenclature* (Stoll, 1964). As Fåhraeus (1983) has pointed out, the name Conodonta used for the single class of the phylum Conodonta recognized in the *Treatise* (Clark et al., 1981) is an improper and unnecessarily corrected transfer (ex order Conodontophorida, Eichenberg, 1930). Fåhraeus's proposal of Conodontata, which he ascribed to Pander, is unacceptable to me on the same grounds.

2. All 7 of the orders in my revised classification of the Conodonta were subsumed in the *Treatise* classification within the order Conodontophorida, a name proposed in an obscure paper by Eichenberg (1930). Although priority does not extend to order-group names generally, *Treatise* editors have consistently operated as if it did. For this reason, it should have been noted that Macfarlane (1923) anticipated Eichenberg by 7 years with proposal (in a similarly obscure place) of an order Conodontes (recte, Conodonta). Macfarlane's "Conodontes" included all conodonts, however, so it cannot be considered a senior synonym of any of the ordinal names used in my revised classification of the Conodonta. The same applies, of course, to the suborders Neurodontiformes and Conodontiformes proposed on structural grounds by Branson and Mehl in 1944. Taken together, those names are junior synonyms of Macfarlane's "Conodontes," but the categories for which they were proposed are no longer recognized as taxonomically significant divisions of the Conodonta.

5. THE MAJOR CONODONT GROUPS

5.1 Introduction

In Chapter 6 I look into general features in the evolutionary history of the Conodonta. Before attention can profitably be paid to such airy matters, however, some attempt must be made to characterize the major conodont groups, illustrate their typical members, and point out the features that provide internal unity for each group and those that indicate their relations to other major groups.

As noted at the end of Chapter 4, my concept of the Conodonta is limited to the group discussed collectively as the order Conodontophorida in the *Treatise* (Clark et al., 1981). Here, however, that group is elevated greatly in taxonomic status and treated as 2 classes, Cavidonti and Conodonti. These are further divided into 7 orders: the Cavidonti into Proconodontida and Belodellida; and the Conodonti into Protopanderodontida, Panderodontida, Prioniodontida, Prioniodinida, and Ozarkodinida.

5.2 Cavidonti and Conodonti

The conodonts I assign to the class Cavidonti are distinguished by thin-walled, smooth-surfaced elements, primarily coniform, that formed unimembrate to quinquimembrate apparatuses, most of which apparently lacked structures in either of the P positions. Cavidonts made their debut in the Late Cambrian (as *Proconodontus*) and persisted (as *Belodella* and *Dvorakia*) into the Devonian.

The Conodonti, established by Branson (1938) as a subclass of the vertebrate class Pisces but here regarded as a class of another phylum, includes the bulk of the Conodonta. This large and important stock apparently originated in Late Cambrian forms assigned by Miller (1980) to *Teridontus*. Typically, these conodonts formed longitudinally striated elements, mostly with relatively short bases and shallow basal cavities, which were assembled by evolved members of most orders into elaborate sexi- to septimembrate apparatuses with complexly denticulate ramiform elements in S and M positions and a varied assortment of pectiniform elements in the two P positions. The stratigraphic range of the Conodonti is from Upper Cambrian to uppermost Triassic.

Although the Cavidonti seem to have appeared a bit before the Conodonti in the Late Cambrian, the latter diversified almost explosively and were clearly the dominant conodonts throughout the history of the phylum. There is currently no evidence that the Conodonti (as *Teridontus*) are descended from cavidont (e.g., *Proconodontus*) ancestors, which might suggest that the Conodonta are polyphyletic. The possibility remains, of course, that the Cavidonti and Conodonti were derived independently from ancestors in the group assigned to the order Paraconodontida by *Treatise* authors. At present, however, there are reasonable structural grounds for being skeptical about affinities between paraconodonts and conodonts and, on those grounds, I have omitted the Paraconodontida from this discussion (and from the Conodonta). Answers to all of these basic questions are surely lurking in as yet undiscovered or undescribed collections from Cambrian rocks.

5.3 The Proconodontida (Cavidonti) and Its Families (Fig. 5.1)

Species of *Proconodontus* (Fig. 5.2), founding stock of the Proconodontidae and the oldest conodonts known, built an apparently unimembrate skeletal apparatus composed of relatively large, deeply excavated, smooth-surfaced, hyaline coniform elements that have subsymmetrically oval transverse sections and keels on the anterior and/or posterior margins. The marginal keels of some elements are minutely and rather irregularly serrate, and this sort of pattern is one of the features that indicates (to me, at least) that *Proconodontus* was also progenitor of the Belodellida, the second order that I recognize among the cavidonts.

Soon after its Late Cambrian debut, *Proconodontus* spawned *Eoconodontus* (Fig. 5.2), which was established by Miller (1980) for proconodontids that built bimembrate apparatuses of albid *Proconodontus*-like coniform elements, one type with a symmetrical cross section, the other with a carina on one side and a laterally bent cusp. *Proconodontus*, the primitive member of the Proconodontidae, is recognized only in Upper Cambrian rocks. *Eoconodontus*, originally included in *Proconodontus* (Miller, 1969), appeared in the latest Cambrian and continued the lineage into the Early Ordovician.

Cambrooistodus cambricus (Miller), the two known species of *Iapetognathus* Landing, and the several species of *Cordylodus* Pander, are assembled in Fig. 5.1 as the family Cordylodontidae. *Cambrooistodus cambricus* (Fig. 5.2) built a bimembrate apparatus in which one element type is deeply excavated and geniculate and the other is closely similar to the asymmetric element of *Eoconodontus*, the presumed ancestral stock.

Cordylodus species (Fig. 5.2) are also distinguished by bimembrate apparatuses, but the posterior margins of both deeply excavated element types are drawn out into short processes that bear several denticles. These dolabrate elements differ primarily in that the posterior process of one joins the cusp in a smooth curve, whereas that of the other joins the cusp with an acute angle. In short, one is the denticulated homologue of the nongeniculate skeletal component of *Cambrooistodus*; the other is a ram-

Fig. 5.1. The Cavidonti. The order Proconodontida includes the Proconodontidae, Cordylodontidae, Fryxellodontidae, and Pygodontidae. The order Belodellida includes the Belodellidae, Ansellidae, and Dapsilodontidae.

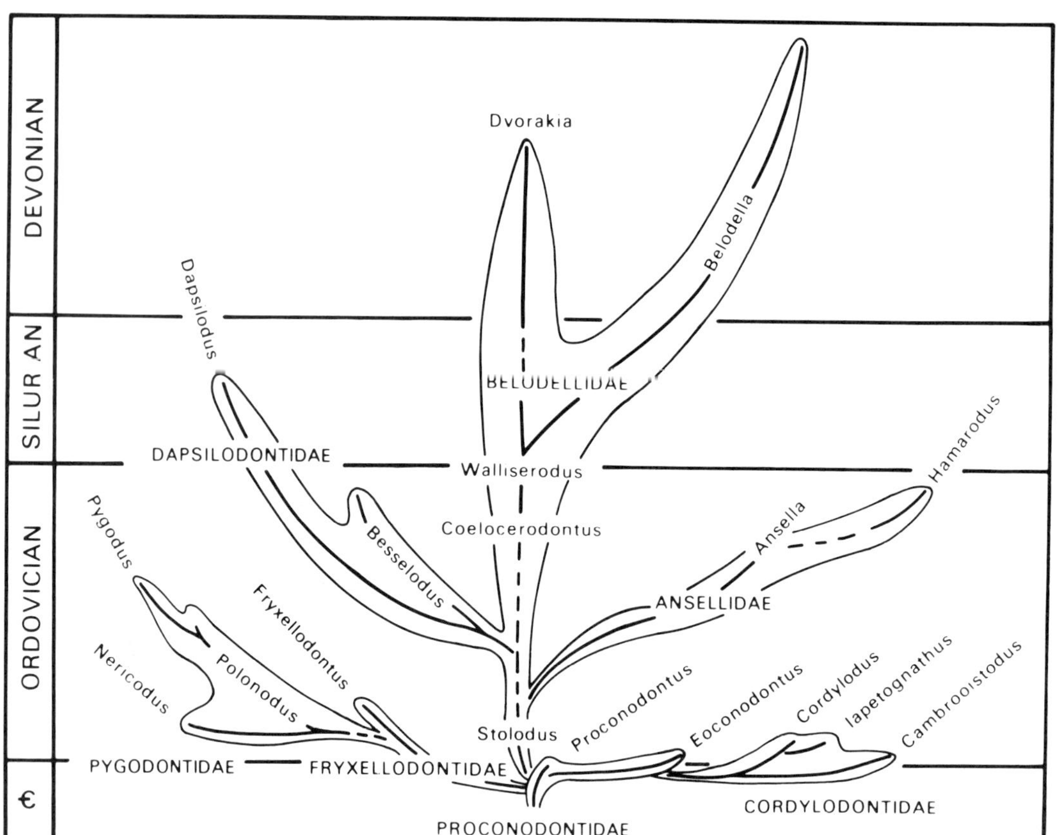

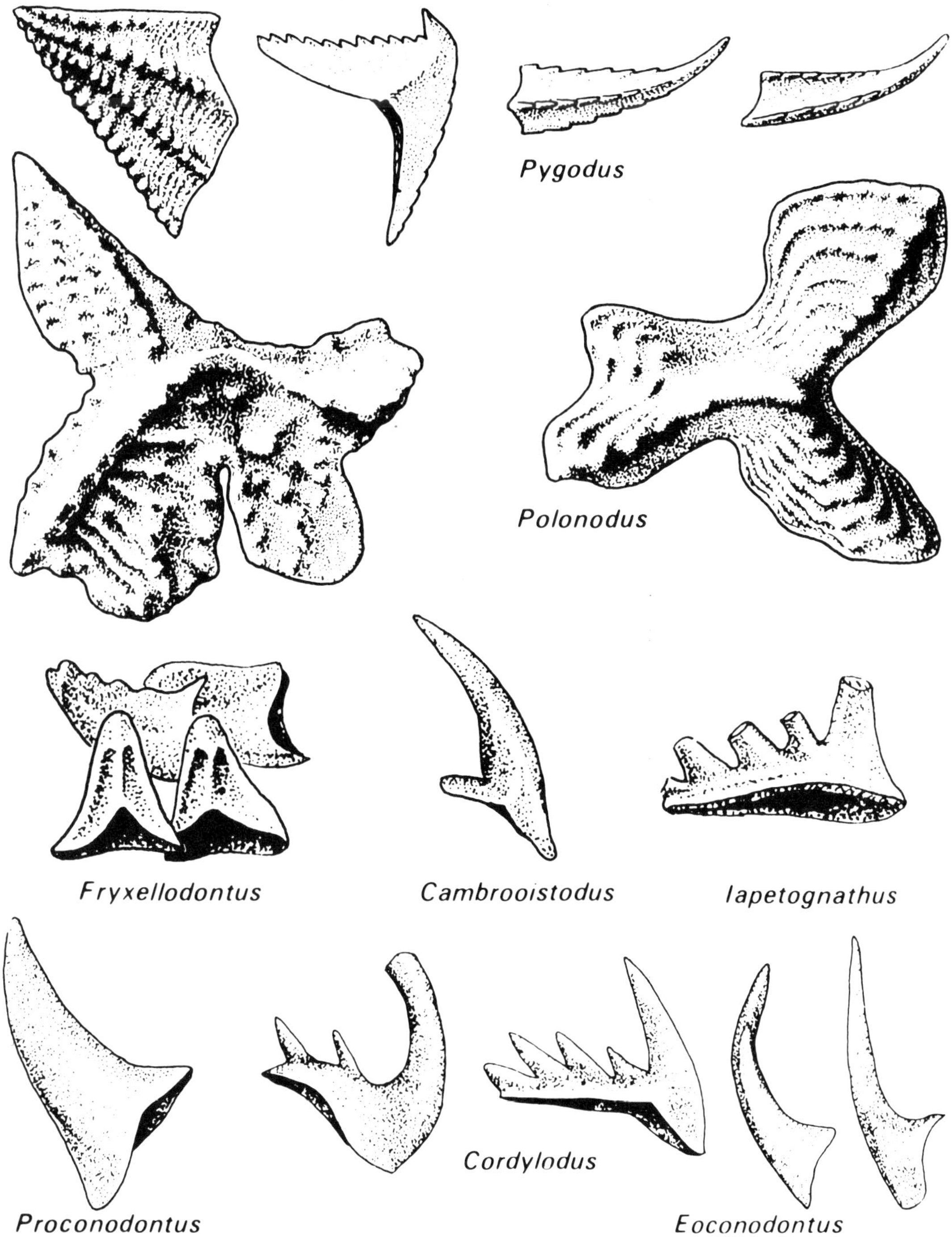

Fig. 5.2. Elements typical of species assigned to various genera of the Proconodontida.

iform version of the geniculate component of *Cambrooistodus* apparatuses. Miller (1980) suggests that *Cambrooistodus* and *Cordylodus* developed independently from *Eoconodontus*. If so, it might be better to include all three genera with *Proconodontus* in a single family, although this would surely complicate diagnosis of the Proconodontidae!

The two known species of *Iapetognathus* (Fig. 5.2) are represented in Early Ordovician

(Tremadocian) strata by ramiform elements that suggest the geniculate dolabrate components of the *Cordylodus* apparatus but apparently differ in orientation. That is, the denticulated process of *Iapetognathus* elements is lateral, not posterior. Landing (Fortey, Landing, and Skevington, 1982) recognizes only a single type of *Iapetognathus* element and thus concludes that the apparatus was unimembrate. The specimens he assigns to *I. preaengensis* exhibit substantial variation, however, so a multimembrate organization cannot be ruled out.

Proconodontus was evidently also the ancestor of *Fryxellodontus* (Fig. 5.2), thus far the sole member of the Fryxellodontidae (Fig. 5.1) and one of the more advanced cavidont conodonts represented in rocks of earliest Ordovician age. Species of *Fryxellodontus* built quadrimembrate skeletal apparatuses of deeply excavated, smooth-surfaced coniform elements that are basically hyaline. Three of the element types form a distinctive symmetry-transition series in which the element I assign to the Sc position is laterally compressed and may have a crudely denticulate posterior margin. A compressed, mitten-shaped element with a broadly rounded apex does not seem to fit in any S position, so it is here assigned to the M slot in the *Fryxellodontus* apparatus. The stratigraphic range of *Fryxellodontus* has not yet been firmly established. Earliest forms are from basal Ordovician strata; however, Serpagli (1974) has tentatively assigned to *Fryxellodontus* a number of specimens from the much younger (but still Early Ordovician) San Juan Formation of western Argentina. These specimens might as well represent an early species of the Pygodontidae, which I regard as probable derivatives of the *Fryxellodontus* stock (perhaps by way of *Nericodus*). Thus the range of *Fryxellodontus* shown diagrammatically in Fig. 5.1 must be taken with a grain of salt.

As indicated in Fig. 5.1, I follow Dzik (1983) and Bergström (1983) in suggesting that Middle Ordovician *Polonodus* and *Pygodus* are related to the *Fryxellodontus* stock of the Cavidonti, perhaps by way of the little-known *Nericodus*. In the *Treatise* (Clark et al., 1981) Bergström created the Pygodontidae for the end members of this stock, to which I also refer *Polonodus* and *Nericodus*.

Nericodus was established by Lindström (1955) for a single incomplete specimen, which Miller's (1980) more recent photographs suggest is a deeply excavated coniform element with a surface ornamented by irregularly distributed nodes that are joined to form equally irregular ridges. Dzik's (1983) illustrations of a couple of additional specimens from the Early Ordovician *Proteus* Zone confirm Miller's interpretation.

Polonodus (Fig. 5.2) includes at least 3 species that formed apparatuses composed of two types of sprawling, deeply excavated coniform elements that lack a distinct cusp and bear surface ornament in the form either of concentric nodose ridges or irregularly distributed ridges and nodes. One element type develops four lobes, two directed anteriorly and two posteriorly, and the pattern of concentric nodose ridges typical of the genus. The other more irregularly ornamented type is basically triradiate, although the three primary lobes may divide into subsidiary ones in elements of various species. There is no evidence as yet that any species of *Polonodus* developed more than a bimembrate apparatus, but none of the species described thus far is based on very many specimens.

Pygodus (Fig. 5.2), which has recently been discussed in detail by Bergström (1983), includes 4 or 5 early Middle Ordovician species that built quadrimembrate apparatuses of specially modified, deeply excavated coniform elements, two of which appear to be homologous with the basically four- and three-lobed elements of *Polonodus* and two of which are deeply excavated pyramidal elements with inconspicuously serrated or minutely denticulated keels on three or four margins. One of these elements, which may be described as quadriramate, is bilaterally symmetrical; the other, termed tertiopedate by Bergström (in Clark et al, 1981), is asymmetric. One of the sprawling, surfically nodose elements in the apparatus of *Pygodus* species is basically trilobate and apparently analogous to just the anterior pair of processes of the quadrilobate element of *Polonodus* species. The other sprawling element of *Pygodus* species is triradiate and possibly related to (although much simpler than) the triradiate element in the apparatuses of *Polonodus* species. Because the pyramidal quadriramate and tertiopedate elements of the *Pygodus* apparatus form a partial symmetry-transition series, they are reasonably assigned to S positions. The sprawling character of the other two elements is like that associated in other conodont species with P-position elements, hence they are assigned to those positions. That also suggests, of course, that the two known elements in the apparatuses of *Polonodus* species might also have been occupants of P positions, although obvious elaboration of the apparatus in the evolution of *Pygodus* from *Polonodus* may merely have made room for additional positions.

5.4 The Belodellida and Its Families (Fig. 5.1)

In Fig. 5.1 (and in the classification outlined in Appendix A), I include three closely related groups of conodonts (Ansellidae, Dapsilodontidae, and Belodellidae) in a newly established order of the Cavidonti for which I propose the name Belodellida. All the conodonts assembled in this new order formed quadri- to quinquimembrate skeletal apparatuses composed of typically thin-walled, smooth-surfaced coniform elements with very deep basal cavities. Those elements commonly have distinct anterior, posterior, and lateral keels or costae that develop fine, needlelike serration or denticulation in the apparatuses of several species.

The central stock of the Belodellida is the Belodellidae. *Treatise* authors assigned *Stolodus, Belodella, Coelocerodontus,* and *Walliserodus* to this family. I add *Dvorakia* (Klapper and Barrick, 1983), which was described and named after the *Treatise* was published. The better-known belodellids formed quadri- to quinquimembrate apparatuses of slender, deeply excavated, smooth-surfaced, basically coniform elements, ornamented by longitudinal keels and costae. The latter bear minute serrations in several species but are undenticulated in others.

The oldest belodellids are included in *Stolodus* (Fig. 5.3), which Lindström (1955) erected for conodonts with deeply excavated coniform elements with three or four asymmetrically disposed longitudinal keels or costae. *Coelocerodontus* (Fig. 5.3), which may not be distinct from *Stolodus,* includes symmetrically costate coniform elements, but the entire apparatus has not yet been reconstructed. *Walliserodus* species (Fig. 5.3) had quinquimembrate apparatuses composed of keeled and costate coniform elements, one a symmetrical form with anterior and posterior keels and interpreted as an occupant of the Sd position. *Dvorakia* (Fig. 5.3) differs from *Walliserodus* primarily in that apparatuses of its two known species apparently lacked elements in the Sd position. *Belodella,* name giver of the family (and the order) (Fig. 5.3), includes species whose skeletal apparatuses are similar to those of *Walliserodus* but are distinguished by S elements with finely denticulated posterior margins.

Species of *Ansella* (Fig. 5.3), principal genus of the Ansellidae, built quinquimembrate skeletal apparatuses with a geniculate coniform element in the M position and an S series of elongate, deeply excavated, surficially smooth ramiform elements with adenticulate or irregularly serrate posterior margins. Elements in the Sa position have three keellike costae and are triangular in cross section; Sb elements have sharp anterior and serrate posterior margins and a planoconvex transverse section; Sc elements are like those in the Sb position in the apparatuses of most known species but are of biconvex cross section and, in a majority of species, have no posterior denticles or serrations.

The apparatus of *Hamarodus* (Fig. 5.3) apparently had a geniculate coniform M element and a symmetry-transition series that included a quadriramate element with a few anterior and lateral denticles in the Sd position, and laterally adenticulate alate, tertiopedate, and dolabrate elements in the Sa, Sb, and Sc positions, respectively. Typical of the *Hamarodus* apparatus, however, are a pair of deeply excavated, laterally compressed elements with serrated postero- and antero-basal margins. Evidently these elements occupied P positions. Affinities of *Hamarodus* are anything but clear. Elements assigned to the S positions are surely not exactly like those in comparable positions in *Ansella,* which also lacks elements in the P positions. But *Ansella* also lacks features that would ally it very obviously with any other major group of conodonts, including the Periodontidae (Prioniodontida), where it was assigned in the *Treatise* (Clark et al., 1981). Thus I follow

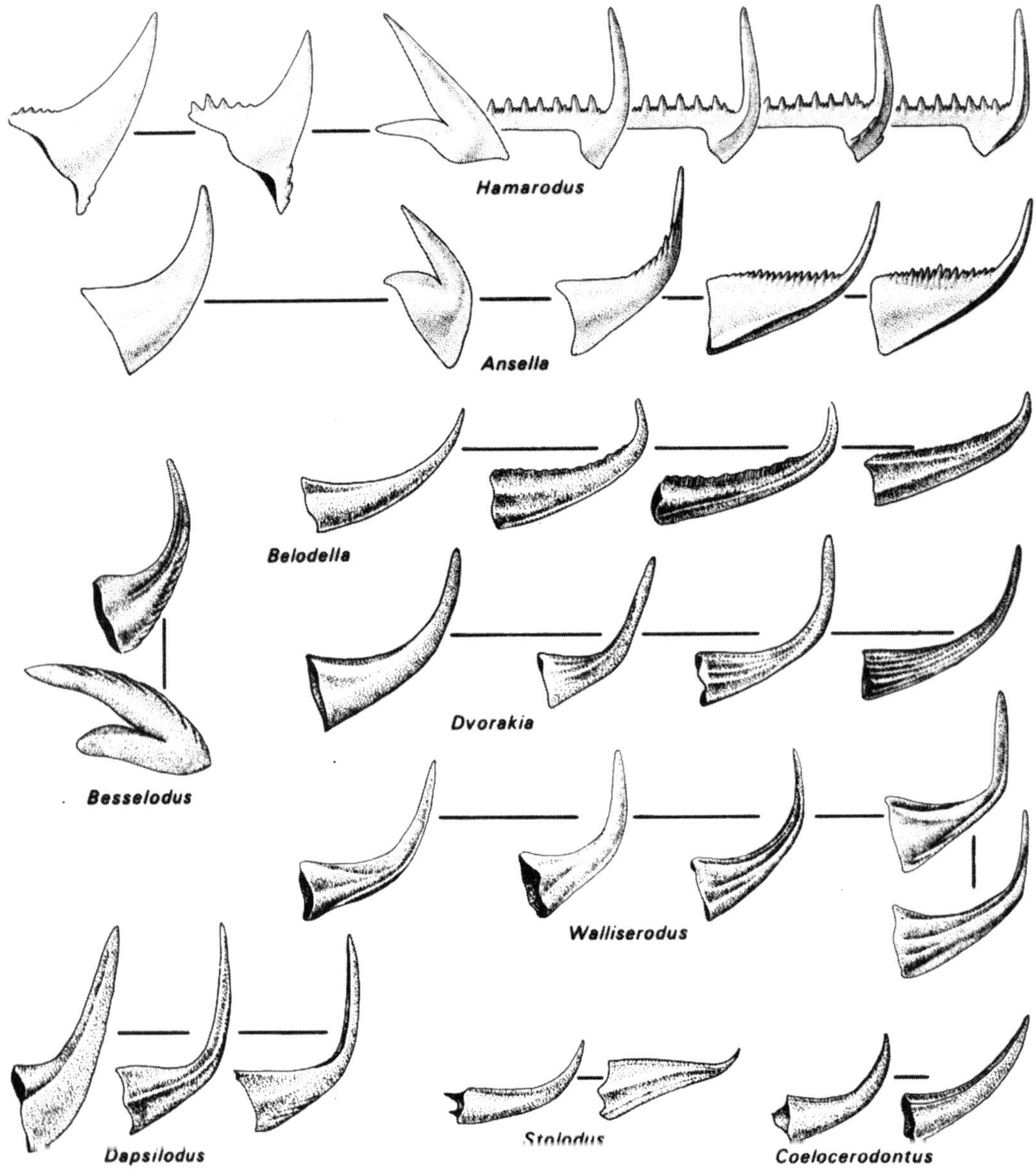

Fig. 5.3. Elements and apparatuses typical of species assigned to various genera of the Belodellida. Components of multimembrate apparatuses joined by heavy horizontal lines; Sa (or Sd) elements on right, followed by Sb, Sc, M, Pb, and Pa elements, if any. Location of elements of *Besselodus* unknown; geniculate coniform element may be M.

Dzik's (1983) vague suggestion that *Hamarodus* might be related to *Ansella* (which, of course, hadn't been named at the time of Dzik's writing).

In Fig. 5.1 I assign *Dapsilodus* and *Besselodus* to a new family, the Dapsilodontidae. Species of the two genera included in this new family formed skeletal apparatuses of deeply excavated, laterally compressed coniform elements that exhibit a distinctive oblique ornamentation adjacent to the anterior margin.

Species of *Dapsilodus* (Fig. 5.3) formed quadrimembrate apparatuses with nongeniculate M elements, whereas M elements in apparatuses of *Besselodus* species (Fig. 5.3) are geniculate. Only two types of elements are known in apparatuses of the latter, but a relationship with *Dapsilodus* is clearly indicated by the distinctive pattern of chevron-shaped ridges adjacent to the anterior margin. At present, *Besselodus* is known only from Ordovician rocks. *Dapsilodus* is also represented in the Ordovician, but it is probably best known and most typically represented in Silurian strata.

5.5 Order Protopanderodontida, New

This division of the Conodonta includes most of the conodonts that built uni- to multimembrate skeletal apparatuses composed of longitudinally striated, laterally unfurrowed coniform elements. Expressly excluded are the panderodontide conodonts, a group that built apparatuses of coniform and rastrate elements with deep lateral furrows; and the oistodontids, whose apparatuses include pastinate coniform (or "acodontiform") elements in an array of coniform types that is otherwise typical of the Prioniodontida.

With the exceptions noted, the Protopanderodontida is the major conodont group identified in the *Treatise* as the superfamily Distacodontacea, which was based on the family Distacodontidae of Bassler (1925). Unfortunately, no one knows much about *Distacodus* Pander, 1856, whose types are presumably lost, or about the Distacodontidae, based on it by Bassler (1925). Consequently, I suggest that *Distacodus* be regarded as a *nomen dubium* and that it not be used as the basis for any supra-generic taxonomic units. The name Protopanderodontida, although a mouthful, is taken

Fig. 5.4. The Protopanderodontida and Panderodontida. The families Protopanderodontidae, Clavohamulidae, Acanthodontidae, and Drepanoistodontidae make up the Protopanderodontida. The Panderodontida consists of the single family Panderodontidae.

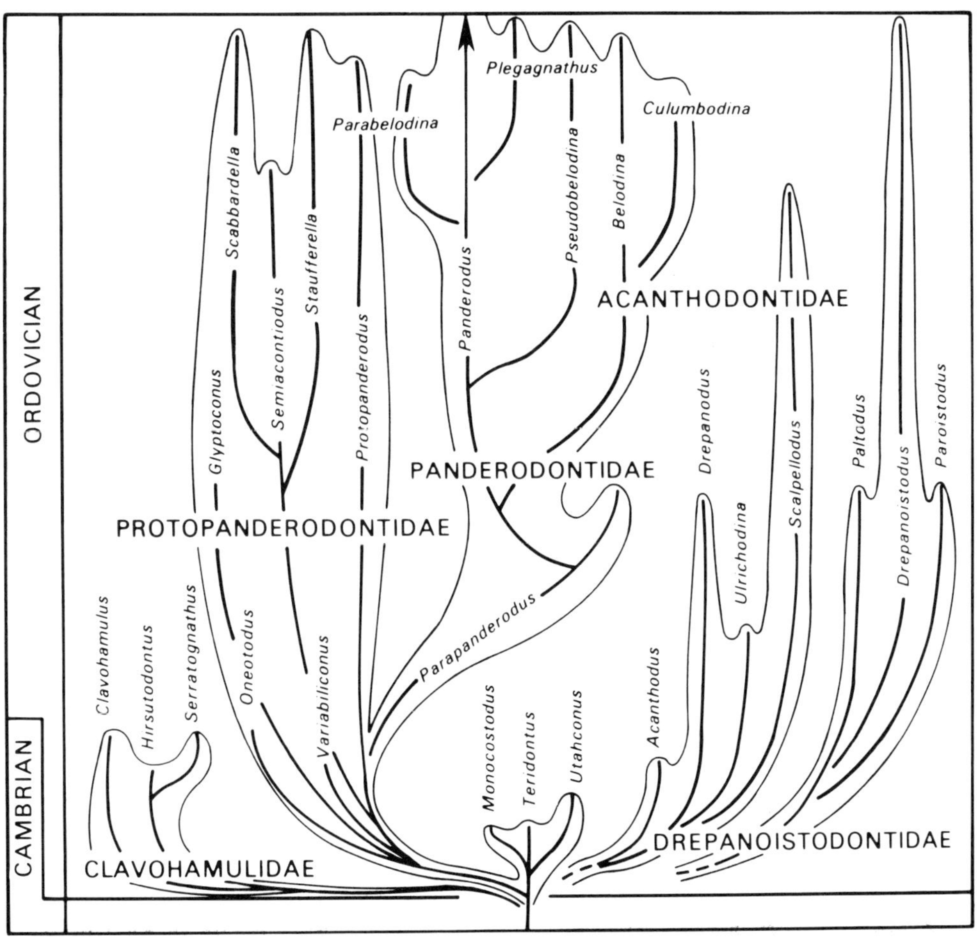

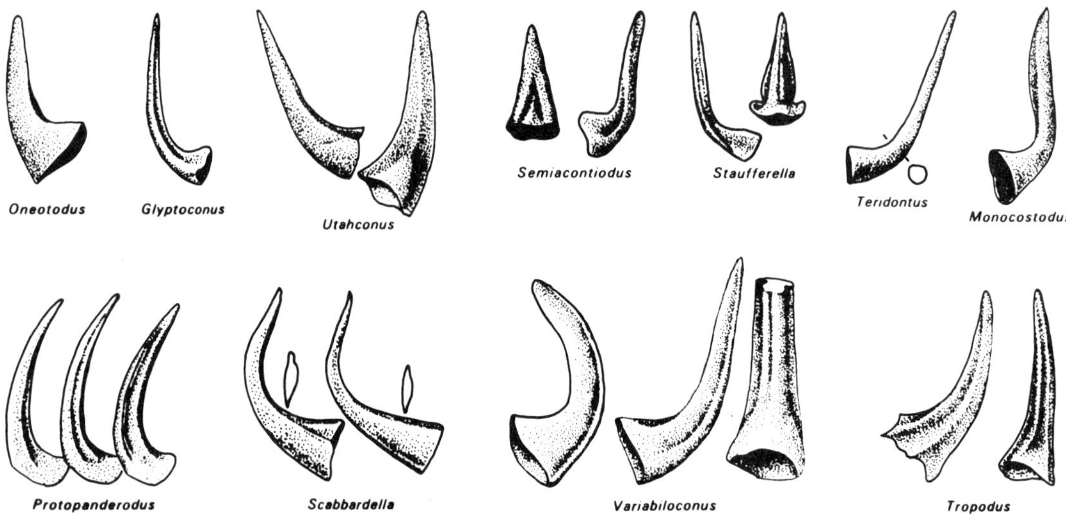

Fig. 5.5. Elements and apparatuses typical of species assigned to genera of the Protopanderodontidae.

from the name of the largest and probably most distinctive family of conodonts with coniform-element apparatuses, the Protopanderodontidae.

The stratigraphic distribution and presumed relationships of the principal genera of the Protopanderodontida are shown diagrammatically in Fig. 5.4.

5.5.1 Family Protopanderodontidae Lindström, 1970

Assigned to this family are conodonts included by *Treatise* authors in the families Teridontidae, Oneotodontidae, Protopanderodontidae, and Scolopodontidae. These conodonts formed uni- to multimembrate skeletal apparatuses of typically (but by no means exclusively) albid, nongeniculate coniform elements with finely striated surfaces. Faces of elements are unornamented in the simpler members of the family but may be marked by longitudinal carinae, costae, or ridges in more advanced or specialized members of the family.

Teridontus Miller, 1980 (Fig. 5.5), which includes the oldest protopanderodontides, formed a unimembrate apparatus of erect or somewhat reclined coniform elements of circular to slightly elliptical cross section that are almost completely white and thus properly described as *albid*. Species of *Monocostodus* Miller, 1980 (Fig. 5.5) built a bimembrate apparatus of albid coniform elements like those of *Teridontus* species but with a sharp, narrow costa on, or on one side or another of, the distal two thirds of the posterior margin.

Oneotodus Lindström, 1955 (as redefined by Ethington and Brand, 1981) (Fig. 5.5) includes conodont species whose apparatuses were composed of stout, albid coniform elements of circular to slightly elliptical cross section. These elements may bear a slight indentation on the posterior margin at the point of maximum cusp curvature, and some form a few low, rather coarse longitudinal costae. As currently defined, *Oneotodus* seems intermediate between *Teridontus* and *Semiacontiodus*; I am not sure the slight differences merit separation at the generic level.

Pseudooneotodus Drygant, 1974 has a bimembrate skeletal apparatus of deeply excavated, albid coniform elements with relatively smooth surfaces. One element type is short and stout and has walls that thicken apically; the other element type is smaller and more slender than the first, but otherwise similar morphologically. The stout elements of *Pseudooneotodus* also develop one to three nodelike apical denticles, and these are especially diagnostic of the genus. Except for an anomalous lower Middle Ordovician occurrence reported by Stouge (1984), *Pseudooneotodus* appears in Upper Ordovician rocks and ranges upward into the Devonian. Its first occurrence is thus separated by a considerable stratigraphic interval from *Oneotodus*, its closest morphologic relative, and this suggests that ancestry and possibly familial assignment of *Pseudooneotodus* may be elsewhere.

Species of *Semiacontiodus* (Fig. 5.5) built bimembrate apparatuses. One element type is bilaterally symmetrical, anteroposteriorly compressed, and has a posterior indentation or costa flanked by posterolateral costae; the other element is asymmetric and has a costa on only one side. Sometime early in the Middle Ordovician, *Semiacontiodus* gave rise to *Staufferella* (Fig. 5.5), whose species formed a bi- or trimembrate apparatus that differed from those formed by *Semiacontiodus* species in that the basal part of the finlike lateral costae is distinctly notched. Dzik (1983) suggests that *Semiacontiodus* was also ancestral in the Middle Ordovician to *Scabbardella* (Fig. 5.5), whose only known species formed a bimembrate apparatus of strongly compressed elements, one type of which is essentially symmetrical and has anterior and posterior costae; the other is asymmetric and more broadly costate on one side than the other.

Shortly after *Semiacontiodus* made its debut in the very early Ordovician, it apparently spawned a stock of protopanderodontide conodonts characterized by a multimembrate apparatus of albid, costate, and noncostate coniform elements. Landing, Barnes, and Stevens (1986) have recently coined the generic name *Variabiloconus* for the best known of these conodonts, *V. bassleri* (Furnish, 1938) (Fig. 5.5). In Fig. 5.4 I suggest that *Variabiloconus* may have been ancestral to *Protopanderodus* (Fig. 5.5), a long-ranging stock of Ordovician conodonts with a bi- to trimembrate skeletal apparatus composed of albid elements with prominently sharp-edged posterior and posterolateral costae. In the very latest phase in the history of the *Protopanderodus* stock, the upper margin of the base was extended posteriorly and came to support a single large denticle.

Species of *Tropodus* Kennedy (1980) (Fig. 5.5), like those of *Protopanderodus*, formed multimembrate apparatuses of deeply excavated symmetrical and asymmetrical coniform elements, all of which bear three to five prominent keellike costae. Repetski (1982) describes the components of multielement *Tropodus comptus*, the type species, as form species of *Paltodus* and *Scolopodus*, which suggests that there is still uncertainty about *Tropodus* in Lower Ordovician circles.

Semiacontiodus, represented in earliest Ordovician rocks by a number of species previously assigned to *Acontiodus*, was probably also ancestral to the long-ranging Early Ordovician stock named *Glyptoconus* by Kennedy (1980). The only species of *Glyptoconus* (Fig. 5.5) that has been adequately described, *G. quadraplicatus* (Branson and Mehl, 1933), has a multimembrate apparatus of hyaline coniform elements that have only slightly expanded bases and slightly recurved cusps that are quadrate in cross section and bear prominent longitudinal grooves on the lateral and posterior faces.

A particularly important event in the history of the Protopanderodontidae was development in the earliest Ordovician of the short-lived stock named *Utahconus* by Miller (1980). Species of *Utahconus* (Fig. 5.5) formed a bimembrate skeletal apparatus of albid elements, one of which is a primitive pastinate (or acodontiform) coniform element. *Utahconus* probably was ancestral to *Rossodus*, the oldest described member of the Prioniodontida. *Utahconus* may also have been the ancestor of the Acanthodontidae, another of the protopanderodontide families. *Parutahconus* Landing (in Fortey et al., 1982) is similar, but its elements may bear small nodes on the base and along the lateral costa.

5.5.2 *Family Clavohamulidae Lindström, 1970*

This family was established for a small group of bizarre protopanderodontide conodonts that formed unimembrate skeletal apparatuses of albid elements that are surficially granulose, nodose, or spinose. Species of *Hirsutodontus* (Fig. 5.6) built apparatuses of *Teridontus*-like coniform elements that developed spines or nodes at the base. The basal part of comparable elements of *Clavohamulus* (Fig. 5.6), however, spread laterally to form a hemispherical mound with a nodose upper surface. Elements of *Serratognathus* (Fig. 5.6), on the other hand, are anteroposteriorly compressed, bilaterally symmetrical structures with a knifelike lower

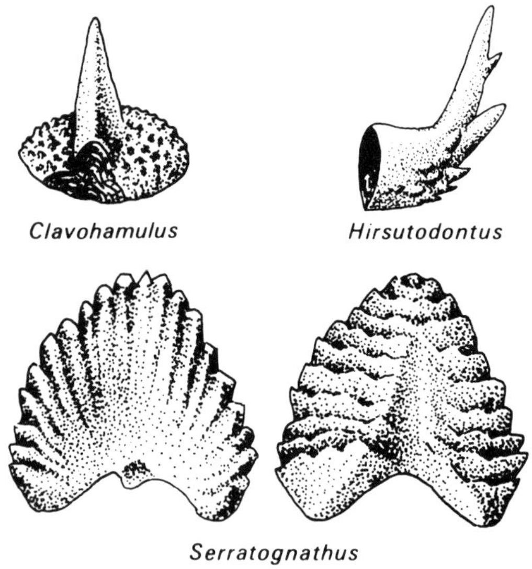

Fig. 5.6. Elements typical of species assigned to genera of the Clavohamulidae.

margin and an anterior face with numerous rows of short closely spaced nodes.

5.5.3 Family Acanthodontidae Lindström, 1970

This division of the Protopanderodontida, for which the name Acanthodontidae has precedence, includes *Drepanodus* (Fig. 5.7), *Cornuodus* (Fig. 5.7), *Ulrichodina* (Fig. 5.7), *Acanthodus* (Fig. 5.7), *Parapaltodus*, and probably *Scalpellodus* (Fig. 5.7). Species of all these genera formed basically bimembrate skeletal apparatuses of nongeniculate coniform elements that are weakly albid or hyaline. One element is compressed, bilaterally symmetrical, and commonly has sharp anterior and posterior margins; the other is similar, but the cusp is twisted laterally with respect to the base, which thus opens to one side. In 1955 Lindström established the form genus *Scandodus* for elements such as the latter, so it is convenient to refer to them as *scandodontiform*. The type form species of *Scandodus*, however, may have been a component of the apparatus of a species of *Ulrichodina* (Dzik, 1983) and hence is not listed separately as a member of this family.

Fig. 5.7. Elements typical of species assigned to genera of the Acanthodontidae.

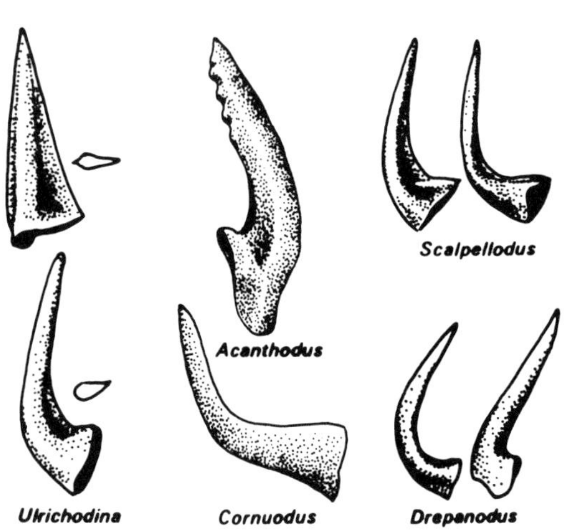

Elements of species assigned here to the Acanthodontidae differ from one another in only minor ways. Those of *Acanthodus* species, for example, commonly develop a row of nodes along the posterior margin, whereas an anterior segment of the basal margin of the bilaterally symmetrical elements of *Ulrichodina* species is crimped inwardly to form a distinctive notch that is characteristic of the genus. The family is confined to the Ordovician.

5.5.4 Family Drepanoistodontidae Fåhraeus and Nowlan, 1978

Species of multielement *Paltodus*, *Paroistodus*, and *Drepanoistodus* form the core of an important group of protopanderodontide conodonts for which many authors have used the name Distacodontidae Bassler, 1925. As noted earlier in this chapter, however, *Distacodus* is best regarded as a *nomen dubium* and thus is not suitable as the type of any suprageneric categories.

Paltodus (Fig. 5.8), which was apparently the rootstock of the Drepanoistodontidae, has a quinquimembrate apparatus with a complete symmetry-transition series, a distinctive geniculate coniform element in the M position, and an undifferentiated group of asymmetrically bicostate nongeniculate elements in the P positions. The anterior segment of the basal margin of alate Sa elements may be flattened, or slightly inturned, and this character suggests *Ulrichodina* or some other member of the Acanthodontidae as the likely ancestor.

Species of *Drepanoistodus* (Fig. 5.8) had a skeletal apparatus that is commonly described as bi- or trimembrate but should probably be

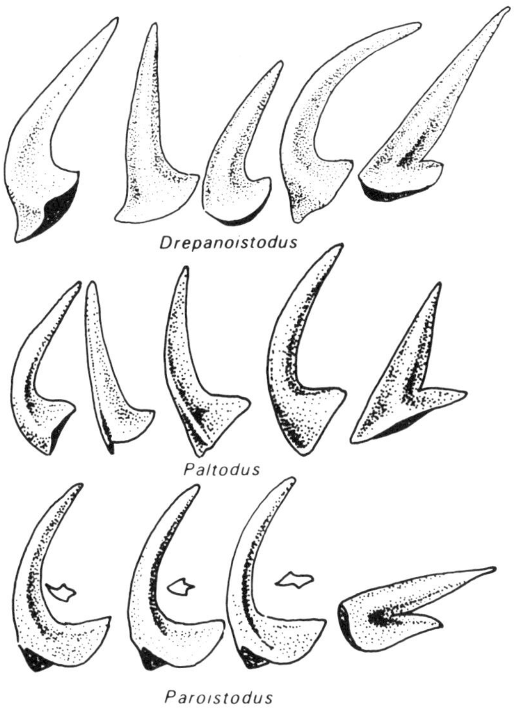

Fig. 5.8. Elements and apparatuses typical of species assigned to genera of the Drepanoistodontidae.

considered quinqui- or even seximembrate. An essentially erect, nongeniculate element with equally concave sides and sharp anterior and posterior margins occupies the Sa position. Sb and Sc positions in the transition series are filled by regularly recurved nongeniculate coniform elements, which are flatter on one side than the other and commonly have basal margins that are different in conformation. M elements are geniculate coniform elements, commonly with long cusps and much shorter bases. P positions were evidently occupied by subsymmetrical nongeniculate coniform elements with an acutely angular anterobasal corner and sides that are almost equally convex. Certain species of *Drepanoistodus* attained an essentially cosmopolitan distribution in the Ordovician, but the genus became extinct at the end of that period.

Paroistodus (Fig. 5.8) species had a bi- to quadrimembrate apparatus that included a symmetry-transition series of laterally costate nongeniculate coniform elements and a geniculate coniform element in the M position that characteristically develops a conspicuous zone of recessive basal margin at the anterobasal corner. Nongeniculate elements that form the symmetry-transition series in the apparatus of *Paroistodus* species tend to develop bases whose upper margins are drawn out posteriorly. Dzik (1983) and Stouge (1984) have both suggested that this process became denticulated in the youngest species of the genus, *P. horridus* (Barnes and Poplawski, 1973).

5.6 The Panderodontida

Species assigned to this distinctive division of the Conodonta are distinguished by tri- to septimembrate skeletal apparatuses composed of longitudinally striated, laterally furrowed coniform or rastrate elements. The latter are basically coniform types but have a few denticles along the posterior margin of the cusp. In only a few apparatuses (e.g., that of *Plegagnathus*) are denticulated parts of any elements drawn out into distinct processes.

In my opinion the major diagnostic feature of the Panderodontida is the distinctive longitudinal fissure or furrow that is developed on one or both sides of every element of the skeletal apparatus, and I have excluded from the order all of the conodonts whose elements lack indication of this feature. This means, of course, that my concept of the Panderodontida is considerably narrower than that adopted in the *Treatise*.

In the narrow sense adopted here, the Panderodontida includes *Panderodus*, *Parapanderodus*, *Pseudobelodina*, *Belodina*, *Culumbodina*, *Plegagnathus*, *Parabelodina*, and *Neopanderodus*. *Pseudopanderodus* (Landing, 1979) is excluded because the ostensibly latest Cambrian coniform elements on which it was based have subsequently been determined to be representatives of *Panderodus* that were somehow mixed into collections from much older rocks.

The oldest panderodontide species known had apparatuses composed of furrowed coniform elements, and it appears likely that they were ancestral to species of genera such as *Pseudobelodina*, *Belodina*, *Culumbodina*, and *Plegagnathus*, whose skeletal apparatuses were formed entirely of rastrate elements. Consequently, at first blush, it would seem feasible to assign panderodontide conodonts with apparatuses of coniform elements to one family, the Panderodontidae, and those with apparatuses of rastrate elements to another, as yet unnamed, family. However, there is reason to believe that species with rastrate-only apparatuses may not make up a monophyletic lineage. That is, species such as *Parabelodina denticulata*, with an apparatus of distinctive rastrate elements, shared the sea briefly in the Late Ordovician with *Panderodus bergstroemi*, which had an apparatus of furrowed coniform elements that are otherwise identical to the rastrate elements of *P. denticulata*. This suggests to me that *Panderodus bergstroemi*, not a species with rastrate skeletal elements, is the likely ancestor of *Parabelodina denticulata*, and that a family created for panderodontide conodonts with rastrate skeletal elements would be polyphyletic. Consequently, I assign all species currently recognized in the order Panderodontida to a single family, the Panderodontidae, which I expect to be subdivided as our knowledge of this puzzling group enlarges.

5.6.1 Family Panderodontidae Lindström, 1970

The oldest conodonts that seem to be referable to the Panderodontidae appeared very early in the Early Ordovician (Deltifer Zone). These conodonts have commonly been regarded as representatives of *Scolopodus gracilis*, a species based by Ethington and Clark (1964) on specimens determined as *Drepanodus striatus* by Graves and Ellison in 1941 but recently transferred to *Parapanderodus* by Stouge (1984). *P. striatus* (Fig. 5.9) had a bimembrate skeletal apparatus of hyaline or only weakly albid coniform elements, one a long, slender, gracefully curved nongeniculate cone of circular cross section, the other an asymmetric nongeniculate cone. Surfaces of both types bear fine longitudinal striations, and forms of circular cross section have a deep furrow in the posterior face.

Parapanderodus striatus was joined later in the Early Ordovician (Evae Zone) by another species with an apparatus composed of at least three types of conspicuously striated, but inconspicuously furrowed, coniform elements, which were made the types of *Protopanderodus asymmetricus* by Barnes and Poplawski (1973). That species was transferred to *Juanognathus*

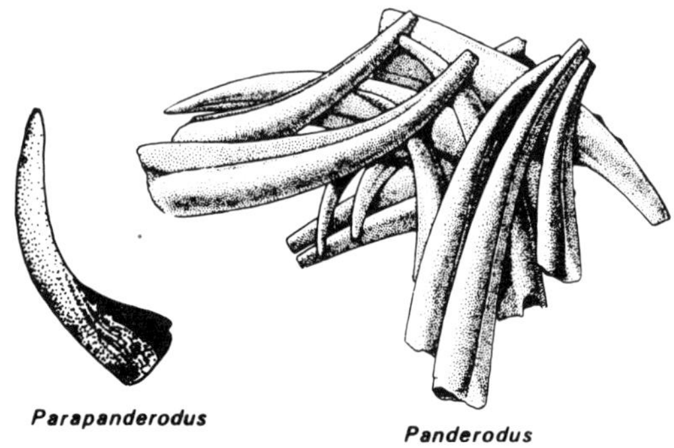

Fig. 5.9. Elements typical of the apparatuses of *Parapanderodus* and *Panderodus*. The fused cluster of *Panderodus* elements is from the Silurian of Podolia. From Dzik and Drygant (1986).

by Bergström (1979), reassigned to *Semiacontiodus* by Stouge (1984) and, most recently, treated as a representative of *Parapanderodus* by Dzik and Drygant (1986). That assignment may also be questioned, but examination of the many specimens of this species in available collections has convinced me that it is surely a panderodont and the fact that it shares some elements in common with species of *Parapanderodus* makes that as good a home for it as any.

Dzik and Drygant (1986) note that *Parapanderodus asymmetricus* (Barnes and Poplawski, 1973) had an apparatus that was at least trimembrate and included a longitudinally striated, alate coniform element with a broad posterior costa bordered on either side by an inconspicuous furrow; a slender, subsymmetrical coniform element with a furrow on only one side of the posterior costa; and an asymmetric coniform element that is conspicuously flattened in a plane normal to the anterior-posterior axis. The latter type of element, which is typical of *P. asymmetricus*, appears to lack a furrow in all published illustrations, but it clearly has one along the left margin of the mid-posterior costa. [It is amusing to note that that side is obscure in all figures of this element; however, a notch in the basal margin in the position of the furrow is clearly visible in figures by Barnes and Poplawski, Stouge, and Ethington and Clark (1982).]

Additional species of *Parapanderodus* are represented in rocks of early Middle Ordovician age. The apparatus of those species apparently contained only two types of slender coniform elements, both of which exhibit a "panderodont furrow" (or a pair of them) on the posterior face. The basal cavity of these elements seems to be somewhat deeper than that in elements of *P. asymmetricus* but much shallower than cavities in elements of younger panderodontid conodonts.

Panderodus (Fig. 5.9) made its debut a bit later in the early Middle Ordovician (Llanvirnian or Whiterockian). According to Löfgren (1978), the apparatus of the earliest species, *P. sulcatus* (Fåhraeus), had a bimembrate apparatus of slender (or "graciliform") and laterally compressed (or "compressiform") coniform elements that are longitudinally striated and have basal cavities about one third of the total element in length. Elements of both types appear from Löfgren's illustrations to be slightly bowed, and the distinctive panderodont furrow is situated on the outer, or convex, side.

Later Ordovician species of *Panderodus* were determined from large collections of discrete elements to have had quinquimembrate skeletal apparatuses (Sweet, 1979), and this is also the case in fused clusters representing *Panderodus unicostatus* (Branson and Mehl) from the Early Silurian of Podolia, which were recently described in detail by Dzik and Drygant (1986). Although certain of the elements in the apparatus I reconstructed in 1979 are nearly bilaterally symmetrical, they are furrowed on only one side and were described only as "similiform." Jeppsson (1983) reports that there are truly symmetrical forms in his collections of *Panderodus* elements from the Silurian of Gotland, and a few have also been discovered by Britt Leatham, one of my students, in the large collections on which I based my 1979 reconstruction. Thus it is appropriate to conclude that the apparatus of fully evolved species of *Panderodus* was probably seximembrate although, as Dzik and Drygant (1986) note, distinctions between elements forming the S series are commonly difficult to make.

Panderodus ranged through the Silurian into the Devonian, but information on the species represented is sketchy. Indeed, specimens are illustrated in only a few reports on Devonian faunas and Silurian forms have received less than adequate taxonomic treatment.

The youngest members of the Panderodontidae are late Early and Middle Devonian species of *Neopanderodus*, which apparently had apparatuses like those of better-known species of *Panderodus* but composed of furrowed coniform elements whose sides are marked by longitudinal striae that are more prominent than those of *Panderodus* elements.

The oldest rastrate elements known are from early Middle Ordovician rocks of about the same age as the ones from which the earliest *Panderodus* elements have been collected. Opinions vary as to the genus represented by these elements, but I regard them as specimens of *Belodina*. Later, better-known species of *Belodina* (Fig. 5.10) formed trimembrate apparatuses characterized by three different types of rastrate elements, which are separated primarily by differences in their radii of curvature and the number of denticles on the posterior margin. One element lacks posterior denticles and is formally coniform.

In 1979 I described a sequence of species of *Belodina* and noted that in younger and younger Ordovician strata denticulate rastrate elements become more numerous and coniform ones proportionately less so. However, one Late Ordovician species, *B. calciprominens*, is known only from large coniform rastrate elements of the sort once included in the form genus *Eobelodina*. Although specimens assignable to *Belodina* are known from rocks of latest Ordovician age, none has yet been recovered from the Silurian.

Pseudobelodina (Fig. 5.10) has a quadrimembrate skeletal apparatus composed of deeply excavated, laterally furrowed, posteriorly "heeled," anterolaterally costate rastrate elements. Three of these element types are bowed to the unfurrowed side and form a symmetry-transition series; the fourth, which may have occupied either the M or one of the P positions, is straight or is bowed slightly toward the furrowed side. The oldest *Pseudobelodina* known to me is an as yet undescribed species from Middle Ordovician (Blackriveran) strata in North America. Although elements of this unnamed species are all denticulate rastrate forms and thus morphologically similar to comparable elements in the apparatus of *Belodina* species, there is no coniform rastrate (or "eobelo-

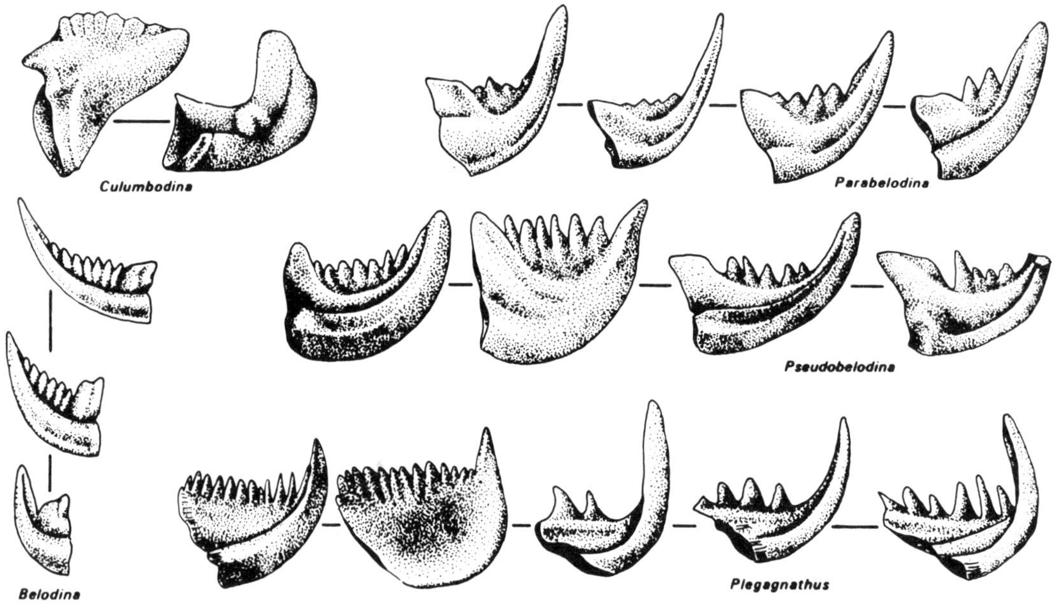

Fig. 5.10. Rastrate-element apparatuses typical of species assigned to various genera of the Panderodontidae.

diniform") element, and I have not seen collections in which there is intergradation in form between elements of this *Pseudobelodina* species and those of *Belodina monitorensis* and *B. compressa*, which are represented in the same rocks. Consequently, I think it likely that *Pseudobelodina* developed directly from *Panderodus* and not from *Belodina*, which may have originated from the *Panderodus* stock much earlier in the Middle Ordovician.

Plegagnathus (Fig. 5.10), which is known thus far only from Upper Ordovician strata in western North America, has a skeletal apparatus of five morphologically distinct rastrate elements, one bowed toward the furrowed side, one bowed to the unfurrowed side, and three laterally costate, discretely denticulated ones that form a symmetry-transition series. *Parabelodina* (Fig. 5.10), also known only from Upper Ordovician rocks in western North America, formed a quinquimembrate apparatus broadly similar to that of *Panderodus* but with components that are heeled rastrate, not adenticulate coniform, elements.

Plegagnathus and *Parabelodina* built the most elaborate rastrate-only apparatuses of all the panderodont conodonts. Their apparatuses are not anticipated in those of species of more long-ranging *Belodina* or *Pseudobelodina* and, as I have already pointed out, the *Parabelodina* apparatus is closely related morphologically to that of contemporaneous *Panderodus bergstroemi*. Hence I suspect that *Plegagnathus* and *Parabelodina* developed directly and independently from *Panderodus*, whose species all have quinqui- or seximembrate skeletal apparatuses.

Culumbodina (Fig. 5.10), as I interpret it, had a bimembrate skeletal apparatus of heeled, smooth-sided, or longitudinally multicostate rastrate elements of two types, both bowed toward the unfurrowed side. Elements that represent early stages of growth have denticles along most of the posterior margin, including the heel; in later growth-stage elements, most of those denticles are overgrown, and they may be completely absent in elements representing the latest ontogenetic stages. Origins of *Culumbodina* are not at all obvious from published information about the genus or its contemporaries. The genus probably appeared a bit earlier in the Siberian Ordovician than in North America, where it made its debut in the early part (Edenian) of the Late Ordovician. Contemporary panderodonts included *Belodina*, which must be considered the most logical ancestral group.

Panderodus and some of the panderodonts with rastrate-element apparatuses seem to have reached the acme of their development and diversity in shallow, tropical, or subtropical seas. *Panderodus* itself has a somewhat more cosmopolitan distribution, but it seems largely to have been replaced in higher-latitude Ordovician faunas by *Drepanoistodus* and, farther poleward, *Walliserodus*. Specialized panderodontids, with rastrate-element apparatuses, were primarily low-latitude forms.

5.7 The Prioniodontida Dzik, 1976

This major division of the Conodonta (Fig. 5.11) includes conodonts in which one or both of the P positions in the fundamentally sexi- or septimembrate cephalic apparatus are occupied by pastinate coniform or pectiniform elements or their platformed equivalents. In a few spe-

Fig. 5.11. Ordovician and earliest Silurian families and genera of the Prioniodontida. Later Silurian and Devonian representatives of the order are shown in Fig. 5.17.

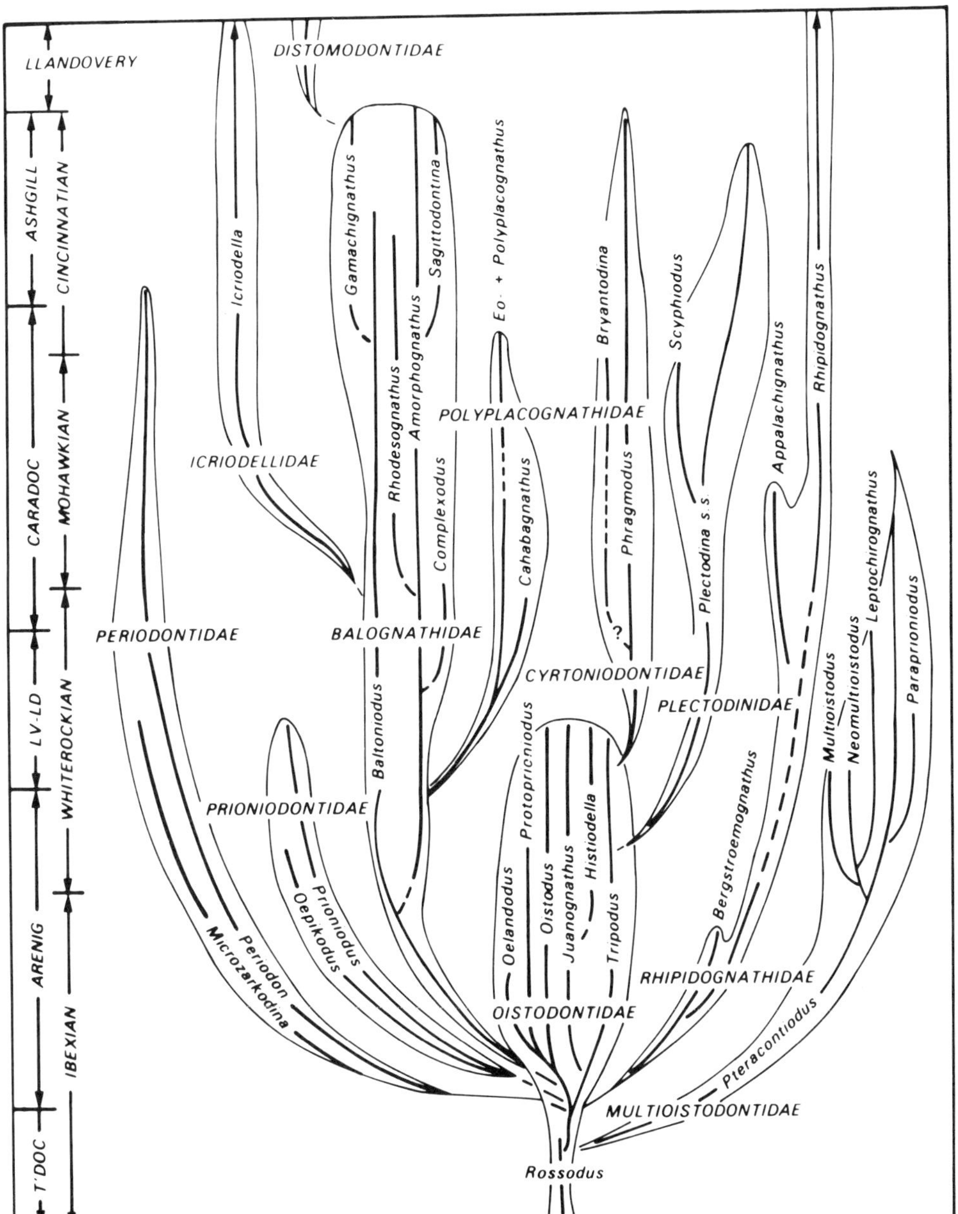

cialized prioniodontides, no mineralized elements seem to have been formed in S or M positions. In several others P positions are occupied, at least at maturity, by pectiniform elements that are formally carminate or angulate but can be shown (or reasonably inferred) to have arisen through ontogenetic or phylogenetic modification of pastinate elements.

The oldest species with a prioniodontide skeletal apparatus are currently the two assigned by Repetski and Ethington (1983) to *Rossodus* (Fig. 5.12), which are commonly represented in rocks of Early Ordovician age at numerous localities in North America, Korea, Australia, Iran, and Siberia. Slightly younger species with closely similar skeletal apparatuses are assembled here in *Tripodus* (Fig. 5.12), but they have been assigned by others to "*Acodus*" or *Diaphorodus*. These species are also represented by widespread, well-documented Lower Ordovician occurrences and appear to be ancestral to a number of distinctive stocks. The youngest Prioniodontida are probably several Late Devonian species of "*Icriodus*" that Sandberg and Dreesen (1984) conclude were derived independently from *Pelekysgnathus* and probably represent an unnamed genus. As noted in my discussion of the Icriodontidae, however, the relationship between Ordovician and Early to Mid-Silurian icriodellids and later Silurian and Devonian icriodontids is anything but clear, and Fåhraeus (1984) has suggested a rather different phylogeny.

5.7.1 *Family Oistodontidae Lindström, 1970*

In this family I include conodonts with a quadri- to quinquimembrate skeletal apparatus

Fig. 5.12. Elements and apparatuses typical of species assigned to various genera of the Oistodontidae. Drawings of *Juanognathus* elements from Serpagli (1974).

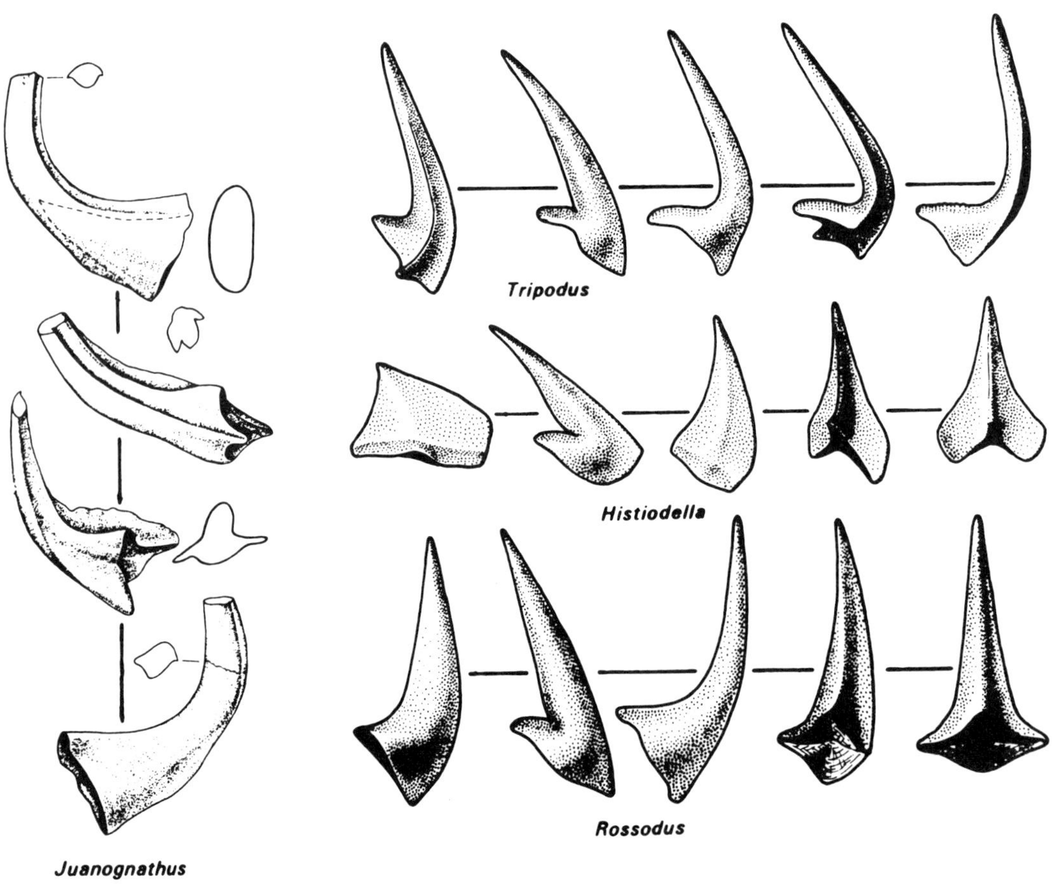

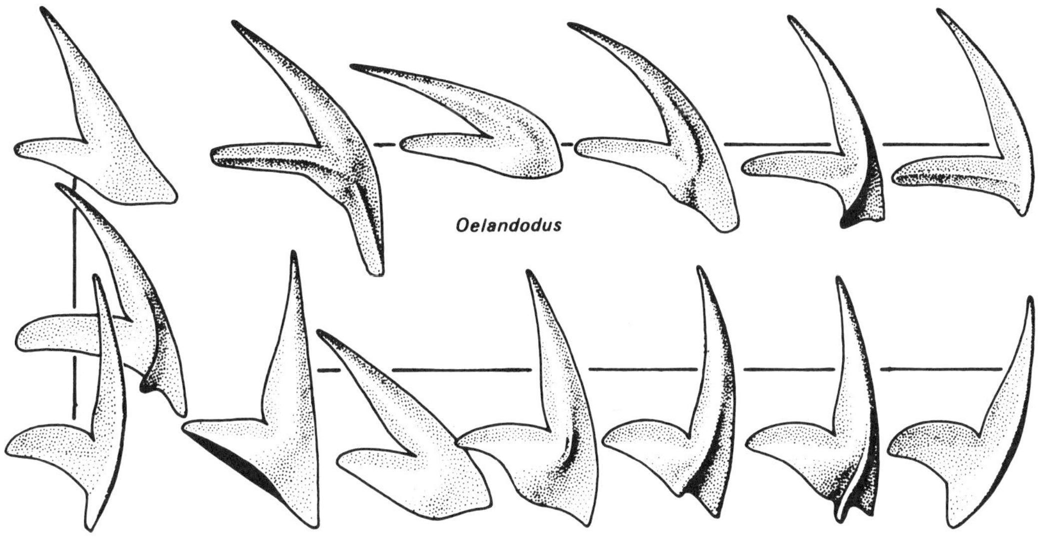

Fig. 5.13. Elements and apparatuses typical of species included in three genera of the Oistodontidae.

of coniform elements, of which one (P) is pastinate (or "acodontiform"), one (M) is geniculate (or "oistodontiform"), and the rest form a symmetry-transition (or S) series, which, when completely differentiated, grades from an alate ("acontiodontiform") element in the Sa position by way of a digyrate (or "paltodontiform") Sb element to a compressed, laterally unornamented nongeniculate ("distacodontiform") element in the Sc position. Conodonts with such apparatuses have been included in *Rossodus* Repetski and Ethington, 1983, *Acodus* Pander, 1856, *Tripodus* Bradshaw, 1969, *Diaphorodus* Kennedy, 1980, *Oelandodus* Van Wamel, 1974, *Protoprioniodus* McTavish, 1973, *Oistodus* Pander, 1856, and probably *Juanognathus* Serpagli, 1974, but it is not clear to me just how many genera are really represented. *Rossodus* (Fig. 5.12) may not have a fully developed symmetry-transition series, hence its two known species seem primitive. (On the other hand, the "acodontiform" P elements and "paltodontiform" Sb elements of *Rossodus* may be so similar that they have not been distinguished). Furthermore, if the skeletal apparatus of the typical species of *Juanognathus* (Fig. 5.12) includes a geniculate coniform element (Serpagli was not sure about this), *Juanognathus* is probably a senior synonym of *Rossodus*, as noted by Repetski and Ethington.

Tripodus (Fig. 5.12) and *Diaphorodus* are based on different, but probably congeneric, species; hence I regard them as synonyms. There will probably always be uncertainty as to how *Acodus* (or many of Pander's genera) should be interpreted in a multielement sense, so I recommend that *Acodus* be consigned to the list of *nomina dubia* and that we go on to more important things.

Oistodus, *Protoprioniodus*, and *Oelandodus*, all shown in Fig. 5.13, include species whose skeletal apparatuses are composed entirely of geniculate coniform elements, which nevertheless exhibit an array of forms closely similar to that in the apparatus of species of *Rossodus* and *Tripodus*. In the *Treatise* Bergström expressed the opinion that further work might show these three genera to be synonyms, and I can see no very good reason (aside from the geniculate nature of their coniform skeletal elements) to refer them to a family separate from *Rossodus* and *Tripodus*. Consequently, in the schematic classification in Appendix A, I group all these genera in the family Oistodontidae (which has priority over Juanognathidae), but refer *Oistodus*, *Protoprioniodus*, and *Oelandodus* to a subfamily Oistodontinae; *Rossodus* and *Tri-*

podus to a subfamily Tripodontinae; and *Juanognathus* and *Histiodella* (a probable derivative of *Juanognathus*) to a subfamily Juanognathinae. This scheme emphasizes the close similarity in skeletal pattern I perceive in species of all these genera which, taken together as the family Oistodontidae, surely represent the rootstock of the Prioniodontida.

For reasons that seemed logical at the time the *Treatise* was written, *Histiodella* Harris (Fig. 5.12) was referred in that volume to the family Rhipidognathidae. Subsequently, McHargue (1982) has published additional information on the ontogeny of *Histiodella* species, on phylogeny within this curious stock, and on the skeletal apparatus of several Middle Ordovician species. In addition, Repetski (1982) has named and described a very early Early Ordovician species, and Stouge (1984) has provided further documentation of the genus in the early Middle Ordovician Table Head Formation of Newfoundland. This new information indicates that the skeletal apparatus of typical species of *Histiodella* is quinquimembrate and exhibits the full array of element types characteristic of the Oistodontidae. The principal defining character of *Histiodella* is the phylogenetic tendency for elements in the P position(s) to develop minute denticles on the margins of their anterior and posterior processes. In Fig. 5.11 I show *Histiodella* as a descendant of *Juanognathus*, but the ancestral stock might just as well be *Tripodus*. It is also possible that Repetski's *H. donnae* represents an independent, early development from either *Tripodus* or *Juanognathus* because it is separated by much of the Early Ordovician from the first representatives of the typical *Histiodella* stock. In addition, Repetski described only a single element of *H. donnae*, although he now admits to another (a geniculate coniform element).

5.7.2 *Family Prioniodontidae Bassler, 1925*

In the Early Ordovician zone just above the one in which *Tripodus* made its debut are conodonts with a seximembrate skeletal apparatus in which the two P positions are occupied by morphologically identical pastinate elements with three denticulate processes; the M position by an element like the geniculate coniform M element of *Tripodus*, but formally a bipennate ramiform element with an elongate, minutely denticulate anterior process; and S positions by a symmetry-transition series that includes bipennate, tertiopedate, alate, and quadriramate elements. It seems clear that these conodonts, referred to *Prioniodus* Pander, are direct descendants of *Tripodus*, from which they differ primarily in a somewhat more elaborate skeletal apparatus in which margins of elements are drawn out into denticulated processes and in which there is also a

Fig. 5.14. Elements and apparatuses of species typical of genera of the Prioniodontidae.

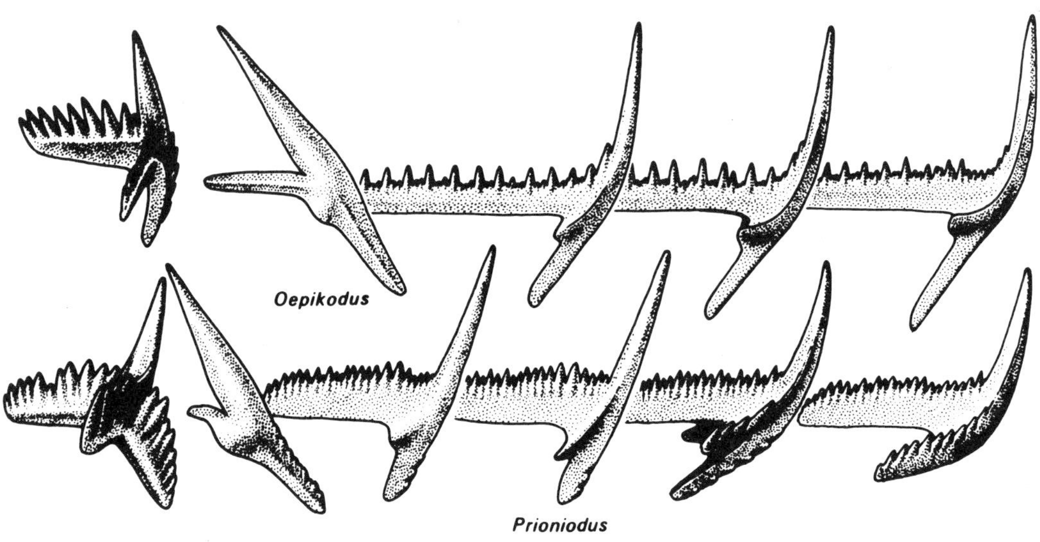

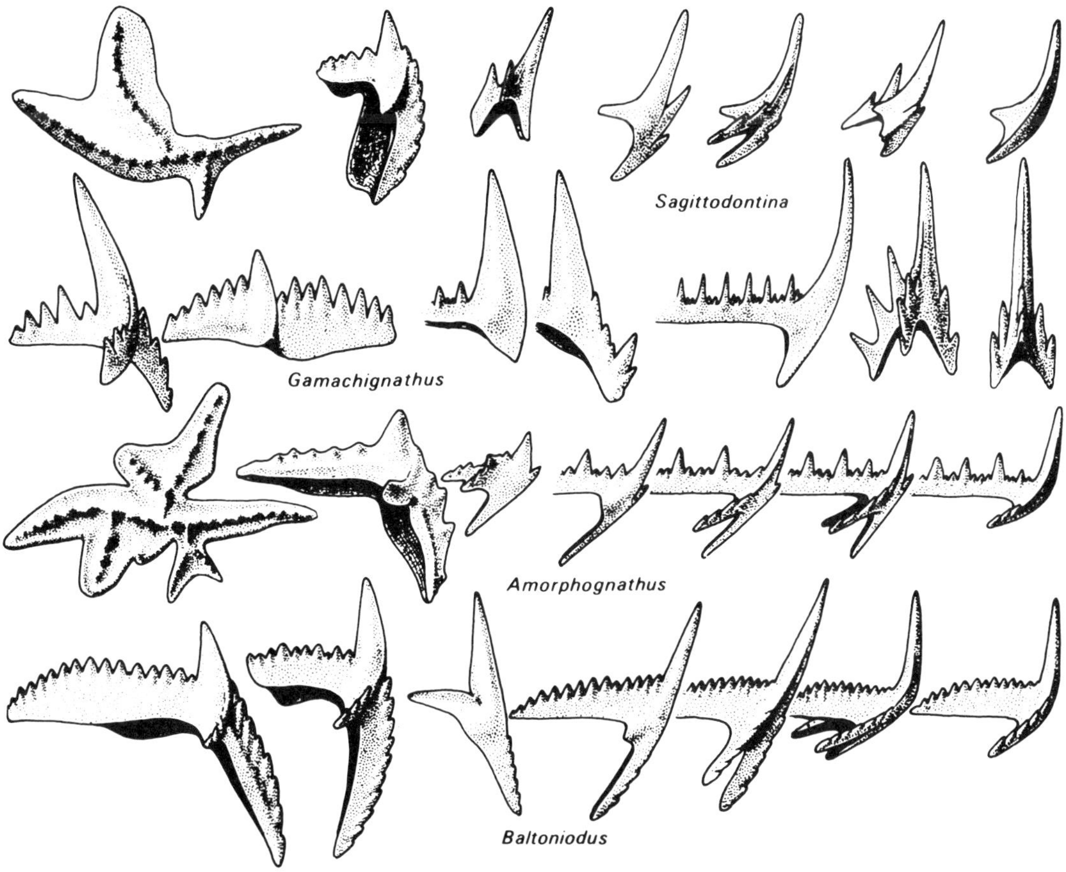

Fig. 5.15. Elements and apparatuses typical of species assigned to genera of the Balognathidae.

greater degree of morphologic differentiation in the S series.

Prioniodus (Fig. 5.14) developed phylogenetically through a succession of similar, biostratigraphically useful species in the Early and early Middle Ordovician, and in rocks a couple of zones above the one in which it first appeared, spawned a new lineage, assigned now to *Oepikodus* (Fig. 5.14). In species of the latter, the skeletal apparatus consists of *Prioniodus*-like pectiniform elements; a geniculate coniform M element; and a symmetry-transition series of quadriramate elements like the one in the Sd position of *Prioniodus*.

5.7.3 Family Balognathidae Hass, 1959

Beginning in Lower Ordovician rocks just above the ones in which *Prioniodus* is first recognized and extending upward into Upper Ordovician strata is a succession of recurrent groups of elements that represent a phylogenetic series of conodont species referred to *Baltoniodus* (Fig. 5.15). Species of *Baltoniodus* are distinguished from those of *Prioniodus* by morphologically different pastinate elements in the two P positions of their skeletal apparatuses. Although this is regarded as of no more than subgeneric significance by several investigators, others have pointed out that *Baltoniodus* appeared a bit later than *Prioniodus* and also probably evolved directly from a younger species of *Tripodus* and not from *Prioniodus*. However this may be, *Baltoniodus* is the universally acknowledged ancestor of *Amorphognathus*, which appeared in a somewhat later part of the Early Ordovician and ranged, through a succession of distincive species, to the end of that period.

Species of *Amorphognathus* (Fig. 5.15), perhaps the best-known and most widely distributed members of the Balognathidae (*Balogna-*

thus is a junior synonym of *Amorphognathus*), had elaborate, septimembrate skeletal apparatuses in which the S positions included ramiform elements comparable to those in similar places in the *Baltoniodus* apparatus, but in which the M position bore bizarre tertiopedate (or "holodontiform") elements and the two P positions were occupied by morphologically distinct, bilaterally asymmetrical pastiniscaphate elements, which in the older multielement literature are termed "amorphognathiform" and "ambalodontiform." *Amorphognathus* species had the most complex apparatuses of all the Prioniodontida and thus represent the pinnacle of differentiation achieved by this important conodont group.

Although Fåhraeus (1984) identifies *Baltoniodus* as the ancestor of *Rhodesognathus* (Fig. 5.16), it is more likely the genus is a descendant of *Amorphognathus*, with which it is allied in the *Treatise* classification prepared by Bergström. When we established *Rhodesognathus* in 1966, Bergström and I believed we were dealing with conodonts that had bimembrate apparatuses of paired, bilaterally symmetrical pastinate (or "ambalodontiform") elements. In collections assembled later by our students, however, the pastinate elements characteristic of *Rhodesognathus* occur in some samples that also include ramiform elements that resemble those of *Amorphognathus*, but lack the platformed elements distinctive of the latter. Thus we now suspect that S (and possibly even M) elements of *Amorphognathus* and *Rhodesognathus* are indistinguishable, and that *Rhodesognathus* has a sexi- or septimembrate apparatus instead of the bimembrate one we visualized 20 years ago when the genus was established. If that is the case, then the close relationship between *Amorphognathus* and *Rhodesognathus* that I show in Fig. 5.11 is even more likely.

Fig. 5.16. Pectiniform elements characteristic of species of genera assigned to the Balognathidae.

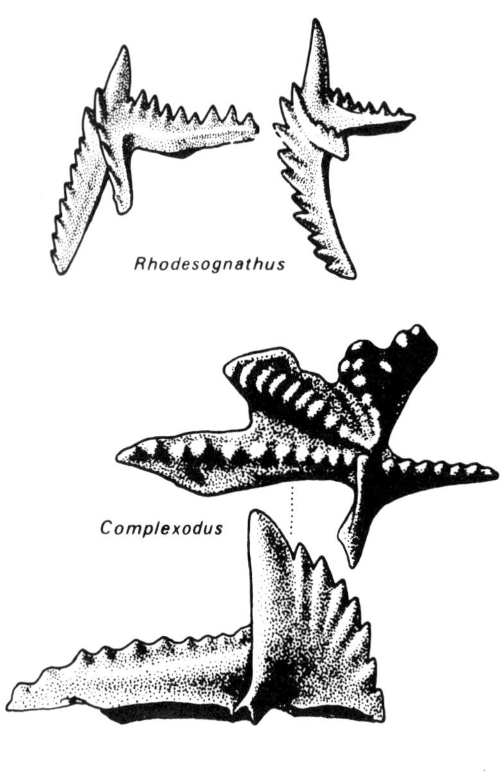

Complexodus pugionifer, whose unimembrate apparatus consists of pastiniscaphate elements that resemble (but differ from) those of *Amorphognathus*, is also included in the Balognathidae of Fig. 5.11 and illustrated in Fig. 5.16. I also refer to the Balognathidae the multielement genus *Sagittodontina* Knüpfer (Fig. 5.15) which, as reconstructed by Bergström (1983), is probably a senior synonym of *Noixodontus* McCracken and Barnes (1982). Neither *Complexodus* nor *Sagittodontina* is well enough known to merit further discussion, although it is certainly of more than passing interest that *Sagittodontina*, which is known from Late Ordovician rocks that were probably deposited in subpolar seas, may be an index to the highest-latitude conodont fauna recorded in rocks of any age.

Recurrent groups of eight distinctive types of Late Ordovician elements, interpreted by McCracken, Nowlan, and Barnes (1980) as two species of *Gamachignathus* (Fig. 5.15), include pastinate elements with three denticuated processes in one P position; bladelike pastinate elements with only two denticulated processes in the second P position; a pair of dolabrate and a pair of bipennate elements in M positions; and an S series of morphologically intergradational alate, tertiopedate, and dolabrate ramiform elements of narrowly pyramidal shape. The apparatus may be dimorphic in the M position and not truly "octimembrate"; even so, clear ancestors are not obvious. Apparatus stucture is, in many respects, more like that of *Prioniodus* than *Baltoniodus* species, but the species most proximate stratigraphically is an unnamed representative of *Baltoniodus* (Savage and Bassett, 1985; Orchard, 1980). Consequently, in Fig. 5.11, *Gamachignathus* (including *Birksfeldia* Orchard, 1980) is included in the Balognathidae and shown tentatively as a descendant of *Baltoniodus*.

5.7.4 Family Icriodellidae, new

I interpret *Icriodella* Rhodes, 1953 as the stem of an important prioniodontide stock that probably also includes *Pedavis* Klapper and Philip, 1971, *Sannemannia* Al-Rawi, 1977, and *Steptotaxis* Uyeno and Klapper, 1980. In the *Treatise* all these genera were included with *Icriodus* and *Pelekysgnathus* in the Icriodontidae, primarily on the basis of obvious similarities in the Pa elements formed by their species. Many of the other elements of the apparatus are different, however, and the differences are sufficient to raise reasonable questions as to the relationship between *Icriodella* and its probable derivatives and the primarily Devonian *Icriodus-Pelekysgnathus* complex. Thus *Icriodella* and its kin are here set aside as a new family, the Icriodellidae, the relationships of which are shown in Fig. 5.17.

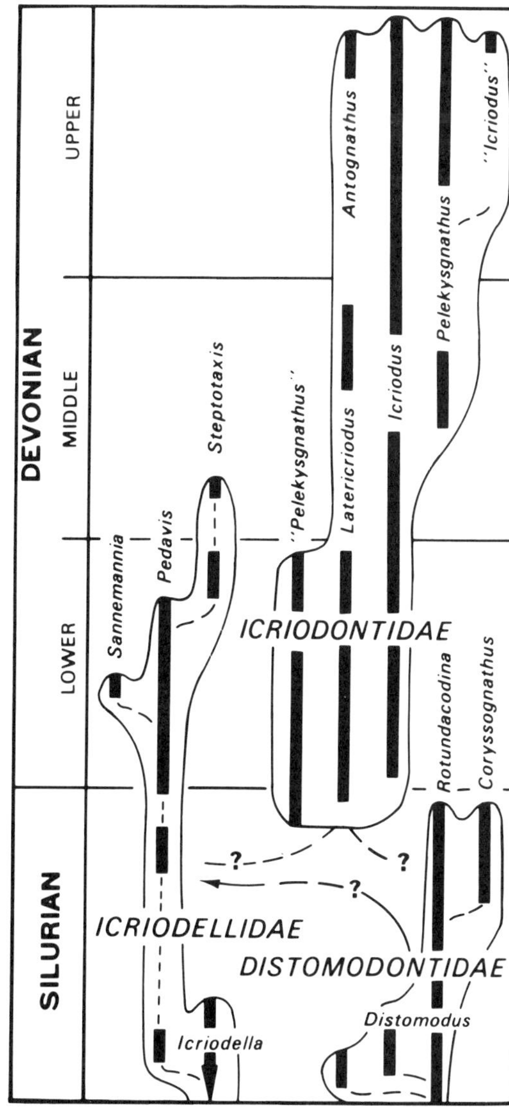

Fig. 5.17. Silurian and Devonian Prioniodontida.

Basically, the icriodellids have a quinquimembrate skeletal apparatus in which the Pa element is pastinate in the apparatus of the oldest species known, pastiniscaphate in those of later forms, and stelliscaphate, with four processes, in apparatuses of the youngest species known. The Pb element, in apparatuses of the *Icriodella* (Fig. 5.18) and *Pedavis* (Fig. 5.18) species for which it is known, is a pyramidal pastinate element with three sharp edges that are adenticulate or only weakly denticulate. I interpret the "coronellan" element in the apparatus of *Steptotaxis* species (Fig. 5.18) as an occupant of the Pb position, primarily because it is related morphologically to the pastinate coniform (or "acodinan") elements assigned to that position in apparatuses of various icriodonts. The apparatus of the only known species of *Sannemannia* is unknown.

M elements of *Icriodella*, *Pedavis*, and probably *Steptotaxis* are bipennate or even pastinate, with adenticulate or faintly serrate edges. In apparatuses of species of *Icriodella* (Fig. 5.18) the occupant of the Sa position is a deeply excavated, pyramidal alate element with minutely serrated lateral processes, whereas elements in other S positions are pyramidal tertiopedate units that are closely similar to the Sa element, but asymmetric with respect to disposition of their lateral processes. S positions in the apparatuses of *Pedavis* species, however, are filled by elongate, prominently costate structures that apparently are nongeniculate coniform elements in the Sb and Sc positions but posteriorly denticulate dolabrate elements in the Sa position. In *Steptotaxis* species (Fig. 5.18) S positions were presumably occupied by acostate, nongeniculate coniform elements of essentially circular cross section.

In the apparatus of *Icriodella praecox* Lindström, Racheboeff, and Henry, 1974, the oldest species of *Icriodella* thus far described, the pastinate Pa element has a stubby, pyramidal cusp, subequal anterior and posterior processes with single rows of denticles, and a short, adenticulate lateral process. In all other species of *Icriodella*, however, the anterior process of Pa elements is somewhat longer than the posterior process and bears a double row of short denticles.

Species of *Pedavis* (Fig. 5.18) are distinguished by apparatuses in which Pa elements are stelliscaphate structures with four denticulate processes, at least one with a double row of denticles and two (the lateral ones) that are commonly deflected somewhat anteriorly. Pa elements of *Sannemannia pesanseris* are similar, but denticulation is evidently in the form of single rows of nodes. Pa positions in the apparatuses of several species from the Lower Silurian of Australia are occupied by stelliscaphate elements that are remarkably similar to those of *Pedavis* (Bischoff, 1986). Other elements in the apparatuses of those species, however, have a conformation typical of the ones formed by various species of the Distomodontidae, and Bischoff (1986) has assigned them to *Distomodus*.

In Fig. 5.17 I indicate a number of possible phylogenetic relationships among the genera I have parceled out among the Icriodellidae, Distomodontidae, and Icriodontidae. *Pedavis* is included with *Icriodella* in the Icriodellidae because of close similarity between elements I assign to the Pa, Pb, and M positions in apparatuses of all its known species. *Sannemannia* is included because the Pa element of its only known species is closely similar to that formed by species of *Pedavis*. Similar reasoning is also behind my decision to assign *Steptotaxis* to this family, even though certain of its more characteristic species were originally assigned to *Pelekysgnathus* of the Icriodontidae. Non-Pa elements of *Steptotaxis* differ considerably from those of *Pedavis*, and S elements of *Pedavis* are quite different from those of *Icriodella*.

I have already noted that Pa elements of species assigned by Bischoff (1986) to *Distomodus* are morphologically very close to those formed by species of *Pedavis*, but that presumed occupants of Pb, M, and S positions are clearly distomodont. In assessing phylogenetic relationships, most students of conodonts have accorded greatest weight to similarities in Pa elements. If that were done in this case, the clear result would be derivation of *Pedavis*

Fig. 5.18. Elements and apparatuses of species typical of the Icriodellidae.

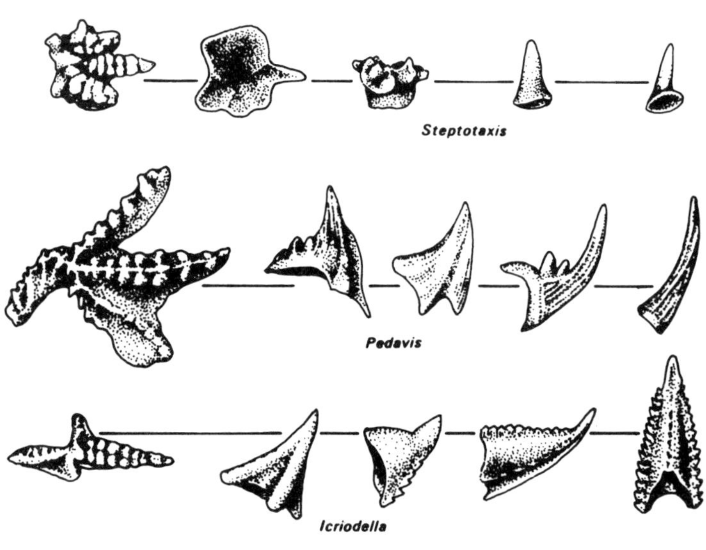

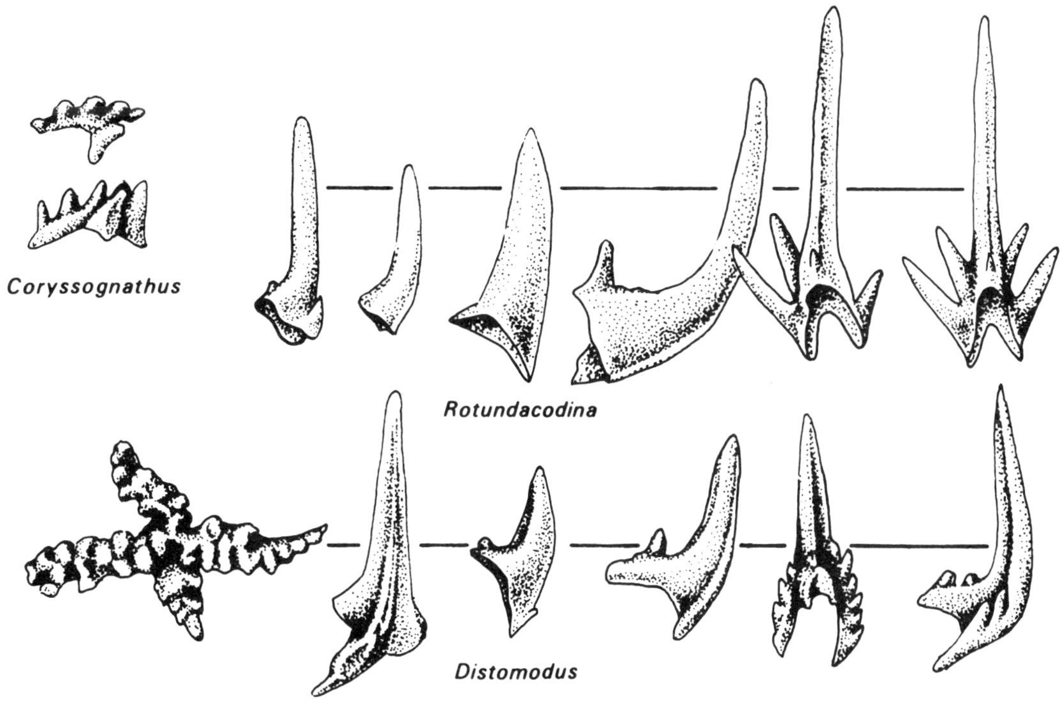

Fig. 5.19. Elements and apparatuses of species typical of genera assigned to the Distomodontidae.

from *Distomodus* or some other genus of the Distomodontidae instead of from *Icriodella.* To my mind, the prominently costate elements that occupy S positions in apparatuses of *Pedavis* species are equally difficult to derive from either the *Icriodella* or *Distomodus* stock and may thus be regarded as the features that distinguish *Pedavis.* Thus, if one considers the remaining elements (Pa, Pb, M), similarity is greatest between *Pedavis* and *Icriodella,* and that has guided me in assigning *Pedavis* to the Icriodellidae.

Broadhead and McComb (1983) argue from comparison of the Pa elements of Late Silurian *Pedavis* and those of latest Silurian-earliest Devonian *Latericriodus woschmidti* that the latter may have developed from the former by neoteny. Although that is an attractive idea, non-P elements in the apparatus of *L. woschmidti* (Fig. 5.20) differ substantially from those in apparatuses of *Pedavis* species and, in the absence of information on the ontogenetic development of the latter, it is difficult for me to see how S elements of the supposed paedomorph (*L. woschmidti*) might be related to earlier ontogenetic stages of ancestors in *Pedavis.* Nevertheless, I indicate in Fig. 5.17 the possibility that *Pedavis,* whose familial assignment is not all that certain, might be the point of origin of *Latericriodus* and the Icriodontidae.

5.7.5 Family Distomodontidae Klapper, 1981

Conodonts assembled here and assigned to *Distomodus, Rotundacodina,* and *Coryssognathus* have a basically quinqui- or seximembrate skeletal apparatus in which ramiform components tend to have short, weakly denticulate or adenticulate processes, and the Pa and Pb positions are occupied by stelliscaphate and pastinate elements or their reduced derivatives. The latter character provides a link with other prioniodontide conodonts and suggests origin of the Distomodontidae in Late Ordovician balognathids such as *Gamachignathus* or *Sagittodontina.*

There is some difference of opinion in the literature as to the apparatus composition of *Distomodus kentuckyensis* Branson and Branson (Fig. 5.19), type-species of the stem stock, and this affects interpretations of familial taxonomy and phylogeny. The apparatus reconstructed by Klapper and Murphy (1974) is quinquimembrate and includes in the P position the bizarre pastinate element described by Nicoll and Rexroad (1968) as the form species *Trichonodella? expansa* (later *Exochognathus expansus*). Cooper (1975), on the other hand, assembled a seximembrate apparatus for *D. kentuckyensis*, which is similar in its major particulars to the one reconstructed by Klapper and Murphy but includes in the second P position a stelliscaphate element like the one described by Branson and Branson (1947) as the form species *Icriodina irregularis*. Klapper's diagnosis of both *Distomodus* and the Distomodontidae in the *Treatise*, however, is clearly dependent on the Barrick and Klapper (1976) reconstruction of multielement *Hadrognathus staurognathoides* Walliser, which has very *Distomodus*-like ramiform elements in the S and M positions and has been referred to *Distomodus* on that basis by most authors dealing with Silurian conodont faunas.

For several reasons, Jeppsson (1979) prefers the apparatus of *Distomodus kentuckyensis* originally reconstructed by Klapper and Murphy, that is, one without a stelliscaphate element in the second P position. Jeppsson notes, for example, that a reconstruction such as the one advocated by Cooper (1975) for *D. kentuckyensis* suggests that *Icriodella*-like P elements developed independently in two different prioniodontide stocks, characterized by otherwise quite different apparatuses. Further, he notes that the apparatus of younger Silurian *Rotundacodina dubius* (Rhodes) (Fig. 5.19), which he includes in *Distomodus*, seems to have lacked any platformed pectiniform elements but apparently included ramiform elements that are morphologically very close to the nonplatformed components of the *D. kentuckyensis* apparatus reconstructed by Klapper and Murphy in 1974.

Jeppsson's arguments make a good deal of sense only if *Rotundacodina dubius* is interpreted as a typical *Distomodus*. However, reconstruction of the apparatus of the type-species advocated by Cooper and supported by others (e.g., Bischoff, 1986) is now based on a substantial and growing body of distributional evidence that I find compelling. Thus I interpret *Distomodus* in the manner illustrated in Fig. 5.19; that is, as a group of species with a seximembrate skeletal apparatus in which the Pa position was occupied by *Icriodina*-like stelliscaphate elements.

Viewed as just described, *Distomodus* made its debut at or close to the beginning of the Silurian and ranged into the early Wenlockian. In the very early history of the stock several lineages developed. I have indicated this by charting *Distomodus* as a "bush" in the lower part of Fig. 5.17. One branch of this bush (the typical branch) developed Pa elements of the type originally identified as *Icriodina*; another built stelliscaphate Pa elements that anticipated the morphology of *Pedavis* (Icriodellidae); and the third, which came along a little later in the Llandovery, formed Pa elements such as those originally identified as *Hadrognathus*.

Very early in the Wenlock, and late in its history, *Distomodus* was apparently ancestral to *Rotundacodina dubius*, the species Jeppsson (1979) apparently took to be typical of *Distomodus*. S and M elements in the apparatus of *R. dubius* (Fig. 5.19) are reminiscent of those of typical *Distomodus*, but those in the Pa and Pb positions are simple pastinate and pastinate coniform elements, respectively. Phylogenetic simplification of elements in these two positions may be an example of paedomorphosis, possibly accomplished by neoteny. *Rotundacodina* ranges to the end of the Silurian and is, in my opinion, the best bet as ancestral stock of the Icriodontidae.

Only the Pa elements of *Coryssognathus* (Fig. 5.19) have been described. These are reduced pastinate pectiniform elements with a poorly developed lateral process of a sort that might easily have been derived from the much simplified Pa elements of *Rotundacodina dubius*. In numerous reports Jeppsson has suggested that these elements represent a species of *Pelekysgnathus*, but Pa elements formed by species of that genus are segminiscaphate, with the cusp at the posterior end. In *Coryssognathus* elements, on the other hand, the cusp is clearly at the anterior end and there appears in several illustrated specimens to be a very short process anterior of it.

5.7.6 Family Icriodontidae Müller and Müller, 1957

Included in this family are *Latericriodus* Müller, 1962; *Icriodus* Branson and Mehl, 1938; two apparently distinct groups of species assigned to *Pelekysgnathus* Thomas, 1949; *Antognathus* Lipniagov, 1978; and an as yet unnamed genus, recognized on phylogenetic grounds by Sandberg and Dreesen (1984) and based on a group of Late Devonian species previously referred to *Icriodus*.

Icriodonts are characterized by a basically multimembrate apparatus that includes bi-, tri- or merely segminiscaphate pectiniform elements in the Pa position. If there are elements in other positions, they are an assortment of rather variable coniform or only weakly denticulated ramiform elements.

Weddige and Ziegler (1979) advocate division of *Icriodus* s. l. into two groups of species, one that formed bi- and trisegminiscaphate Pa elements, and one with segminiscaphate Pa elements. For the first group they used the generic name *Latericriodus* (Müller, 1962), for the latter, *Icriodus* (Branson and Mehl, 1938). I follow Weddige and Ziegler in the summary that follows but note that *Icriodus* in this sense is almost surely polyphyletic. That is, *I. angustus* and *I. angustoides*, both with the segminiscaphate Pa elements typical of *Icriodus*, have been shown by Klapper and Ziegler (1967) and Carls and Gandl (1969) to have evolved at different times from different ancestors in *Latericriodus*.

Serpagli (1983) has demonstrated that the apparatus of *Latericriodus woschmidti*, founder of the genus, is seximembrate (Fig. 5.20). A bisegminiscaphate pectiniform element occupies the Pa position; a pastinate, low-pyramidal coniform element the Pb position; a more compressed pastinate coniform (or "acodiniform") element the M position; and a morphologically intergradational series of alate, tertiopedate, and dolabrate elements the Sa, Sb, and Sc positions, respectively. Klapper and Philip (1971), however, did not recognize homologues of the S-series elements of *L. woschmidti* in the apparatus of *L. latericrescens* (Fig. 5.20), which they concluded was bi- or trimembrate. Thus, if *L. woschmidti* and *L. latericrescens* did, indeed, form part of a single lineage, a feature in the evolution of that lineage may have been the loss of S-series elements. Al-

Fig. 5.20. Elements and apparatuses of species typical of the Icriodontidae.

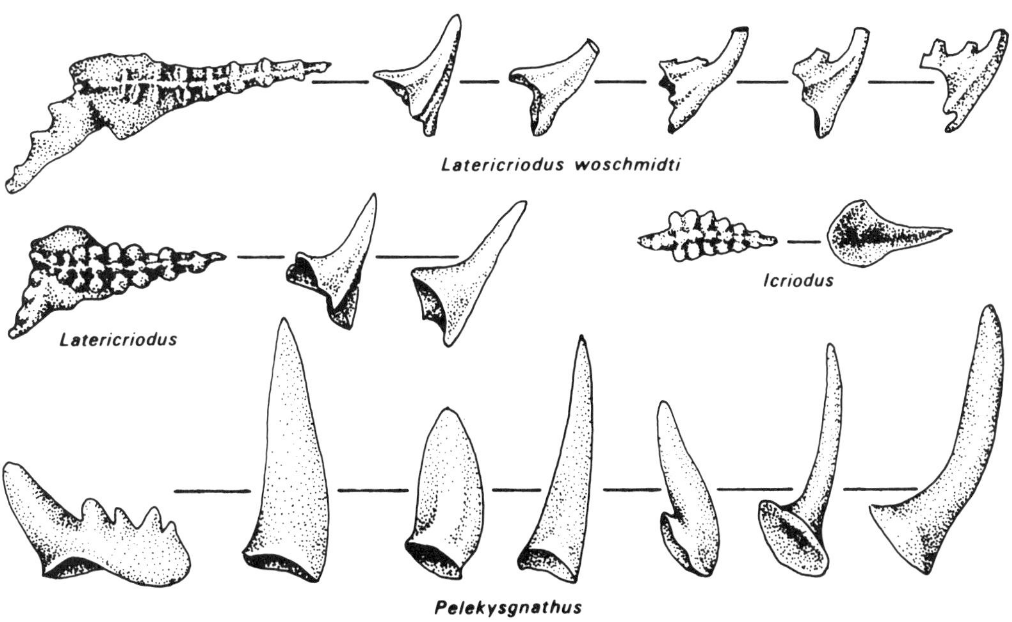

ternatively, it might be preferable to recognize these differences in apparatus architecture by assigning the two species to different genera.

Although Bultynck (1972) advocated an *Icriodus* apparatus (Fig. 5.20) formed entirely of the segminiscaphate elements on which Branson and Mehl (1938) based the form genus *Icriodus*, most other authors agree that the apparatus of *Icriodus* species, like that of *Latericriodus latericrescens*, consisted of segminiscaphate Pa elements and an assortment of mostly tiny pastinate or pyramidal coniform elements. However, Chatterton and Perry (1977) included in the apparatus of *I. hadnagyi*, the oldest *Icriodus* known, a group of highly variable coniform elements, not all of which are "acodiniform." They noted that *Pelekysgnathus csakyi*, represented in somewhat younger rocks, has a similar array of non-P elements. This is intriguing, of course, because it suggests that at least some species of *Pelekysgnathus* may have been derived from *Icriodus*, not from *Pelekysgnathus* ancestors. The plot became even thicker in the Late Devonian when, according to Sandberg and Dreesen (1984), *Pelekysgnathus* spawned several species in whose apparatuses Pa positions are occupied by *Icriodus*-like segminiscaphate elements! Thus *Icriodus* may not only be polyphyletic, but it may be its own "grandpa."

Pelekysgnathus (Fig. 5.20) appeared in the latest Silurian (in the form of *P. index*) a little bit ahead of the first representative of *Latericriodus* (*L. woschmidti*) and well ahead of the first *Icriodus* (*I. hadnagyi*). *P. index* and Early Devonian species assigned to "*Pelekysgnathus*" in Fig. 5.17 formed apparatuses with single-rowed segminate elements in the Pa position, pastinate coniform (or "acodinan") elements in the Pb and (probably) M positions, and tiny nongeniculate coniform elements of essentially circular cross section in S positions. These Early Devonian species are separated by an appreciable stratigraphic gap from a Middle and Late Devonian group that includes *P. inclinatus*, the type-species. Because of the stratigraphic gap, several authors have suggested referring the Early and Middle-Late Devonian groups of *Pelekysgnathus* species to different genera but, until recently, there has been little morphologic evidence in support of such a separation. In 1984, however, Sandberg and Dreesen reported that the apparatus of *P. inclinatus* (Fig. 5.20), the type species and central member of the Middle-Late Devonian group, also includes a geniculate coniform (or "oistodontiform") element, one of the few to be reported from rocks younger than the Ordovician.

Sandberg and Dreesen (1984) have also identified a geniculate coniform (or "oistodontiform") element in the apparatus of one member of a group of five Late Devonian species currently assigned to *Icriodus* but thought instead to be end members in two different lineages within *Pelekysgnathus*. Pa elements, the only ones known for most of these species, are *Pelekysgnathus*-like, but have three denticle rows, not just one. Should it ultimately be desirable to create new generic names for these late-stage icriodont stocks, *Bouckaertodus* (Gagiev, 1979) is apparently available for one of them.

Segminiscaphate Pa elements of the poorly known *Antognathus* have stubby posterolateral processes and an asymmetrically T-shaped outline as viewed from above or below. Origin and relations of *Antognathus* are not known, but its two Late Devonian species were apparently restricted in their distribution to hypersaline lagoonal environments.

Carls and Gandl (1969) have worked out basic phylogenetic patterns in earliest Devonian *Latericriodus* and two derivative *Icriodus* branches; Klapper and Ziegler (1967) have considered somewhat later events in the history of *Latericriodus*; Weddige (1977) and Weddige and Ziegler (1979) describe evolutionary patterns in Middle Devonian lineages of *Icriodus*; and Sandberg and Dreesen (1984) have recently considered late events in the history of the Icriodontidae.

The Icriodontidae are important stratigraphically in shallow-water Devonian biofacies, and icriodonts have become famous as Devonian "index fossils." However, as prior remarks may indicate, numerous problems in icriodont taxonomy and phylogeny remain to be solved. For example, it now seems clear that either the three-rowed segminiscaphate elements by which species of *Icriodus* have customarily been identified or the single-rowed segminiscaphate elements taken as typical of *Pelekysgnathus* may have been produced independently at different times in different lineages. Also, origins of the family are by no means certain.

Klapper and Murphy (1974) suggested that *Icriodus sensu lato* might have evolved from *Pelekysgnathus index*, but they found no specimens that would bridge the gap in morphology that separates Pa elements of *Latericriodus woschmidti* and *P. index*. Nicoll (1982) regarded *Rotundacodina dubius* (Distomodontidae) as the most likely ancestor, and Broadhead and McComb (1983) regard *Latericriodus woschmidti* as a neotenous derivative of *Pedavis*. Fåhraeus (1984) suggests derivation of *Pedavis* from *Distomodus*, in which presumably he included *Rotundacodina dubius*. I regard the neotenous route suggested by Broadhead and McComb (1983) as unlikely because it was devised primarily to explain the origin of the Pa elements of *I. woschmidti* from those of *Pedavis latialatus*, and it thus ignored other elements in the apparatus, which are surely not the same in the two species. On the other hand, there is substantial similarity between non-P elements in the apparatuses of *R. dubius* and *I. woschmidti*, as Serpagli (1983) has shown, and the remote ancestors of *R. dubius* certainly built scaphate elements comparable to those of *L. woschmidti*. (Of course, the scaphate elements some of them built were even closer to those of *Pedavis*, which I ally to *Icriodella* because of greater similarities in M and Pb elements.)

While I do not claim to have special insight into the ancestry of the icriodonts—or even the clairvoyance that would be required to make a monophyletic unit out of this family—I suggest that a derivation of *Latericriodus woschmidti* from *Rotundacodina dubius* explains the characters of more of the *Latericriodus* apparatus than does a derivation from *Pedavis*. Pending badly needed additional information on Late Silurian conodont faunas, I submit that this is a more straightforward route than is the one by way of *Pedavis*, whose own origins are yet to be amply documented.

5.7.7 *Family Polyplacognathidae Bergström, 1981*

Late in the Early Ordovician a species of the *Amorphognathus* group (Balognathidae) was apparently the progenitor of a stock of prioniodontide conodonts characterized by a reduced, bimembrate apparatus of bilaterally asymmetrical stelliplanate and pastiniplanate elements. Recurrent groups of such elements are the basis for recognizing a succession of species of *Eoplacognathus* (Fig. 5.21). These species have been of considerable biostratigraphic utility in rocks primarily of early Middle Ordovician age. The stock seems to have died out, however, early in the late Middle Ordovician.

During its early Middle Ordovician heyday, *Eoplacognathus* seems to have given rise to an important line of species that Bergström (1983) refers to the genus *Cahabagnathus* (Fig. 5.21). The bimembrate apparatus of *Cahabagnathus*, like that of *Eoplacognathus*, is composed of a pair of stelliplanate and a pair of pastiniplanate pectiniform elements. However, those of the *Cahabagnathus* apparatus form mirror-image pairs, whereas those of *Eoplacognathus* are bilaterally asymmetrical and have more simply ornamented upper sides.

Fig. 5.21. Elements typical of the apparatuses of species assigned to genera of the Polyplacognathidae.

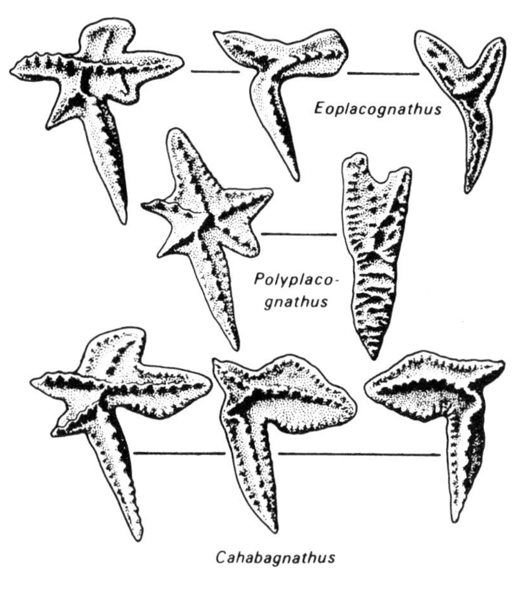

Eoplacognathus was probably also ancestral to *Polyplacognathus* (Fig. 5.21), whose only well-known species, *P. ramosus*, is distinctive of the North American Midcontinent Province in the late Middle Ordovician and is recognized by a bimembrate skeletal apparatus that includes mirror-image pairs of stelliplanate and pastiniplanate pectiniform elements. *P. ramosus* is currently somewhat isolated, both stratigraphically and geographically, from any potential ancestor in *Eoplacognathus*; however, the gap may be filled by poorly known forms from the Middle Ordovician of Siberia.

5.7.8 *Family Multioistodontidae Harris, 1964*

The stem of this important family is a group of species that has been referred by various authors to *Triangulodus*, or *Eoneoprioniodus*, or *Pteracontiodus*. In this discussion, I use *Pteracontiodus* of Harris and Harris (1965) (Fig. 5.22), which was proposed in the same year as *Eoneoprioniodus* Mound, but almost a decade earlier than *Triangulodus* Van Wamel, 1974. *Trigonodus* Nieper, 1969, is almost certainly a junior synonym.

The skeletal apparatus of species in the *Pteracontiodus* plexus is a quinqui- to seximembrate array of dominantly hyaline coniform elements, which includes two pairs of morphologically similar pastinate elements (Pa, Pb), a pair of geniculate elements (M), and pairs of bipennate and tertiopedate coniform elements that join a presumably unpaired alate coniform element to form a completely differentiated symmetry-transition (S) series. This pattern is obviously very much like that of *Tripodus*, which suggests a close relationship. *Pteracontiodus* and *Tripodus* appear at about the same time, however (in the form of *P. brevibasis* and *T. deltatus*), so it is not clear which (if either) is ancestor and which (if either) is descendant.

In the latest part of the North American Early Ordovician (Ibexian), or the very earliest part of the Middle Ordovician (Whiterockian), *Pteracontiodus* was joined briefly by a succession of species with skeletons of largely hyaline elements identical in plan to those of the several species of *Pteracontiodus* but with more distinctly developed processes that bear one to several simple denticles. These species have been grouped in *Multioistodus* by most authors but are referred here to *Neomultioistodus* (Fig. 5.22) because the skeletal apparatus of the type-species of *Multioistodus* apparently lacked an element in the M position. It may be a reduced derivative of the *Pteracontiodus-Neomultioistodus* stock. *Leptochirognathus* (Fig. 5.23),

Fig. 5.22. Apparatuses typical of species of genera included in the Multioistodontidae. Sa elements at top of each row, followed below by Sb, Sc, M, and P elements.

Fig. 5.23. Elements typical of the skeletal apparatus of *Leptochirognathus*, a member of the Multioistodontidae. From Bauer (1987).

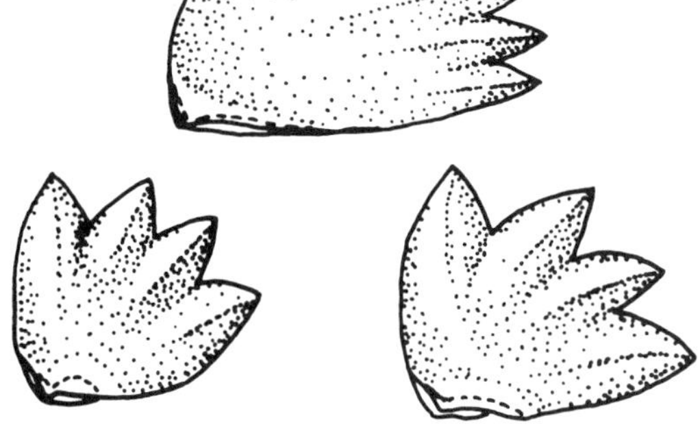

whose best-known species has a bimembrate apparatus, was probably also derived from an early species of *Neomultioistodus* (e.g., *N. clypeus*) by simplification of elements in the P positions and by loss of elements in the M and S positions.

Lower Middle Ordovician (Whiterockian) strata in North America also yield two recurrent groups of dominantly hyaline elements that are identical in plan to those by which I recognize species of *Pteracontiodus*, *Neomultioistodus*, and *Multioistodus*, but with shorter cusps and longer processes, which bear numerous white denticles. These groups record two closely related species of *Paraprioniodus* (Fig. 5.22). The older species (*P. costatus*) apparently had a geniculate coniform element in the M position of its skeletal apparatus, whereas the M position in the apparatus of the currently unnamed younger species is occupied by a dolabrate (or "cyrtoniodontiform") ramiform element.

Pteracontiodus ranges, in the form of at least one unnamed species, into Middle Ordovician rocks well above those dominated by the other members of this family, but apparently was not ancestral to other conodonts. Dzik (1983) reported that, in a manuscript then (and still) in preparation, he and I might consider *Paraprioniodus*, a *Pteracontiodus* derivative, as the ancestor of the important Ordovician *Phragmodus*. For various reasons, however, that now seems unlikely, and *Phragmodus* is shown in Fig. 5.17 as a possible descendant of an as yet unidentified species in the *Tripodus* complex.

5.7.9 *Family Plectodinidae, new*

In this new family I include prioniodontide conodonts with a sexi- or septimembrate skeletal apparatus that includes a pastinate (or "dichognathiform") element in the Pa position, an angulate pectiniform element in Pb, a dolabrate or bipennate element in M, and alate, digyrate, and bipennate ramiform elements in the Sa, Sb, and Sc positions, respectively.

Since about 1969 I have assigned conodonts with such apparatuses to *Plectodina* Stauffer and *Aphelognathus* Branson, Mehl, and Branson, taken in a multielement sense. However, I have also assigned to those genera a number of species with apparatuses in which the Pa position is occupied by an angulate pectiniform element (e.g., *P. tenuis*, *P. florida*) because of the often repeated observation that, in large samples from a succession of late Middle Ordovician populations, the denticulated lateral process of the 'dichognathiform' Pa elements rotates gradually anteriorly and ultimately fuses with the short, adenticulate anterior process to produce an element in younger samples that is formally angulate. As Bergström and I noted in 1972, this sequence of events results in a type of apparatus that is basic to the major group of dominantly post-Ordovician conodonts herein identified as the order Ozarkodinida.

Because it is unfortunate to have a major evolutionary step such as development of the Ozarkodinida buried taxonomically within a genus, I suggest that *Plectodina* be reserved for species like *P. aculeata* (Fig. 5.24) with pastinate Pa elements, and that those species with angulate Pa elements be transferred to another genus, which I will identify here only as "*Plectodina*."

The problem with *Aphelognathus* is just the reverse. That is, the type-species of *Aphelognathus* has an angulate Pa element and is thus appropriately an ozarkodinide (instead of a prioniodontide) conodont; the two species with pastinate Pa elements now assigned to this genus (*A. gigas*, *A. kimmswickensis*) will need a new generic assignment, or can be accommodated temporarily in *Plectodina*. Thus, in Fig. 5.11, I show the range only of *Plectodina sensu stricto*. Species of "*Plectodina*" and *Aphelognathus* with angulate Pa elements are included in the ozarkodinide family Spathognathodontidae, whose content and stratigraphic distribution are shown digrammatically in Fig. 5.36.

Plectodina s.s. (in the form of *P. aculeatoides*) almost made it to the end of the Ordovician, and specimens representing a host of variable (and largely undescribed) species are common in samples from low- and intermediate-latitude Middle Ordovician rocks.

The Plectodinidae seems also to be the best home for *Scyphiodus* (Fig. 5.25), whose only known species has a unimembrate skeletal apparatus that consists of an anguliscaphate pec-

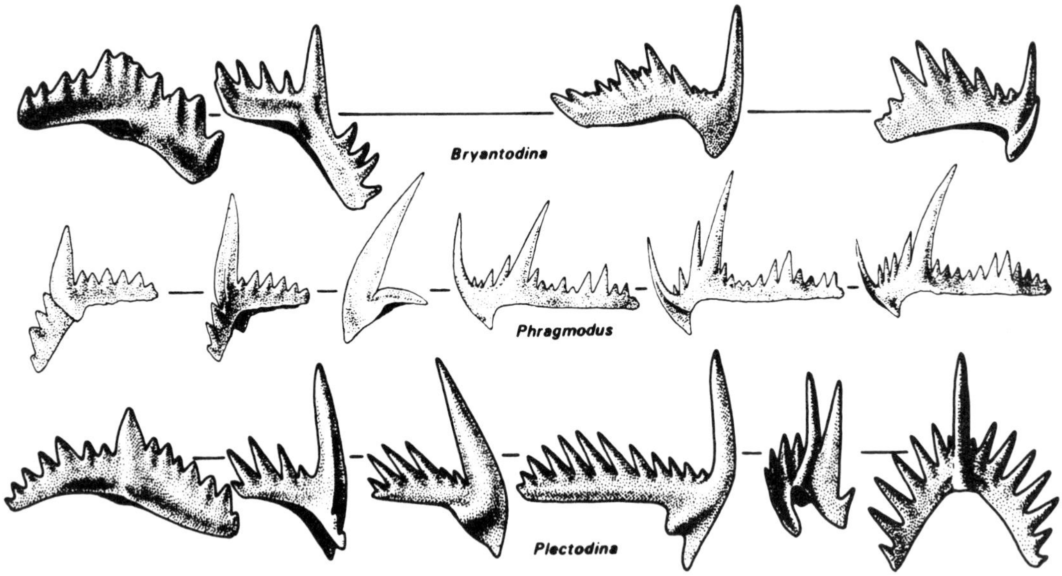

Fig. 5.24. Elements and apparatuses typical of species assigned to *Plectodina* (Plectodinidae), and *Phragmodus* and *Bryantodina* (Cyrtoniodontidae). Sa elements on right, followed by Sb, Sc, M, Pb, and Pa elements.

tiniform element with a bladelike anterior process that is slightly offset longitudinally from a posterior process of about the same length that bears three parallel rows of denticles connected transversely by ribs. In a recent study of a large collection of *Scyphiodus* elements from the Winnipeg Formation of South Dakota (Sweet, 1982), I discovered that specimens representing very early growth stages are clearly pastinate (or "dichognathiform"), but that lateral and anterior processes coalesce in elements representing successively later ontogenetic stages. Thus mature *Scyphiodus* elements retain only the slight longitudinal offset to mark their pastinate origins. *Scyphiodus primus* was apparently one of only two prioniodontide conodonts of North American origin (the other was *Polyplacognathus ramosus*) to develop a platformed pectiniform element and the only one to develop a reduced, unimembrate apparatus.

Fig. 5.25. Pa elements typical of *Scyphiodus primus* (Plectodinidae). Lower figure is of an element representing a very early ontogenetic stage. The pastinate nature of elements at this developmental stage is the principal clue to suprageneric affinities. The upper two figures are lateral and upper views of Pa elements representing a late stage in ontogeny.

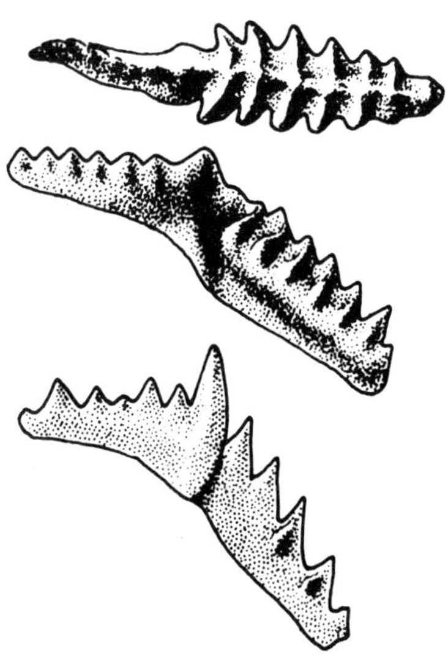

5.7.10 *Family Cyrtoniodontidae Hass, 1959*

Somewhat mysteriously, I here refer the genera *Phragmodus* and *Bryantodina* to the family Cyrtoniodontidae. This is necessary because the type-species of Stauffer's form genus *Cyrtoniodus* was probably based on specimens that represent the M elements of *Phragmodus cognitus*. *Cyrtoniodus* is thus a probable synonym of *Phragmodus*, but the family Cyroniodontidae based on it by Hass (1959, as subfamily Cyrtoniodontinae) has clear priority over the family Phragmodontidae established by Bergström in 1981.

Phragmodus (Fig. 5.24), characteristic of low-latitude Middle and Late Ordovician faunas, has a distinctive seximembrate apparatus in which the P positions are occupied by pastinate (or "dichognathiform") elements, the M position by a dolabrate (or "cyrtoniodontiform") or geniculate coniform element, and the S positions by the "phragmodontiform" ramiform elements distictive of the genus. The latter, which include alate, tertiopedate, and bipennate types, all have a terminal or subterminal cusp with smooth sides or with a sharp costa on one or both sides and a long, arched, denticulated posterior process at the crest of which is a major denticle that may rival or exceed the cusp in size.

In 1972 Bergström and I included *Phragmodus* and *Plectodina* in the family Cyrtoniodontidae and speculated that these genera, which dominate Middle and Late Ordovician conodont faunas in the North American Midcontinent, originated in *Prioniodus*. It now seems more likely that the prioniodontide plan of the skeletal apparatus in species of these three genera is indicative of a common, but probably not direct, origin. Thus, in Fig. 5.11 I show *Prioniodus*, *Phragmodus*, and *Plectodina* as heterochronous derivatives of *Tripodus*.

With some reservations I also include *Bryantodina* (Fig. 5.24) with *Phragmodus* in the Cyrtoniodontidae. *Bryantodina* has an array of ramiform elements in S positions that look very much like the distinctive "phragmodontiform" elements of *Phragmodus*. However, P positions are occupied in the apparatuses of *Bryantodina* species by carminate and angulate pectiniform elements that are quite different from the pastinate P elements of *Phragmodus* and cannot be shown at this time to be either ontogenetic or phylogenetic modifications of such elements. In favor of an association with *Phragmodus*, I note that in the apparatus of *P. cognitus* one of the pastinate elements is replaced phylogenetically by an angulate pectiniform one that retains a clue to its pastinate ancestry in the form of a riblike offset on the outer face of the cusp. No such offset ornaments the outer face of either P element in the apparatus of *Bryantodina typicalis*, however; hence, if either developed by phylgenetic reduction of an orginally pastinate form, that reduction is not documented by extant collections.

5.7.11 *Family Rhipidognathidae Lindström, 1970*

In the *Treatise* Bergström assembled *Bergstroemognathus*, *Appalachignathus*, *Rhipidognathus*, *Carniodus*, and *Histiodella* in the Rhipidognathidae, which was assigned to the Prionìodontacea (the Prioniodontida of my taxonomy). Following publication of new information on *Histiodella*, which I have cited in Section 5.7.1, I feel more comfortable assigning that genus to the Oistodontidae. Also, *Carniodus* is out of place here, because the seximembrate skeletal apparatus of its only known species has a pastinate element with three distinctly denticulated processes in one P position and an alate element with a long, denticulated posterior process in the Sa position. Because neither of these characters is a feature of the apparatus of any other rhipidognathid species, I feel that *Carniodus* should be reassigned. The only prioniodontide group in which *Carniodus* could be accommodated is the Balognathidae, but there is little morphologic resemblance between most of its skeletal elements and those formed by typical balognathids. Furthermore, Silurian *Carniodus* is separated from the Ordovician balognathids by a considerable stratigraphic gap. Consequently, I follow Bischoff (1986) in assigning *Carniodus* to the Pterospathodontidae (Ozarkodinida), whose numerous Silurian species built apparatuses that included strikingly similar P elements.

Genera that remain in the Rhipidognathidae include species with tri- to septimembrate skeletal apparatuses in which P position(s) are occupied by structures that are formally angulate or carminate pectiniform elements but are allied morphologically to the "acodontiform" P elements of *Tripodus* and other stem prioniodontides through possession of a distinct rib or costa on the outer face of the cusp, which may be marked on the basal margin by a conspicuous boss or lappet. In short, those elements may be regarded as very simple pastinate ones, in which no sharp edge or denticulate process ever developed along the anterior edge of the

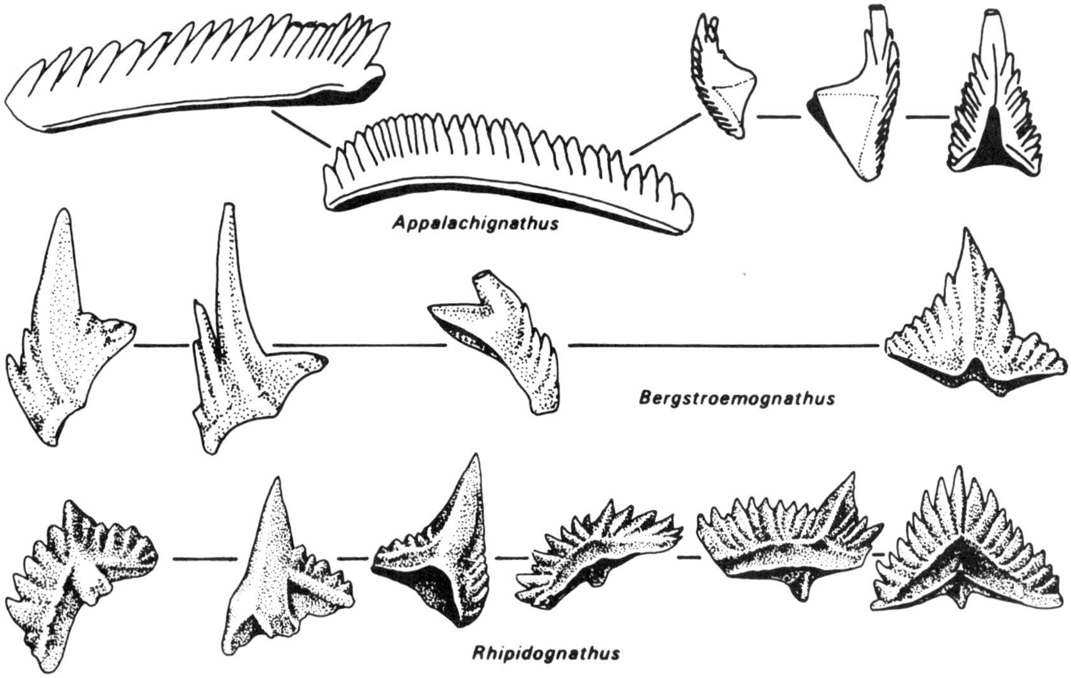

Fig. 5.26. Elements and apparatuses typical of species of the three principal genera of the Rhipidognathidae. Figures of *Bergstroemognathus* from Serpagli (1974); those of *Appalachignathus* from Bergström et al. (1974). Sa elements on right, followed to left by Sb, Sc, M, Pb, and Pa elements.

cusp. (A similar situation, of course, exists in the family Periodontidae.)

Bergstroemognathus (Fig. 5.26), known now also from the Lower Ordovician of China, has a basically trimembrate apparatus, with an anteroposteriorly compressed, palmate, alate Sa element that lacks a posterior process; a bipennate M element that lacks posterior denticles but has a long, denticulated anterior process; and a bowed element (or P and Sb elements) with a distinct cusp and short, sparingly denticulated processes.

Rhipidognathus, which has been reported now from the Lower Ordovician of China, from much lower in the Middle Ordovician of North America than was the case at the time *Treatise* copy was written, and from mid-Silurian rocks, has also been discovered to have a seximembrate, not a trimembrate apparatus, the components of which are illustrated in Fig. 5.26.

The type and only known species of *Appalachignathus* has a skeletal apparatus that is more elaborate than that of *Bergstroemognathus extensus*, but one in which there are also clear homologues (Fig. 5.26). Ancestors of *Appalachignathus* are not obvious in any known part of the *Rhipidognathus* stock, but that is the genus stratigraphically most proximate, so it is the one identified as ancestral in Fig. 5.11.

5.7.12 Family Periodontidae Lindström, 1970

Periodon (Fig. 5.27) and *Microzarkodina* (Fig. 5.27), which were assigned in the *Treatise* to the Periodontidae and Cyrtoniodontidae, respectively, are here included in the same family, not because I have great experience with, or basic new information on, either genus, but because it seems likely from recent studies (Löfgren, 1978; Dzik, 1983) that species of the two genera represent branches of the same Early Ordovician stock that evolved subsequently along somewhat different lines.

The type-species of *Periodon*, *P. aculeata*, has a seximembrate skeletal apparatus, with angulate pectiniform P elements, one of which is conspicuously twisted, or "tortiform"; dolabrate (or falodontiform) M elements, with an adenticulate posterior process; and an S series in which the ramiform components have short, minutely denticulated lateral processes and a long posterior process that is arched, laterally compressed, and surmounted at the point of maximum geniculation by a denticle that is wider and conspicuously longer than the others. The stratigraphic (and probably phylogenetic) predecessor of *P. aculeata*, *P. flabellum*, had geniculate coniform, not anteriorly denticulate dolabrate elements in the M position, and the ramiform S elements were apparently neither so profusely denticulated nor so conspicuously arched as those in comparable positions in the apparatus of *P. aculeata*. An even older species of *Periodon*, which has not yet been named, is represented in the Proteus and Elegans zones of Jämtland (Löfgren, 1978) by adenticulate alate elements interpreted as occupants of the Sa position and by adenticulate pastinate (or acodontiform) elements. The latter, and the "tortiform" angulate elements that evidently replace them in the apparatuses of younger species, are the elements that indicate that *Periodon* and the Periodontidae belong in the Prioniodontida. Of course, the distinctive S elements of *P. aculeata* and *P. grandis*, the species of *Periodon* that I know best, are also generally similar to those of *Phragmodus*, and this led me in the heyday of form taxonomy to suggest that *Periodon* might be an older name for *Phragmodus*. I no longer hold such a view.

Fig. 5.27. Elements typical of the skeletal apparatuses of species assigned to the Periodontidae. Top elements in each row Pa, followed downward by Pb, Sa, Sb, Sc, and M elements.

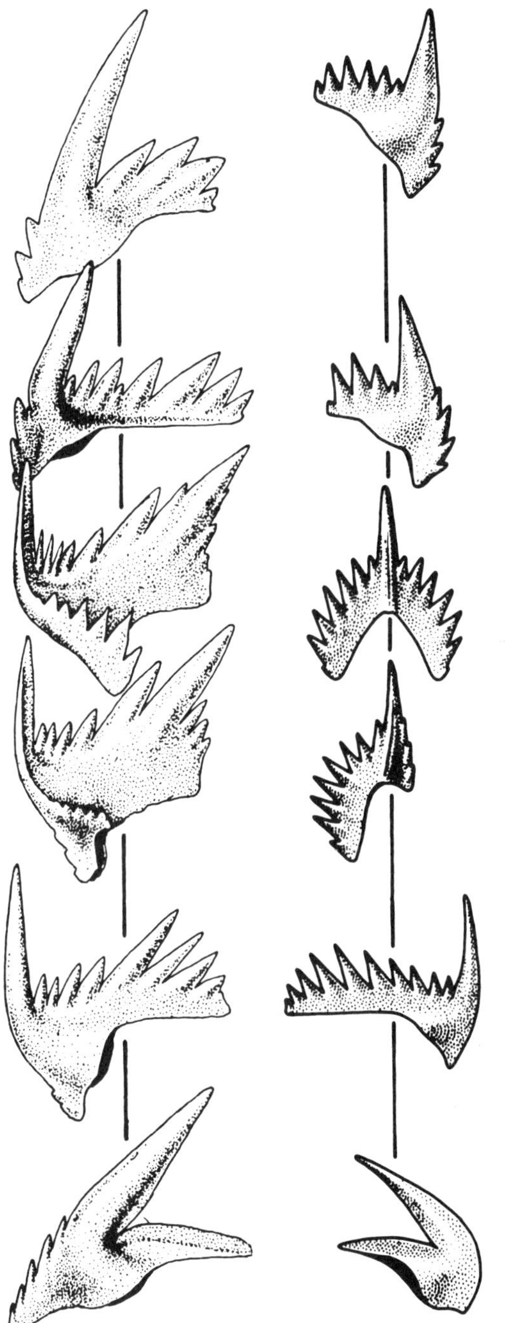

Species of *Microzarkodina*, such as *M. flabellum*, the type-species, have quinqui- or seximembrate skeletal apparatuses in which the rounded anterior margin of pastinate (?) P elements is not developed as a process and the short, anteriorly deflected lateral process commonly bears no more than a single denticle. M elements are geniculate coniform structures, and S positions are occupied by alate, tertiopedate, and bipennate elements with a long, straight posterior process.

According to the views just summarized, *Periodon* emerged from roots in early oistodontid populations by development of a highly differentiated skeletal apparatus in which processes of component elements gradually became longer and more conspicuously denticulated. As in the presumably ancestral Oistodontidae, P elements never differentiated into conspicuously pastinate structures, but S elements became distinctively arched and denticulated. In *Microzarkodina* development was apparently similar, but the posterior processes of S elements are not arched, M elements develop no anterior denticles, and the feature I interpret as a lateral process in P elements remained short and bore only a single denticle.

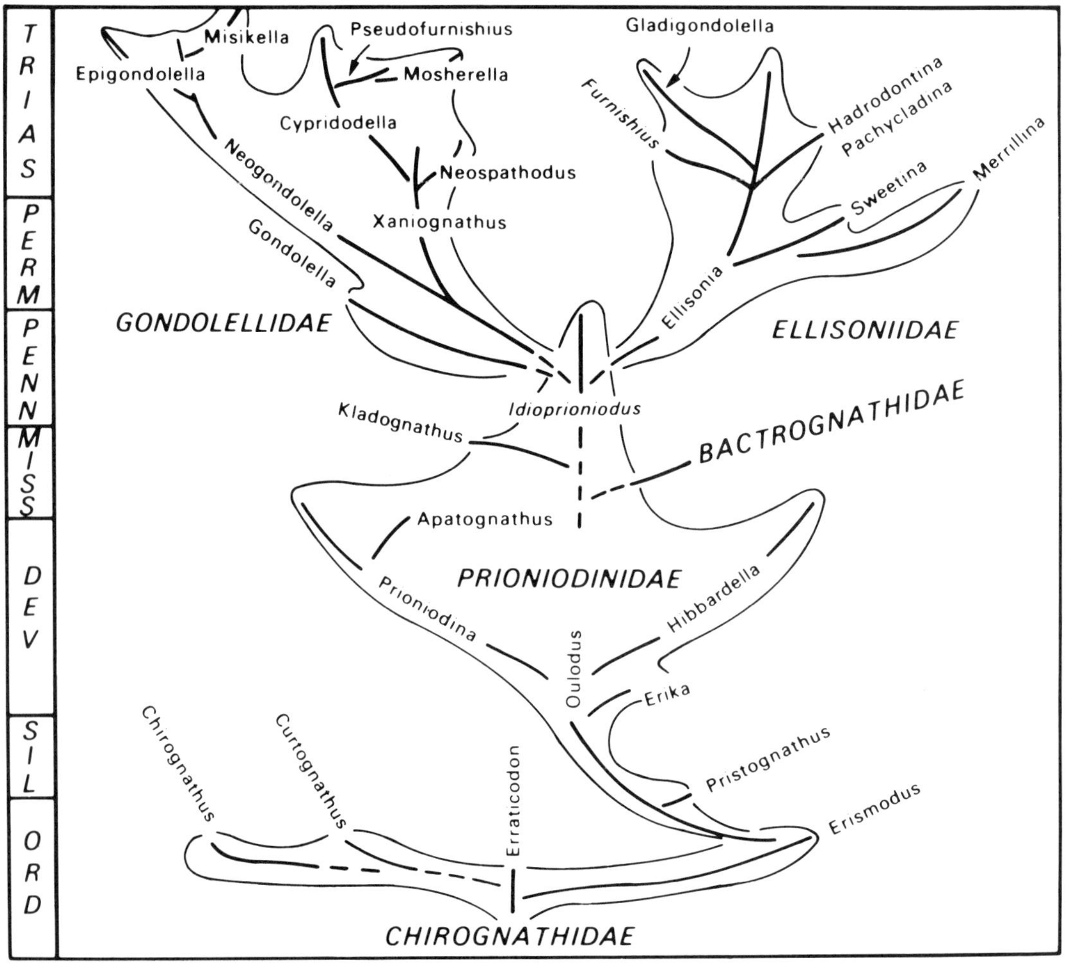

Fig. 5.28. Genera and families of the Prioniodinida.

5.8 The Prioniodinida, New

A few of the conodonts included here were assigned by *Treatise* authors to the Prioniodontida, but a majority were included in the superfamilies Hibbardellacea and Gondolellacea, which are here joined to form the order Prioniodinida. As so conceived, the Prioniodinida includes 5 families and 45 genera. Its oldest known representatives are Early Ordovician forms assigned to *Erraticodon* (Chirognathidae); its youngest are the last survivors of the Conodonta (*Misikella*), which are included in the Gondolellidae and are found as fossils in rocks of latest Triassic age. Temporal distribution and inferred relations among major prioniodinide generic stocks are summarized in Fig. 5.28.

Except as their skeletal apparatuses have been specially modified, adapted, or reduced, conodonts of the order Prioniodinida are characterized by a mineralized cephalic apparatus that is sexi- or septimembrate and includes highly distinctive extensiform digyrate elements in the two P positions. All six or seven types of elements in the apparatus tend to be stout, of the same size, and intergradational morphologically. Denticles are characteristically discrete and peglike; and white matter is commonly of minor amount and, in elements of several species, confined either to the axial regions of cusp and denticles or to diffuse, fuzzy-edged clouds that are irregularly distributed through cusp and denticles. Species of a few early genera (*Chirognathus*, *Curtognathus*) and of at least one Triassic one (*Parachirognathus*) formed elements with little white matter and thick lamellae that are mineralized by apatite crystallites whose prism surfaces are oriented parallel to the direction of growth. Such elements tend to break as if they were internally fibrous, and Branson and Mehl used this feature to distinguish the order Neurodontiformes.

It is also a characteristic of the Prioniodinida that there is little difference in size or basal morphology between the distinctive digyrate elements assigned to the two P positions, the ramiform elements that occupy M positions, and those that make up the morphologically conservative symmetry-transition series. For this reason, different authors have assigned comparable elements of various species to different positions in the apparatus, with the result that homologies are not always obvious. I will attempt to be uniform in the following discussion (and in the classification in Appendix A), but my views do not always coincide with those of other authors.

5.8.1 Family Chirognathidae Branson and Mehl, 1944

Erraticodon, the oldest known chirognathid and the most ancient member of the Prioniodinida distinguished thus far, appeared late in the Early Ordovician, when it achieved a very wide biogeographic distribution. *E. patu* Cooper (Fig. 5.29), the oldest (and best-known) species, is distinguished from other chirognathids primarily by features of the curious element I assign to the Pa position. At first glance that element appears to be pastinate (or "dichognathiform"), and its association in the same apparatus with a Pb element that is an extensiform digyrate (or "oulodontiform") element, and thus typically prioniodinide, might be taken to suggest that *E. patu* straddles the boundary between the Prioniodontida and the Prioniodinida. Closer inspection, however, reveals that the distinctive Pa element of *E. patu* is also basically an extensiform digyrate element, but the anterobasal part of the cusp extends downward to form a distinct anticusp that bears several denticles on its upper margin and is thus a true anterior process. It seems unnecessary to invent a new morphologic term for such an element, which (if need be) might be described as a quadriramate structure with a short, adenticulate posterior process.

Ancestors of *Erraticodon* (and thus the Prioniodinida) are not known, but it is surely reasonable to regard *Erraticodon* as the plexus from which *Erismodus* evolved in the shallow Middle Ordovician seas of North America. Species of *Erismodus* (Fig. 5.29) differ from those of *Erraticodon* primarily in that the Pa components of their seximembrate apparatuses lack a denticulated anterior process (but retain a bosslike anticusp), and bipennate M elements lack denticles on the short posterior process (and thus have sometimes been described as "falodontiform"). The oldest species of *Erismodus* with which I am acquainted are represented by typically "fibrous" or "neurodontiform" elements. However, the skeletal elements of later species (e.g., *E. quadridactylus*) have more white matter and commonly break like other nonfibrous types. This suggests to me that "fibrous" histology may somehow be related to the probably quite unusual conditions that obtained in the shallow, very warm Midcontinent waters in which early *Erismodus* developed. Later species, such as *E. quadridactylus*, may represent forms adapted to the more normal conditions that probably characterized Midcontinent seas later in the Middle Ordovician.

The bizarre *Chirognathus* (Fig. 5.29), represented in the shallow, tropical waters of the North American Midcontinent by a single species, is allied morphologically to *Erismodus*, from which it differs primarily in a skeletal apparatus composed of elements with a scarlike attachment surface skewed to one face, an alate Sa element without a posterior process, and an M element with denticulated anterior and posterior processes. *Curtognathus* (in which I include *Trucherognathus* and *Polycaulodus*) is also a likely derivative of *Erismodus*, but its apparatus has not yet been competely reconstructed and, like *Chirognathus*, it is not likely to be on the "main line" of prioniodinide evolution. These two genera are significant, how-

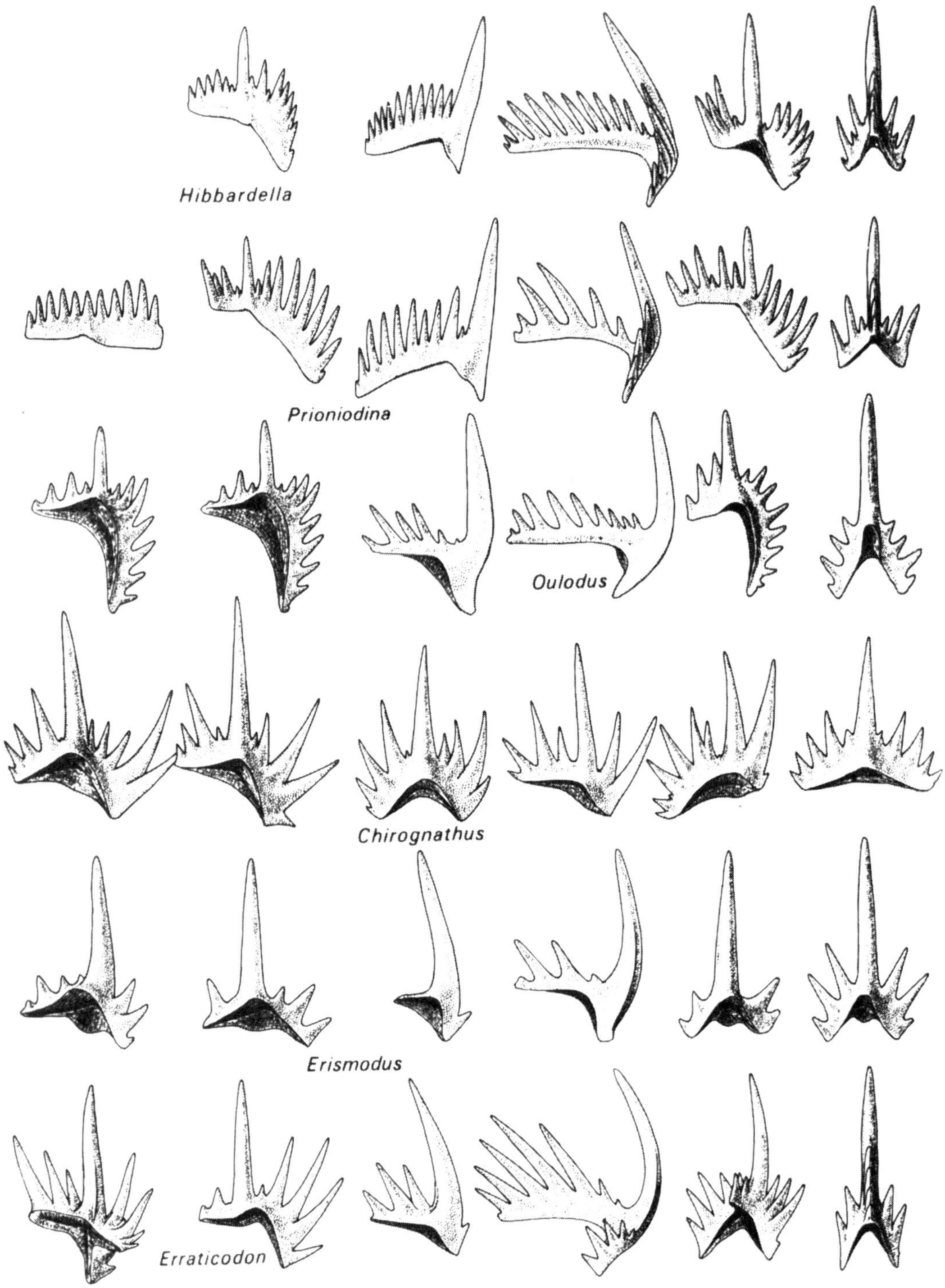

Fig. 5.29. Apparatuses of species typical of the Chirognathidae (*Erraticodon, Erismodus, Chirognathus*) and Prioniodinidae (*Oulodus, Prioniodina, Hibbardella*). In all apparatuses Sa elements are at the right, followed toward the left by Sb, Sc, M, Pb, and Pa elements. The Pa element of *Hibbardella* has not been described.

ever, in being the primary "fibrous" conodonts and thus the nucleus of Branson and Mehl's Neurodontiformes.

5.8.2 *Family Prioniodinidae Bassler, 1925*

Oulodus (Fig. 5.29), the oldest prioniodinid, seems clearly to have evolved from *Erismodus*, as Schönlaub and I suggested in 1975. In the apparatus of *O. serratus*, the type species, there are extensiform digyrate (or "oulodontiform") elements in both positions, only minor patches of white matter in any element, and a dolabrate (instead of an anteriorly denticulate bipennate) element in the M position. These features, and scrappy evidence from Ordovician rocks in northeast Oklahoma (Amsden and Sweet, 1983) of a possibly slightly older species, suggest that *O. serratus* may have been somewhat specialized, since all potential ancestors in *Erismodus* have an extensiform digyrate (or "prioniodiniform") element in the Pa position, and this pattern is repeated in all younger species of *Oulodus*. However this may be, the plan of the sexi- or septimembrate apparatus of *Oulodus* is clearly anticipated in those of *Erismodus* and *Erraticodon*, its ancestor, and there can be little doubt about close relationship.

Oulodus quickly adapted to the probably rigorous conditions in shallow subtidal parts of late Middle Ordovician seas, in which its species were soon established as ecologic hallmarks. With enlargement of those marginal zones in the later Ordovician, *Oulodus* also increased in diversity, and its evolutionary development is recorded in Late Ordovician and Early Silurian strata by a succession of chronospecies that is of considerable utility in Late Ordovician biostratigraphy (Sweet, 1984). *Pristognathus*, represented by a single, short-ranging Late Ordovician species, is closely related in apparatus architecture to *Oulodus* and clearly belongs with that genus in the Prioniodinidae.

Several Silurian species assigned either to *Oulodus* or *Delotaxis* (which I include in *Oulodus*) carry the Prioniodinida through the Silurian into the Devonian but provide evidence of very little change in basic plan or diversification. In the Early Devonian, when some of the dominant ozarkodinides began to develop platformed pectiniform elements in P positions and started to experiment with ontogenetic or phylogenetic reduction in the development or mineralization of elements in various apparatus positions, the Prioniodinida remained conservative and development produced no startling modifications.

Devonian workers have not yet dealt in detail with the Prioniodinida, but Klapper and Philip (1972) have demonstrated that the apparatus of *Hibbardella angulata* (Fig. 5.29), type-species of *Hibbardella*, differs from that of *Oulodus* primarily in that the alate Sa element of the former has a denticulate posterior process, whereas that of *Oulodus* has a stubby posterior process that lacks denticles. For this reason the assembly of elements from the upper Middle Devonian of Manitoba that Uyeno (Norris, Uyeno and McCabe 1982) has recently referred to *Oulodus* may represent a species of *Hibbardella*, instead.

Murphy and Matti (1982) have established the genus *Erika* (Fig. 5.30) for a single Early Devonian species, which had a seximembrate apparatus with extensiform digyrate elements in both P positions, a breviform digyrate element in the M position, and a symmetry-transition series that includes alate, extensiform digyrate, and bipennate ramiform elements. The apparatus of *Erika divarica*, the only species known, is similar in many ways to that of both *Oulodus* and *Hibbardella*; however, the M element is not much like that formed by species of either genus, and the alate Sa element lacks a posterior process and thus differs from any other prioniodinide conodont with which I am familiar.

According to Nicoll (1980), the apparatus of Late Devonian species of *Apatognathus* (Fig. 5.30) is also prioniodinide, and I include the genus with the Prioniodinidae in Fig. 5.28. The element typical of *Apatognathus* in form taxonomy was apparently an occupant of the M position.

Sparling (1981) has also concluded that the apparatus of at least one Middle Devonian species of *Prioniodina* (*P. tortoides*) (Fig. 5.29) is closely comparable to those of *Oulodus* and *Hibbardella* in element type and number and in morphologic features such as style of denticulation. Sparling's discussion suggests that

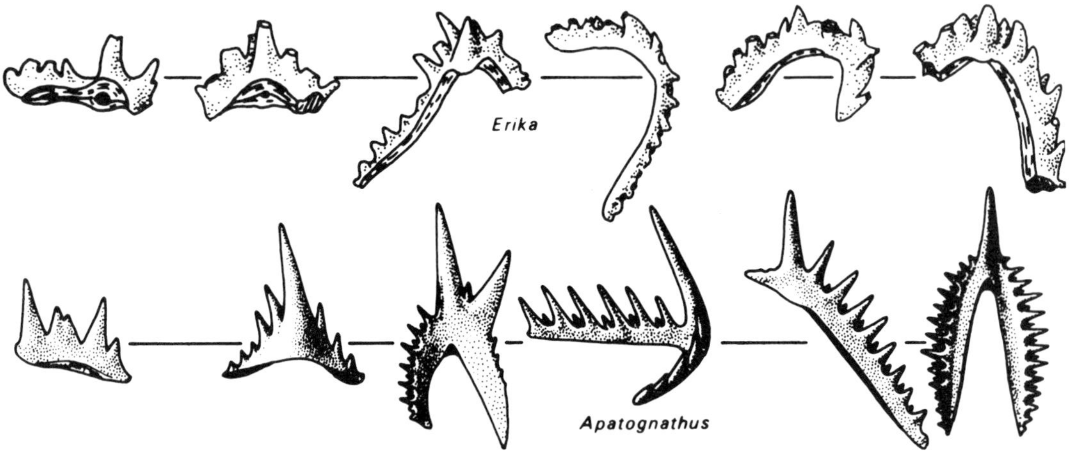

Fig. 5.30. Elements and apparatuses typical of species assigned to *Apatognathus* and *Erika* (Prioniodinidae). Sa elements on right, followed to left by Sb, Sc, M, Pb, and Pa elements. Sketches of *Erika* elements from Murphy and Matti (1982).

other species of *Prioniodina* await description in a multielement sense, and I have anticipated that this might be the case by using the name Prioniodinida for the order and Prioniodinidae (not Hibbardellidae) for the family.

Collections from pre-Chesterian Mississippian rocks contain a variety of large, discretely denticulated ramiform and pectiniform elements that probably represent various species of the Prioniodinidae, but no one has yet worked out the recurrent groups recorded or dealt with them taxonomically. These groups almost certainly include the skeletal record of prioniodinid species ancestral to the specialized Bactrognathidae, which will be considered in Section 5.8.3. They probably also record early representatives of *Idioprioniodus*, a distinctive prioniodinid that has a documented stratigraphic range from Chesterian through Pennsylvanian and clearly played an important role in the late Paleozoic history of the Prioniodinida.

Rexroad (1981) reports that Chesterian (Mississippian) elements previously identified as form species of *Kladognathus*, *Magnilaterella*, *Hibbardella*, *Neoprioniodus*, and *Ligonodina* form recurrent groups that represent congeneric species of a multielement genus for which *Kladognathus* Rexroad, 1958, seems to be the oldest available name. Elements representing form species of *Lambdagnathus* Rexroad, 1958, have been cited by Norby (1976) and Rexroad (1981) as potential additions to the recurrent groups on which *Kladognathus* species are now based. This seems eminently logical to me, despite the fact that these relatively rare elements do not group with other components of the recurrent groups at a statistically significant level. An imaginative reconstruction of *Kladognathus* is included in Fig. 5.31.

The Sa position in the multielement *Kladognathus* apparatus was apparently occupied by alate elements with a denticulate posterior process referred previously to *Hibbardella*; the Sb position was filled (in my opinion, not Rexroad's) by tertiopedate elements typical of the form genera *Kladognathus* and (probably) *Lambdagnathus*; and the Sc position by bipennate elements formerly assigned in form taxonomy to *Ligonodina*. M elements of multielement *Kladognathus* are pick-shaped dolabrate elements of the sort long identified as form species of *Neoprioniodus*. Rexroad (1981) did not recognize P elements in his reconstruction of the apparatuses of *Kladognathus* species, but I suggest that elements previously recognized as form species of *Magnilaterella* are more likely to have occupied P positions in the apparatuses of *Kladognathus* species than the Sb positions to which Rexroad (1981) assigned them.

As just reconstructed, multielement *Kladognathus* fits the general model of the Prioniodinida, and I assign it to this group here. There is little in the published record to suggest the immediate ancestry of *Kladognathus*, but species of the genus appear to have been members of nearshore, shallow-water associations that also included species of *Cavusgnathus*. With only a few exceptions, this ecologic preference is also characteristic of the Prioniodinida and may help confirm the assignment.

Idioprioniodus (Fig. 5.31) was diagnosed in a multielement sense by Merrill and Merrill (1974), and their reconstruction has been confirmed in a number of subsequent studies. Elements of the skeletal apparatus of the two species known are large, mostly hyaline, and include distinctive digyrate pectiniform elements in both the P positions; dolabrate M elements (including the type form species of *Neoprioniodus*); and a symmetry-transition series of four ramiform elements with large cusps and very short, sparsely denticulated processes. Especially significant in this array of skeletal components is the digyrate element that I assign to the Pb position. Lateral processes of this element tend to be quite short (although they may be longer than those of other elements of the *Idioprioniodus* apparatus), and they project downwardly and anteriorly so as to form a distinct V with the inflated, but undenticulated, base of the cusp. A closely similar type of digyrate element is common in collections of Permian and Triassic conodonts and was referred by Clark and Ethington (1962) to the form genus *Enantiognathus*. I have for some time regarded such "enantiognathiform" digyrate elements as occupants of the Pb position in the skeletal apparatus of *Ellisonia* and in those of a succession of species of multielement *Xaniognathus* and *Cypridodella*. Hence, occurrence in the Pb position of the *Idioprioniodus* apparatus of elements reminiscent of these later Paleozoic and Triassic specimens provides strong support for the origin of *Ellisonia* and *Xaniognathus* in *Idioprioniodus* or a closely related genus. And, because I regard *Xaniognathus* as the most generalized component of the family Gondolellidae and *Ellisonia* as the central radical of the Ellisoniidae, structural similarity of Pb elements plays an important role in my ideas concerning the origin of those lineages.

5.8.3 *Family Bactrognathidae Lindström, 1970*

In the *Treatise* Austin and Rhodes assembled the distinctive Mississippian genera *Bactrognathus*, *Doliognathus*, *Dollymae*, *Eotaphrus*, *Scaliognathus*, and *Staurognathus* in a family Bactrognathidae. However, they were unable to suggest the relationship of this obviously interrelated stock to any of the other groups of conodonts described in the *Treatise*.

In 1985 Karl Chauff demonstrated (to my satisfaction, at least) that *Bactrognathus*, *Doliognathus*, and *Staurognathus*, the central elements of the Bactrognathidae, have skeletal ap-

Fig. 5.31. Elements and apparatuses typical of species assigned to *Kladognathus* and *Idioprioniodus* (Prioniodinidae). Sa elements on right, followed to left by Sb, Sc, M, and P (or Pb and Pa) elements.

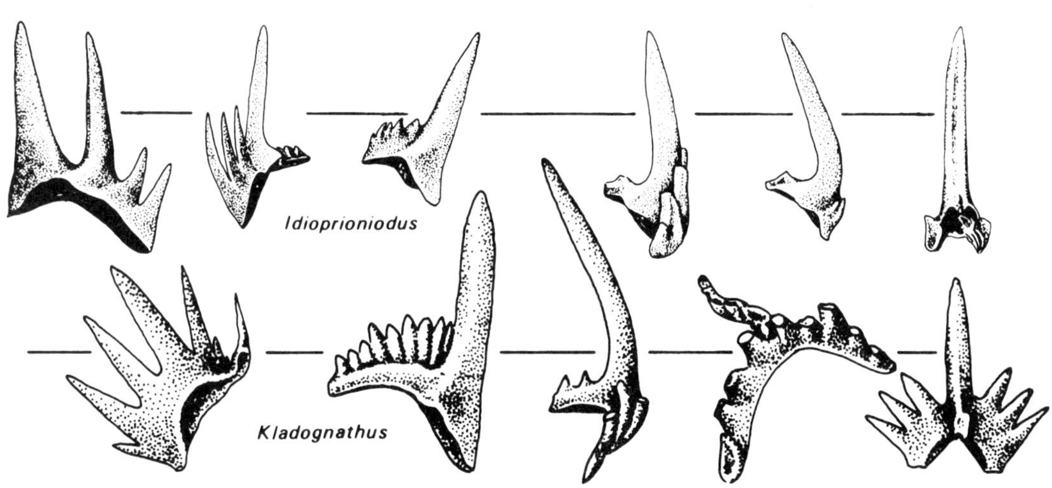

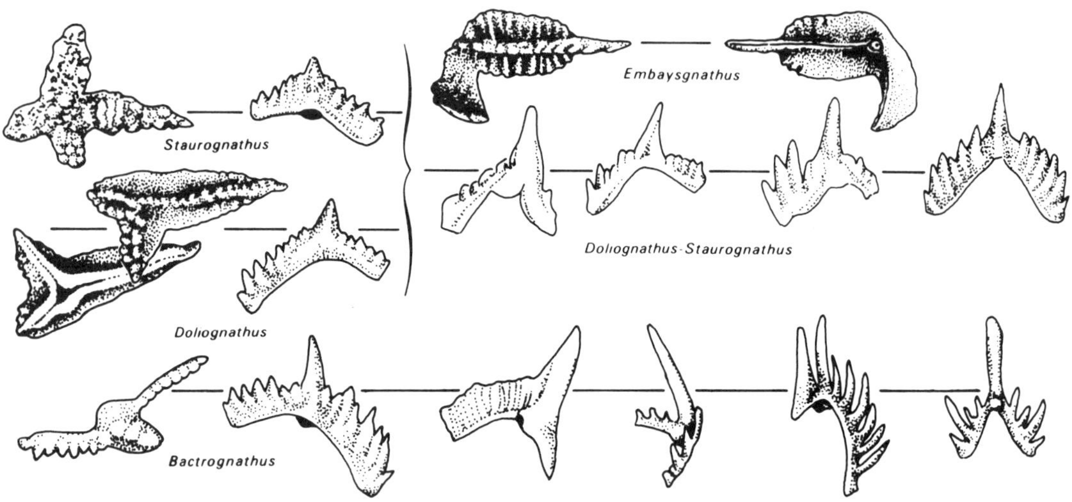

Fig. 5.32. Elements and apparatuses of species assigned to genera of the Bactrognathidae. Complete apparatus of *Embaysgnathus* unknown; *Doliognathus* and *Staurognathus* species have S and M elements that are indistinguishable. *Embaysgnathus* figures original; others are redrawn from Chauff (1985).

paratuses with so many prioniodinide features that association with this order seems quite likely. Chauff showed that species of both *Bactrognathus* (Fig. 5.32) and *Doliognathus* (Fig. 5.32) have an array of ramiform elements in the symmetry-transition (or S) and M series that is closely similar to those in the apparatuses of species of the prioniodinids *Hibbardella* and *Prioniodina*, and that P positions in the apparatuses of species in each genus are occupied by extensiform digyrate elements, which are also analogous to elements assigned a comparable position in *Prioniodina* and *Hibbardella* apparatuses. In the apparatus of the oldest-known species of *Bactrognathus* (*B. hamatus*) the short lateral process bends slightly outward and the long process curves slightly inward distally; otherwise, the structure might be described as a carminate pectiniform element with bladelike anterior and posterior processes. As indicated in Fig. 5.32, the short lateral process of the Pb element in younger species bends more distinctly outward and, in one group (*B. excavatus* and *B. anchorarius*), broad, platformlike processes develop on both the anterior and posterior sides of the cusp. Similar phylogenetic modification of the Pb element characterizes the *Doliognathus-Scaliognathus* lineage.

Ancestors of the Bactrognathidae were not identified by Chauff, who concluded that the plexus might have originated in a group of conodonts perhaps related to *Cryptotaxis* that lived far offshore and is thus not represented in the known Mississippian record. While this may be a reasonable interpretation, I wonder if Chauff might not have been looking in the wrong direction? From the time of their inception in the Early Ordovician, prioniodinide conodonts seem to have been most diverse taxonomically and most widespread biogeographically in very shallow-water environments. Episodically, denizens of those biotopes seem to have spawned stocks that successfully, if briefly, invaded more normal offshore habitats. The Bactrognathidae (and at least some species of *Idioprioniodus*) may represent such a stock.

The Bactrognathidae are of additional significance in a discussion of the Prioniodinida because, if I have interpreted them correctly, they represent the first attempt at platform making by this major group of conodonts. As in other major conodont groups, the platformed elements arose in *Bactrognathus*, *Doliognathus*, *Staurognathus*, and probably in the curious *Embaysgnathus* (Fig. 5.32) by modification of elements in one of the P positions. It may not be entirely fortuitous that, despite their origin in extensiform digyrate structures, these platformed pectiniform elements came to be stelliplanate and thus similar in their general organization to those developed at various times in unrelated prioniodontide and ozarkodinide stocks.

5.8.4 Family Ellisoniidae Clark, 1972

The first representatives of the Ellisoniidae, interpreted by von Bitter and Merrill (1983) as two species of *Ellisonia*, appeared in the Atokan (Pennsylvanian). The youngest species known are of Late Triassic (mid-Carnian) age, and the type-species (*E. triassica*) is Early Triassic. In architecture of their skeletal apparatuses, species of *Ellisonia* (Fig. 5.33) are similar to species of *Idioprioniodus*, from which they almost certainly developed. The apparatus is a sexi- or septimembrate complex, with morphologically conservative alate, digyrate, and bipennate ramiform elements in the S and M positions, an extensiform digyrate element in the Pa position and, in the Pb position, another digyrate element, with processes reflected sharply downward and anteriorly. I have earlier commented on the significance of these "enantiognathiform" digyrate elements in establishing relationships within the Prioniodinida.

Obviously only minor morphologic differences separate the apparatuses of *Ellisonia* species from those of partly contemporaneous species of *Idioprioniodus*. Such features include the tendency for processes of *Ellisonia* elements to be longer and more profusely denticulated than those of *Idioprioniodus* and for the attachment surfaces of *Ellisonia* elements to be composed of small basal pits bordered by prominent zones of recessive basal margin. In some *Idioprioniodus* elements the attachment surface exhibits features similar to those just described, but in none that I have seen are pits so small or recessive-margin zones so broad and conspicuous. It might also be noted that the M element in apparatuses of *Idioprioniodus* species is a distinctive dolabrate structure, whereas the one in *Ellisonia* apparatses is bipennate and readily confused (by me, among others) with elements common in Sb positions in the apparatuses of other species.

Early in the Permian a species of *Ellisonia* was probably the ancestor of a species of *Sweetina* (Fig. 5.33), known thus far only from a distinctive Pb element from the Leonardian Hess Limestone of Texas. Later Permian collections, described in detail by Wardlaw and Collinson (1986) and Croft (1978), include the complete array of elements, which indicates that *Sweetina* had a seximembrate apparatus like that of *Ellisonia* but had an element in the Pb position that is reminiscent of the triradiate extensiform digyrate element of Ordovician *Erraticodon*, the earliest prioniodinide known. That element has one long and one rather short lateral process and a stubby posterior process that bears only one or two denticles.

Species of *Merrillina* (Fig. 5.33), another Permian ellisoniid, developed skeletal apparatuses that were closely similar in their conformation to those of *Sweetina* species but had extensiform digyrate Pb elements that lack a posterior process. In successively younger Permian populations, the short lateral process of Pb components becomes even shorter, and in the youngest Permian collections I have seen, the short process is largely vestigial. Species of both *Sweetina* and *Merrillina* were evidently restricted to the Permian and were apparently best adapted to very shallow, agitated, probably hypersaline water. Commonly, their large elements are the dominant (or only) components of collections in which they occur; however, those elements also tend to be fragmentary, so reconstruction of apparatuses is difficult and only a few species are known in detail.

In the Early Triassic (Scythian) *Ellisonia* was joined (or replaced) in shallow-water environments by species of *Furnishius*, *Hadrodontina*, and *Pachycladina* (all illustrated in Fig. 5.33), which formed seximembrate apparatuses whose component elements are distinguished by wide zones of recessive basal margin and whose alate Sa elements lack a posterior process. Pa elements are extensiform digyrate structures in apparatuses of species of both *Furnishius* and *Hadrodontina*, but extensiform Pb elements develop an additional row of denticles on the anterior face. In Pb elements of *Hadrodontina* the supplemental row joins the main denticle row at both ends. In comparable elements of *Furnishius* the supplemental row joins at only one end and additional denticles

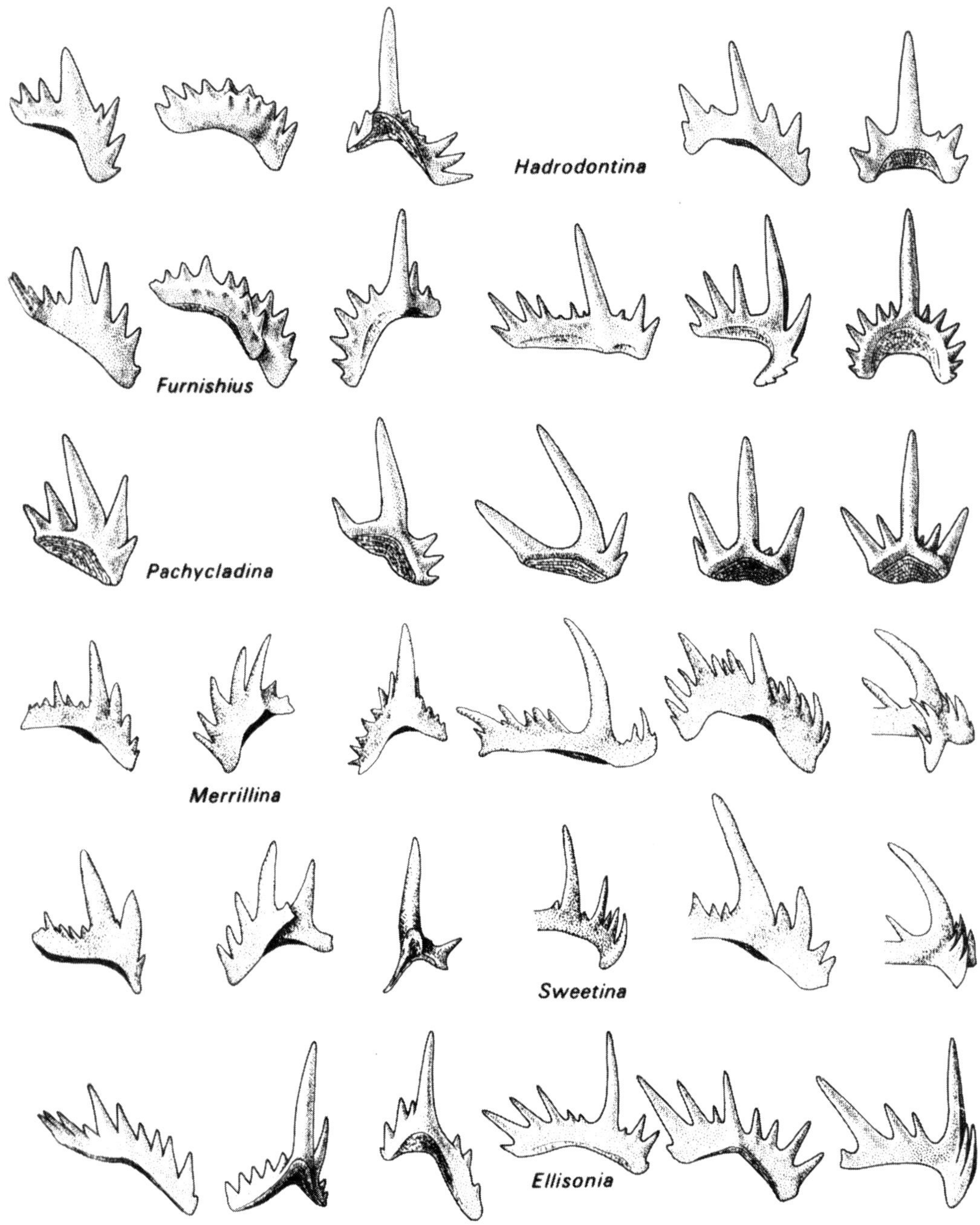

Fig. 5.33. Apparatuses of species assigned to genera typical of the Ellisoniidae. In all apparatuses the Sa element is on the far right, followed to the left by Sb, Sc, M, Pb, and Pa elements. Gaps in the horizontal rows indicate that no element assignable to that position is known.

may develop in the space between that row and the main denticle row. The only known species of *Pachycladina* had an apparatus that appears to be an extreme version of the ones formed by species of *Furnishius* and *Hadrodontina*, but it lacks supplementary rows of denticles in either P element.

The type-species of *Ellisonia*, *E. triassica*, is abundantly represented in collections I made many years ago from the Lower Triassic

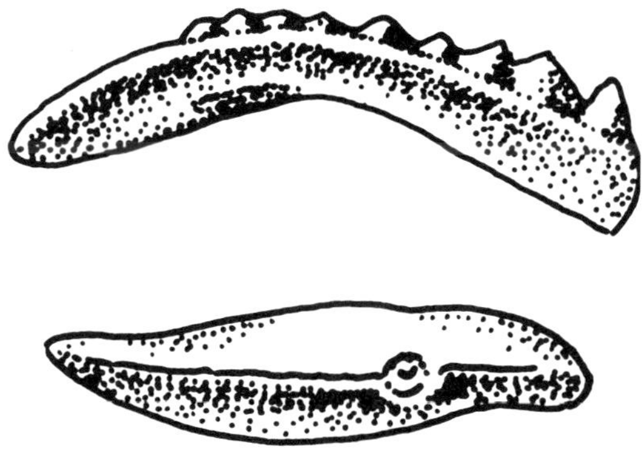

Fig. 5.34. Lateral and underside views of a Pa element of *Gladigondolella* (Ellisoniidae).

Thaynes Formation of Utah, and one of those collections also contains a number of platformed pectiniform elements that are gradational in morphology with the unplatformed Pa elements of *E. triassica*, and are also identical in form with the elements on which Huckriede (1958) based *Polygnathus tethydis*, the type-species of *Gladigondolella* Müller (1962) (Fig. 5.34). Paull (1983) noted the same features in her large Triassic collections and described the Lower Triassic specimens as representatives of *G. meeki*.

A distinctive feature of gladigondolellid Pa elements is the fact that narrow platform brims completely surround both anterior and posterior processes, hence the elements are formally anguliplanate. Comparable elements of the gondolellids, which also occur in Triassic rocks, are segminiplanate and thus are readily separated from those of *Gladigondolella*.

5.8.5 Family Gondolellidae Lindström, 1970

The late Paleozoic and Triassic conodonts assembled by *Treatise* authors in the families Gondolellidae and Xaniognathidae form an obviously interrelated stock that is best considered as a single family. The name Gondolellidae, coined by Lindström in 1970, is the oldest one available for the combined family.

Gondolellid conodonts formed a basically seximembrate skeletal apparatus distinguished primarily by angulate Pa elements, breviform digyrate Pb and M elements (the former "enantiognathiform"), and a symmetry-transition series that includes alate Sa elements with a long, profusely denticulated posterior process; extensiform digyrate Sb elements; and bipennate Sc elements, commonly with long, delicate anterior and posterior processes. In the apparatuses of generalized gondolellids, the posterior process of Pa elements tends to be short and fragile, whereas the anterior process is stout and is commonly surrounded at mid-height by a well-developed midlateral rib. In specialized gondolellids, however, the posterior process of Pa elements disappears and the segminate structures that result may also develop platform segments that are confined to the sides or joined around the posterior end of the cusp to form a continuous brim.

For many years it has been my contention that when gondolellid conodonts developed segminate or segminiplanate elements in the Pa position they also ceased mineralization of elements in other positions in the apparatus. According to this view, the Gondolellidae would include species with both unimembrate and seximembrate apparatuses. This interpretation of gondolellid organization is based on my observation that collections made from rocks within the range of the Gondolellidae commonly contain either the segminate Pa elements or an assortment of ramiform and pectiniform elements from which it is possible to assemble a complete seximembrate apparatus. Only rarely do collections include a balanced assortment of platformed and ramiform elements.

For a number of gondolellid species, particularly Triassic ones referred to *Neospathodus*, *Platyvillosus*, *Epigondolella*, *Mosherella*, and *Misikella* (see Fig. 5.35), I believe my original idea holds. That is, those species formed unimembrate apparatuses—or at least only those elements in the Pa positions were mineralized. However, evidence from a couple of natural assemblages [one almost certainly coprolitic (Rieber, 1980)] indicates that other gondolellid species with segminate or segminiplanate Pa elements may have had complete seximembrate apparatuses. For example, *Pseudofurnishius* (Fig. 5.35) and at least some species of *Gondolella* (Fig. 5.35) and *Neogondolella* (Fig. 5.35) appear to have had apparatuses composed of segminiplanate elements in Pa positions and ramiform elements like those typical of *Xan-*

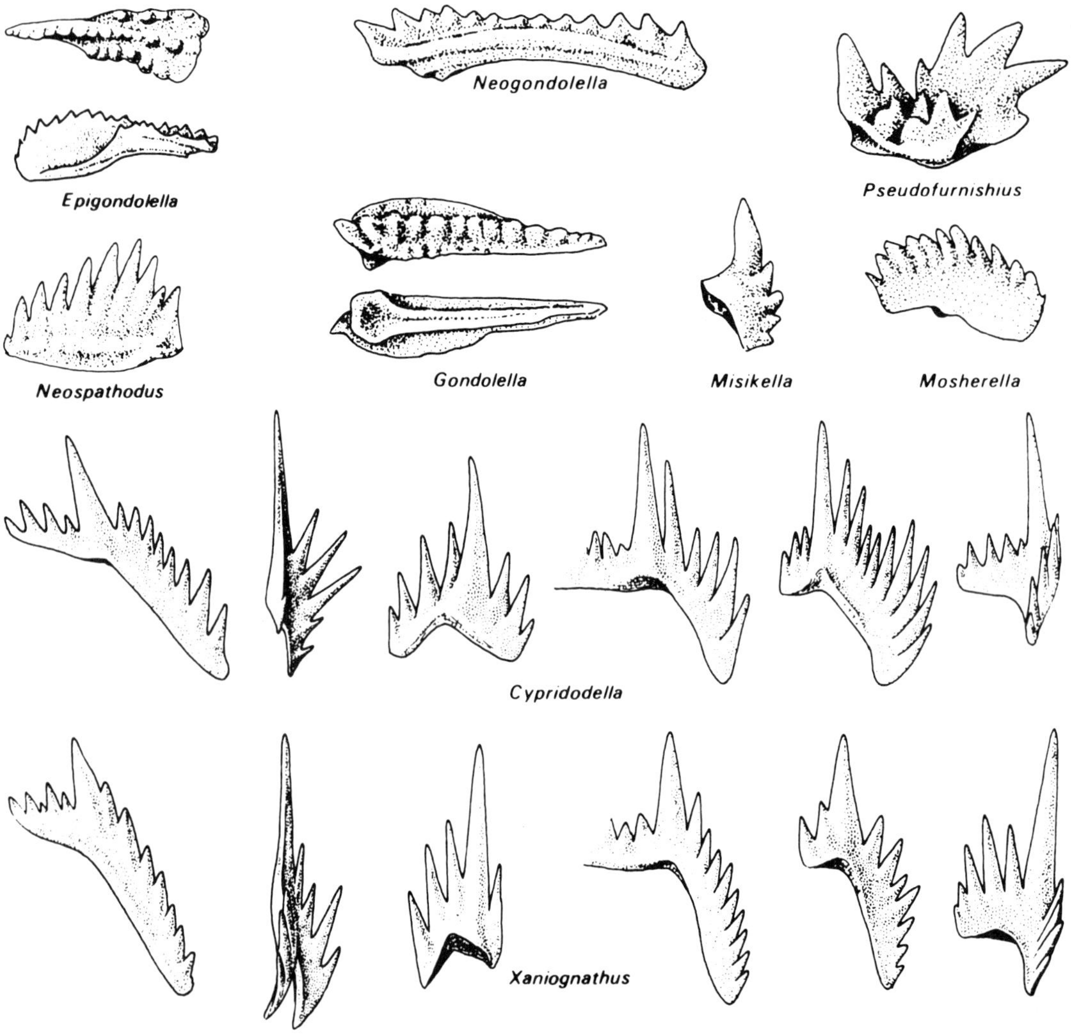

Fig. 5.35. Elements and apparatuses of species typical of genera assigned to the Gondolellidae. In the apparatuses of *Cypridodella* and *Xaniognathus* Sa elements are on the right, followed toward the left by Sb, Sc, M, Pb, and Pa elements. Only Pa elements are illustrated for the 7 genera at the top. Some of them may have had multimembrate apparatuses in which non-Pa positions were occupied by ramiform elements indistinguishable from those in the apparatuses of *Xaniognathus* and *Cypridodella*.

iognathus or *Cypridodella* (Fig. 5.35) in M and S positions.

In citing this evidence I do not mean to suggest (as some authors have done recently) that all elements previously assigned to multielement *Xaniognathus* or *Cypridodella* are merely ramiform components of the apparatuses of species of *Gondolella* or *Neogondolella*. The matter is clearly not so simple. Instead, I contend that species of *Xaniognathus* and *Cypridodella* made up the long-ranging, conservative stem-stock of the Gondolellidae from which more specialized forms with segminate or segminiplanate Pa elements developed from time to time. In some of those more specialized species the apparatus was modified only by substitution of a segminate (or segminiplanate) element for the more conservative angulate element in the Pa position. In others, development of segminate or segminiplanate Pa elements seems to have been accompanied either by almost complete reduction of the apparatus or by a failure to mineralize elements in M and S positions.

Origin of the Gondolellidae is obscure. Apparatuses of *Xaniognathus* species, the most generalized gondolellids, are very close in structure to those characteristic of *Ellisonia* species, and it is likely that both stocks appeared at about the same time in Atokan (Pennsylvanian) seas. I have already commented on the likelihood that *Ellisonia* developed from the prioniodinid *Idiognathodus*, and the gondolellids may have had a similar ancestry. This suggestion receives some support from the fact that Pennsylvanian *Xaniognathus* and *Gondolella* characterize a relatively deepwater biofacies that overlaps one in which *Idioprioniodus* is also very common. *Ellisonia* and its kin, of course, are especially common in rocks that represent very shallow-water environments.

My views on relationships between major genera of the Gondolellidae are summarized diagrammatically in Fig. 5.28, and apparatuses characteristic of typical gondolellid species are illustrated in Fig. 5.35. It is my suspicion, however, that formation of segminate (or segminiplanate) elements in Pa positions (and commonly concurrent apparatus reduction) happened repeatedly in the history of the Gondolellidae, perhaps as an expression of iterative adaptation by successive members of the *Xaniognathus-Cypridodella* stock to similar environmental conditions. If that were the pattern, the taxonomy summarized in Fig. 5.28 is vastly oversimplified, of course, although it is difficult to imagine a defensible alternative.

5.9 The Ozarkodinida Dzik, 1976

The basic skeletal apparatus of ozarkodinide conodonts is sexi- to septimembrate, with P positions occupied by carminate and angulate pectiniform elements or their platformed analogues. In ozarkodinide conodonts with fully developed apparatuses, elements in S and M positions are commonly generalized in morphology and, from published descriptions at least, are much the same from one species to the next. Pa elements, on the other hand, vary greatly and have provided the principal means of taxonomic differentiation. The oldest ozarkodinide conodonts are from rocks of late Middle Ordovician age; the youngest are from earliest Triassic strata.

Except for a few genera transferred from the Prioniodontida, my concept of the Ozarkodinida is essentially that of the superfamily Polygnathacea in the *Treatise* classification. In Fig. 5.36 (and in Appendix A) I divide the Ozarkodinida into 12 families. *Treatise* authors recognized only 6.

The central family in the taxonomic scheme of Fig. 5.36, the Spathognathodontidae, includes an assortment of ozarkodinide genera that is superficially quite heterogeneous. It could be made much less so by including *Eognathodus* with the Polygnathidae; by erecting separate families for the compact but distinctive stocks now represented by *Ancyrodella*, by *Amydrotaxis* and *Ancyrodelloides*, and by *Polygnathoides*; and by further revision of *Bispathodus* and reassignment of the resulting parts. In effect, that type of taxonomic trimming was begun in the *Treatise* by creating separate families for *Kockelella* and *Ancoradella* and for *Pterospathodus* and its kin.

I have resisted the temptation to erect the several new genera and families just suggested on the grounds that such a task is best undertaken by someone far more familiar than I with all the species involved. On the other hand, I have not hesitated to refer *Alternognathus* and *Siphonodella* to the family Elictognathidae, because species of both are well set off from those retained in the Spathognathodontidae and authors who have contributed most to unraveling the evolutionary history of the two generic stocks agree that, despite resemblance in some morphologic particulars, this stock originated in the Spathognathodontidae independently of the Polygnathidae and at a much later time in the Devonian. Those authors may not, however, be too happy about my choice of Elictognathidae.

I have also divided the Idiognathodontidae of the *Treatise* classification into Gnathodontidae and Idiognathodontidae, because such a division recognizes a natural and highly significant interval in the mid-Carboniferous history of both groups. The Palmatolepidae is newly established for a phyletic lineage previously submerged in the Polygnathidae, and the *Diplognathodus-Sweetognathus* lineage is re-

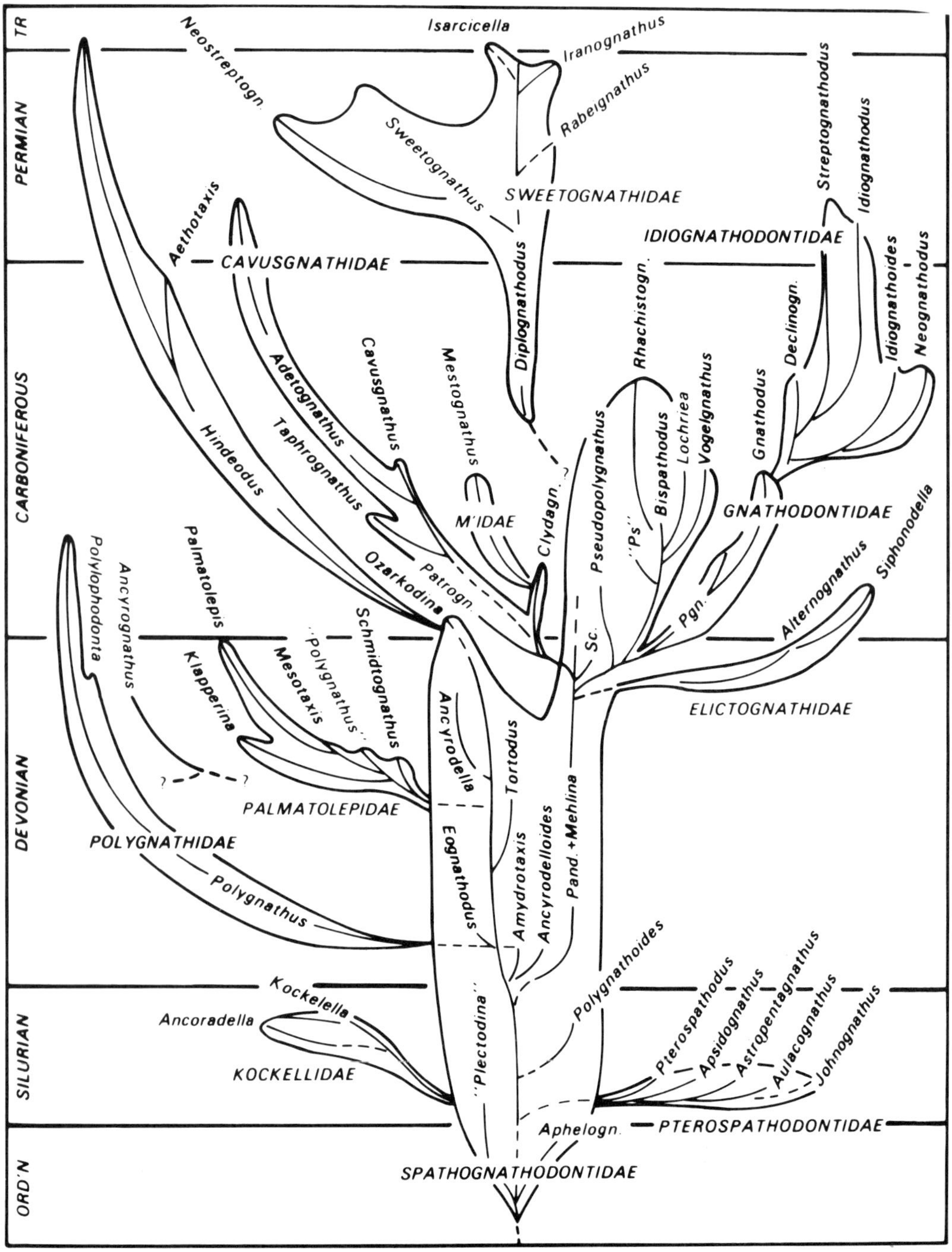

Fig. 5.36. Families and genera of the Ozarkodinida. *Pand.* = *Pandorinellina*; *Sc.* = *Scaphignathus*; *Pgn.* = *Protognathodus*; *Ps.* = *Pseudopolygnathus*; M'IDAE = Mestognathidae; other abbreviations are obvious.

moved from the Anchignathodontidae and treated as the ozarkodinide family Sweetognathidae, which was independently derived from *Bispathodus* or its relatives sometime in the Mississippian.

5.9.1 *Family Spathognathodontidae Hass, 1959*

Spathognathodontid conodonts are a variable, plastic, and long-ranging group, as befits their status as rootstock of the Ozarkodinida. The sexi- or septimembrate skeletal apparatus is basically simple, with bladelike carminate and angulate pectiniform elements in both P positions; a dolabrate or bipennate M element; and an undistinguished symmetry-transition series that includes alate, digyrate, and bipennate elements in the Sa, Sb, and Sc positions, respectively.

The alate Sa elements of most of the earliest spathognathodontids (e.g., "*Plectodina*," *Aphelognathus*) lack a well-developed posterior process, and breviform digyrate elements occupy the Sb position. A few, however, have Sa elements with long, profusely denticulated posterior processes and at least one species had developed extensiform digyrate Sb elements by early in Cincinnatian (Late Ordovician) time.

The Spathognathodontidae are recorded first near the base of the North American upper Mohawkian (i.e., the Rocklandian) by conodonts with several distinctive types of apparatus. One, typified by the specimens Bergström and I (1966) referred to "*Bryantodina*" *abrupta* (but probably referable now to *Yaoxianognathus abruptus*), has a simple carminate Pa element, an angulate Pb element, a dolabrate (or "cyrtoniodontiform") M element, and a symmetry-transition series with a bipennate Sc element, an extensiform digyrate Sb element, and an alate Sa element with a long, denticulated posterior process. Bergström and I (1966) referred the ramiform assembly to a species we named *Plectodina*? *posterocostata*; I transferred it to "*Bryantodina*" *abrupta* in 1979.

A second type of early spathognathodontid is represented by the Lexington Limestone specimens from Kentucky and Ohio that Bergström and I (1966) made the types of *Bryantodina*? *staufferi*. The apparatus of *B.*? *staufferi* includes only slightly arched, angulate Pa elements, conspicuously arched Pb elements, and an alate Sa element that lacks a denticulated posterior process. Other components of the skeletal apparatus of *B.*? *staufferi* have yet to be identified. *B.*? *staufferi* ranges into only the early part of the Late Ordovician (Cincinnatian), but a species with morphologically similar P elements, "*Ctenognathus*" *pseudofissilis* of Lindström (1959), may have continued this line into even younger parts of the Ordovician.

The third and by far the most common type of early spathognathodontid is represented by species that have been included in *Aphelognathus* and in *Plectodina*. The type-species of *Plectodina*, however, has a pastinate and not an angulate or carminate pectiniform element in the Pa position and is thus a typical prioniodontide. The type-species of *Aphelognathus*, and all other speces of the genus except Middle Ordovician *A. kimmswickensis* and *A. gigas*, has a pair of typically spathognathodontid P elements. Thus, in this volume I will assign species such as *Plectodina tenuis*, *P. florida*, and *P. bullhillensis* to "*Plectodina*," for their P elements are angulate and carminate pectiniform elements, not the pastinate and angulate structures typical of *Plectodina* s. s. Similarly, *Aphelognathus gigas* and *A. kimmswickensis* will have to be included for now in "*Aphelognathus*," in recognition of the prioniodontide structure of their Pa elements.

"*Plectodina*" (Fig. 5.37) evolved from *Plectodina*, in the late Middle Ordovician, an event documented in great detail in sequential collections from the Lexington Limestone of the Cincinnati Region. *Aphelognathus* (Fig. 5.37), which was also common in the late Middle Ordovician, came to be even more widespread in marginal parts of the low-latitude seas that covered North America in the Late Ordovician. The youngest "*Plectodina*" species are evidently the Early Silurian "*P.*" *hassi* and "*P.*" *oldhamensis*, which have been referred in most studies to *Ozarkodina* but evidently represent "*Plectodina*," because their skeletal apparatuses seem to have included breviform digyrate (or "zygognathiform") Sb elements, instead of the extensiform digyrate (or "plectospathodontiform") Sb elements typical of *Ozarkodina*.

The three types of spathognathodontid conodonts described thus far from Ordovician rocks differ from one another in several particulars and probably represent different lines of development. That there were also other lines is indicated by discovery a few years ago of distinctive but as yet undescribed forms with extensiform digyrate Sb elements in early Late Ordovician rocks in both Kentucky and Oklahoma and by the sporadic occurrence in Late Ordovician strata in Europe of more or less typical representatives of *Ozarkodina* (e.g., *Ozarkodina* n. sp. of Bergström), whose P, M, and Sc elements have been known for more

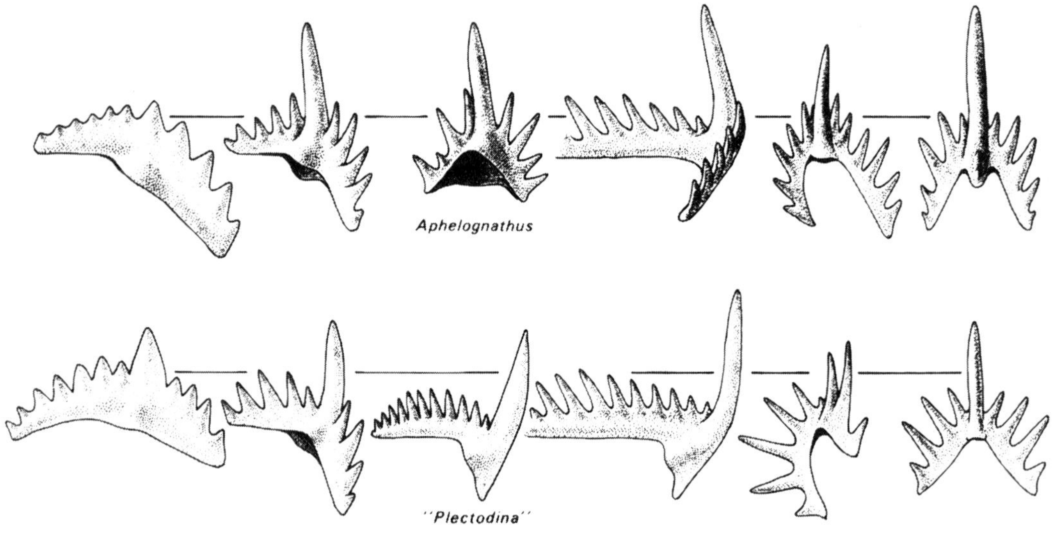

Fig. 5.37. Elements and apparatuses typical of species of *Aphelognathus* and "*Plectodina*," the most widespread and characteristic of the Ordovician ozarkodinides. Sa elements on right, followed to left by Sb, Sc, M, Pb, and Pa elements.

than a decade (Sweet and Bergström, 1972) but whose Sa and Sb elements have yet to be described.

Certain representatives of *Ozarkodina* (Fig. 5.38) appear (in the guise of *O. inclinata*) a bit above the base of the Silurian. These forms and their successors in later Silurian and Devonian strata have a conservative seximembrate skeletal apparatus, with carminate (or "spathognathodontiform") Pa elements, angulate (or "ozarkodiniform") Pb elements, dolabrate or bipennate M elements, and a transition series of alate, extensiform digyrate (or "plectospathodontiform"), and bipennate (or "hindeodelliform") elements in the Sa, Sb, and Sc positions, respectively. The alate Sa element of *Ozarkodina* species lacks a denticulated posterior process. Many authors refer Devonian and a few early Carboniferous species with triramous alate elements in the Sa position to *Pan-*

Fig. 5.38. Elements and apparatuses of species typical of the important ozarkodinide stem forms *Ozarkodina* and *Pandorinellina* (Spathognathodontidae). Sa elements on right, followed to left by Sb, Sc, M, Pb, and Pa elements. Note that the Sa element of *Ozarkodina* lacks a posterior process, whereas that of *Pandorinellina* has a well-developed posterior process.

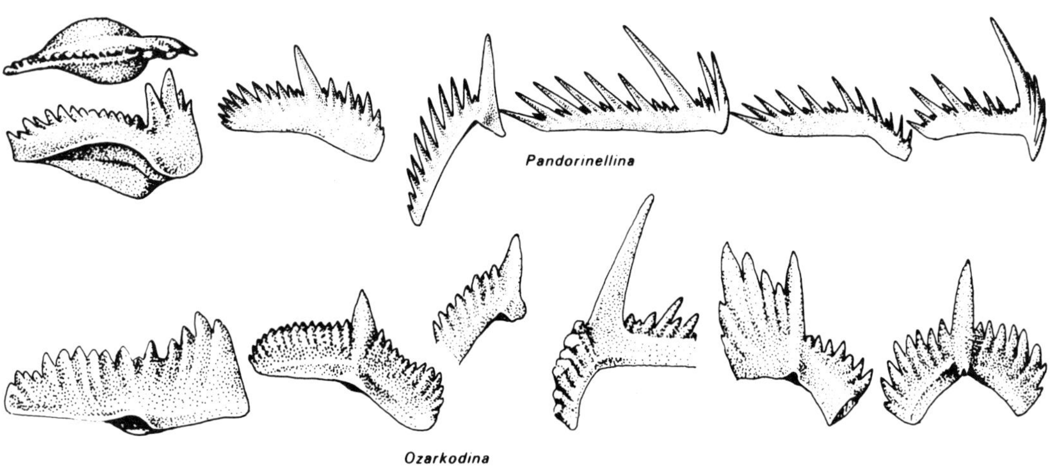

dorinellina Müller and Müller, 1957 (Fig. 5.38), which Uyeno (in Norris, Uyeno, and McCabe, 1982) has shown to be closely similar in apparatus construction to *Mehlina* Youngquist, 1945 (Fig. 5.40).

Unfortunately, the complete apparatuses of only a handful of spathognathodontid species have been worked out, so it is not always possible to decide if *Ozarkodina* or *Mehlina* or *Pandorinellina* is the appropriate generic assignment for many Devonian and early Carboniferous forms or which generic stock is involved at various evolutionary junctures. Sandberg and Ziegler (1979) have somewhat simplified the problem by restricting *Pandorinellina* to those spathognathodontids whose Pa elements are morphologically close to those of *P. insita* (Stauffer), the type-species. Carminate Pa elements of the latter have a high finlike anterior process that exhibits a slight right-lateral offset at its junction with the posterior process, which is built of much shorter denticles.

Although there are certainly well-marked differences in other elements of the spathognathodontid apparatus, dividing the family into subordinate taxonomic groups and tracing developmental history has been accomplished primarily through studies of Pa elements. This structure exhibits great plasticity, as I attempt to show diagrammatically in Fig. 5.39.

In typical representatives of *Ozarkodina* (Fig. 5.39A), the basal cavity of Pa elements is a groovelike indentation along the lower margin that expands laterally to form a cuplike structure at about midlength of the element and tapers toward the process extremities. From time to time in the long history of the *Ozarkodina-Mehlina-Pandorinellina* plexus, however, and possibly also in the still-obscure Early Silurian history of "*Plectodina*" and species derived from *Aphelognathus*, the laterally expanded portion of the basal cavity came to occupy a longer and longer segment of the underside of the posterior process. Elements like this (e.g., Fig. 5.39.G and 5.39H) have a ladle- or scooplike appearance, with the unmodified anterior process serving as handle and the posteriorly widened basal cavity as scoop. An exercise in basal-cavity modification like this apparently preceded development in the early Silurian of *Kockelella* and the Kockelellidae (Fig. 5.39B, 5.39C, and 5.39D); in the Early Devonian of *Ozarkodina selfi* (Fig. 5.39G), the short-lived species from which Lane and Ormiston (1979) derive both *Eognathodus* (Fig. 5.39H) and *Polygnathus* (Fig. 5.39I) of the Polygnathidae; and in the late Middle Devonian, when *Ozarkodina semialternans* (Wirth) spawned "*Polygnathus*" *latifossatus* Wirth and the Palmatolepidae. Evidently, as Branson and Mehl noted long ago, the notoriously unstable basal cavity of spathognathodontid Pa elements expanded and migrated toward the posterior end for the last time in the early Mississippian. Species produced in that episode and their successors are assembled here in the families Anchignathodontidae, Sweetognathidae, Cavusgnathidae, and Mestognathidae.

In a second type of development, sides of Pa elements expanded at or near midlength to form lobulate lappets on one or both sides above the central part of the basal cavity. Not uncommonly, the upper sides of these lateral extensions grew longer phylogenetically and came to support one to several nodelike denticles, commonly arranged in single rows. Pa elements modified in this way characterize early stages of *Pterospathodus* (Fig. 5.39F) in the early Silurian; *Amydrotaxis* and *Ancyrodelloides* (Fig. 5.39E, 5.39P, and 5.39Q) in the early Devonian; and *Ancyrodella* (Fig. 5.39K, and 5.39L) much later in the Devonian. Each of these modification episodes represents a geologically short excursion by spathognathodontid ozarkodinides into the never-never land of platform building in the Pa position. Although *Treatise* authors identified the platform-building excursion that led to *Pterospathodus* and its allies as a stock worthy of recognition at the family level, the others were not so identified and I will not tamper with this situation.

In species of Silurian *Polygnathoides* and mid-Devonian *Tortodus*, Pa elements were modified in yet a third way. That is, riblike thickenings of the sides of basically carminate spathognathodontid elements developed into smooth-surfaced brim- or shelflike platforms along a segment of the element, commonly the posterior process (e.g., *Tortodus*, Figs. 5.39J and 5.41), or around the entire element. The latter type of development characterizes the bizarre Silurian *Polygnathoides* (Fig. 5.40), which

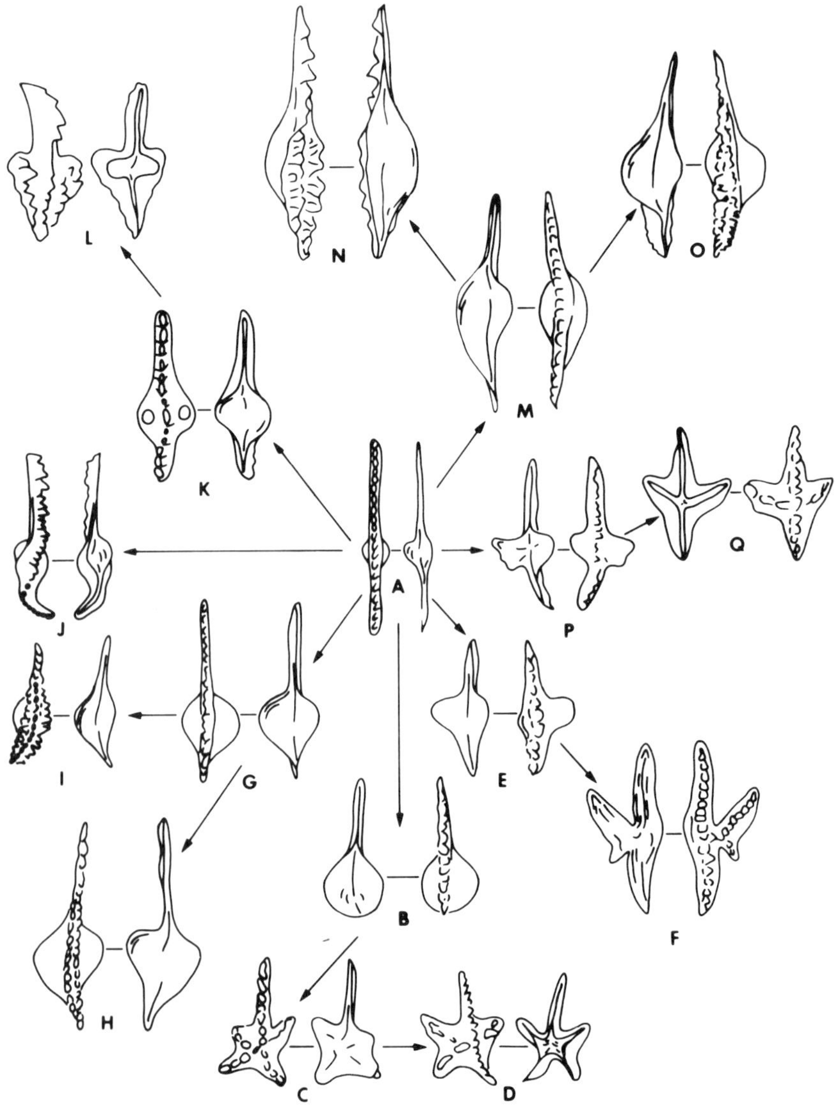

Fig. 5.39. Diagram to illustrate the various ways in which the sides of Pa elements in the Spathognathodontidae were produced to form elements typical of different lineages. (A) Conservative *Ozarkodina*; (B), (C), (D) various kockelellids; (E) *Amydrotaxis*; (F) *Pterospathodus*; (G) *Ozarkodina*; (H) *Eognathodus*; (I) *Polygnathus*; (J) *Tortodus*; (K), (L) *Ancyrodella* lineage; (M) *Bispathodus*, conservative forms; (N) *Rhachistognathus*; (O) advanced *Bispathodus*; (P), (Q) *Ancyrodelloides*.

Jeppsson (1983) has recently determined to have a very ordinary, indeed primitive, spathognathodontid apparatus. Although this type of Pa-element modification yields highly distinctive types of structures, neither *Tortodus* nor *Polygnathoides* appears to have been on the "main line" to anywhere; hence both are retained in the Spathognathodontidae instead of being made types of new families.

Finally, a fourth type of Pa-element modification characterizes skeletal apparatuses in species of Early Devonian *Eognathodus* (Figs. 5.39H and 5.41) and later Devonian and Carboniferous species of *Bispathodus* and its numerous derivatives (Figs. 5.39M, 5.39N, 5.39O, and 5.41). Pa elements of these species have basal-cavity configurations similar to those of either *Ozarkodina inclinata* (Fig. 5.39A) or *O. selfi* (Fig. 5.39G) but differ in gradual development of double, subparallel rows of denticles on the upper surface. One denticle row, the primary one, extends from one end of the element to the other. At the anterior end it forms a single-rowed "blade," which is bordered on either its right or left side by the shorter secondary row. Numerous modifications of this basic pattern lead to ozarkodinide groups identified here as the families Gnathodontidae and Cavusgnathidae.

History of the Spathognathodontidae was marked by several intervals of accelerated innovation that were expressed primarily in modifications of Pa elements along lines just discussed. In the late Llandovery (Silurian), for example, migration of the expanded part of the basal cavity toward the posterior end initiated morphology that came to be typical of Pa elements in the apparatuses of species of *Kockelella* and the Kockelellidae (Fig. 5.39B, 5.39C, and 5.39D). At about the same time, lateral expansion of the central portion of the basal cavity, eventually to form denticulated platform elements, established the pattern characteristic of Pa elements in apparatuses of species of *Pterospathodus* (Fig. 5.39F) and other members of the short-lived Pterospathodontidae.

The Early Devonian included two closely spaced episodes of innovation in the spathognathodontid stock. In the first, shortly after the beginning of the Devonian, *Pandorinellina* (Fig. 5.38) separated from contemporary *Ozarkodina* populations (Fig. 5.38) by developing Sa elements with a denticulated posterior process and distinctively finned Pa elements. At about the same time, initiation of *Amydrotaxis* (Fig. 5.40) and *Ancyrodelloides* (Fig. 5.40) was heralded by rapid, lobulate expansion of the central parts of basal cavities in Pa elements in segments of widespread populations of *Ozarkodina remscheidensis*. In the second Early Devonian episode, posterior widening of the basal cavity in Pa elements of somewhat later populations of the same *Ozarkodina* stock led to *O. selfi* (Fig. 5.39G) then, promptly, to early representatives of *Eognathodus*, with double-rowed Pa elements (Figs. 5.39H and 5.41), and *Polygnathus*, with triple-rowed Pa elements (Fig. 5.39I). Although Klapper and Johnson

Fig. 5.40. Elements and apparatuses of species typical of genera of the Spathognathodontidae. The complete apparatuses of *Amydrotaxis*, *Polygnathoides*, and *Mehlina* are shown, but only the P elements of *Ancyrodelloides* are illustrated. The M and S elements of the latter are not known.

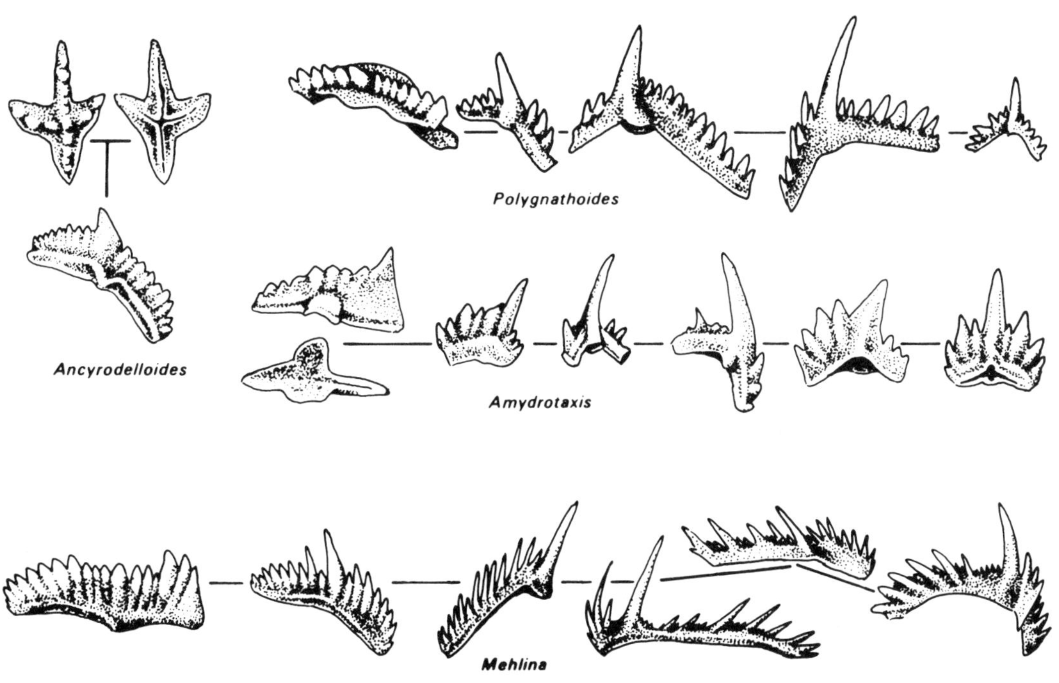

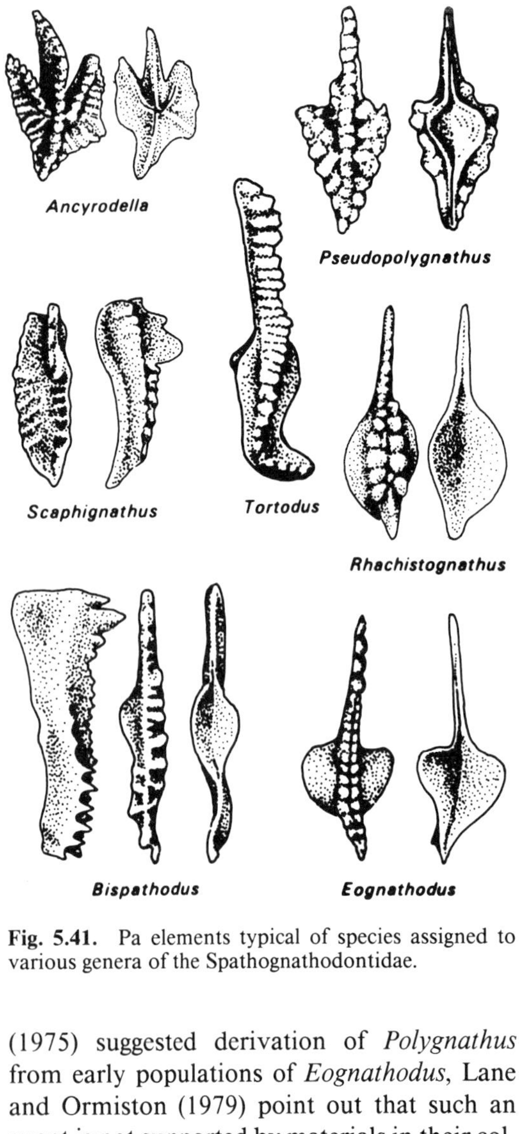

Fig. 5.41. Pa elements typical of species assigned to various genera of the Spathognathodontidae.

(1975) suggested derivation of *Polygnathus* from early populations of *Eognathodus*, Lane and Ormiston (1979) point out that such an event is not supported by materials in their collections from Alaska, which lack Pa elements transitional in morphology between those of *E. sulcatus*, the supposedly ancestral population, and *Polygnathus pireneae*, the one that initiated radiation of *Polygnathus*. For the latter reason, I retain *Eognathodus* (Fig. 5.41) in the Spathognathodontidae and, in Fig. 5.36, show *Polygnathus* as an independent development from *Ozarkodina* and as the stem of a different family, the Polygnathidae.

In the latest Middle Devonian a brief episode of Pa-element modification in populations of either *Ozarkodina sannemanni* or *Pandorinellina insita* (or contemporaneously in both) records initiation of the important Late Devonian *Ancyrodella* (Fig. 5.41) stock. Morphologically, this episode recalls the Early Devonian one that led to development of *Ancyrodelloides* (Fig. 5.40) and, because the stock that arose is readily included in a single, monogeneric lineage that appears to have expired without issue near the end of the Frasnian, I include it also in the Spathognathodontidae.

Finally, modifications of Pa elements in populations at several stages in the Late Devonian history of *Pandorinellina* (Fig. 5.38) led to a confusing welter of the double-rowed forms by which one recognizes early species of *Scaphignathus* (Fig. 5.41) and, a bit later, various species of *Patrognathus* and *Clydagnathus*. *Mehlina* (Fig. 5.40) gave rise to *Bispathodus* (Fig. 5.41), the ancestor of the Gnathodontidae and Sweetognathidae, and *Alternognathus* initiated the lineage that led, in the latest Devonian, to *Siphonodella*, which is of great stratigraphic importance in the earliest Carboniferous. Hence *Alternognathus* and *Siphonodella* are shown in Fig. 5.36 (and in Appendix A) as parts of a separate family, the Elictognathidae. Similarly, *Patrognathus* and *Clydagnathus* are commonly grouped with later representatives of the Cavusgnathidae, and I follow that practice in Fig. 5.36 and in Appendix A.

I retain Devonian-Early Carboniferous *Bispathodus* (Fig. 5.41) in the Spathognathodontidae, however, because species with both single- and double-rowed Pa elements are included in the genus in authoritative reports [e.g., Ziegler, Sandberg, and Austin (1974); Sandberg and Gutschick (1984)]. Even if *B. stabilis* and *B. utahensis*, which have single-rowed Pa elements (and triramous alate Sa elements) and are now considered to be the long-ranging stem forms of *Bispathodus*, were reassigned to *Mehlina* of the Spathognathodontidae, *Bispathodus* would require additional surgery. That is, studies summarized by Ziegler, Sandberg, and Austin (1974) clearly indicate that the "*B. bispathodus*" and "*B. aculeatus*" branches of the *Bispathodus* lineage developed from *B. stabilis* at different times. Because the "aculeatus" branch includes *B. spinulicostatus*, the type-species, only it could be retained in *Bispathodus*. Another generic name would be needed for species of the "bispathodus" branch. There would then be the question of familial assignment for *Bispathodus* and the new genus. Since the situation with Early Devonian *Eognathodus* (Fig. 5.41) seems almost precisely parallel, it seems to me that the best solution is to assign broadly conceived and unrevised *Bispathodus* [and its obvious later Carboniferous derivative, *Rhachistognathus* (Fig. 5.41)], to the Spathognathodontidae.

Finally, *Bispathodus stabilis*, with conservative, single-rowed Pa elements, was apparently the stock from which species assigned to *Pseudopolygnathus* (Fig. 5.41) developed at least twice, once in the late Devonian and a second time in the early Carboniferous. The complete apparatus of *Pseudopolygnathus* has yet to be described, but Pa elements of primitive species in both the Devonian and Carboniferous iterations of the genus combine open, *Bispathodus*-like basal cavities with at least rudiments of three (not 2) denticle rows. Later species in both lineages are distinguished by carminiplanate Pa elements, with a large basal pit that is surrounded by a zone of recessive basal margin. In this respect, Pa elements of *Pseudopolygnathus* are homeomorphic with those of *Polygnathus*, and the two heterochronous lineages now included in the genus will probably merit separate generic and familial identity sometime in the future. At present, it seems best to retain *Pseudopolynathus* and "*Pseudopolygnathus*" in the Spathognathodontidae with the ancestral *Bispathodus stabilis* stock.

5.9.2 Family Kockelellidae Klapper, 1981

Toward the middle of the Llandovery (early Silurian), the basal cavity of Pa elements in a spathognathodontid species such as *Ozarkodina abrupta* (Aldridge, 1972) began to widen toward the posterior and also to expand conspicuously to the sides, giving rise to carminiscaphate elements with a relatively long, even-crested anterior process and a rather short posterior process surmounted by small denticles that decline in size toward the process tip. Near the end of the Llandovery, continued modification resulted in Pa elements with an even shorter posterior process and a large, almost posterior basal cavity whose basal margin has an almost circular profile in superior view (Fig. 5.39B). Pa elements of this sort characterize the apparatus of *Kockelella ranuliformis* (Walliser), which also includes angulate Pb and bipennate (but nevertheless pick-shaped) M elements and a symmetry-transition series composed of bipennate Sc, asymmetrically alate Sb, and symmetrically alate Sa elements, the latter lacking a denticulated posterior process.

Subsequent development of the *Kockelella* lineage involved little apparent modification of elements in the M or S positions but substantial change in those in the Pa position. That change is marked first by development of short processes on one or both sides of the cusp, with no conspicuous modification in profile of the basal margin. The basal cavity of Pa elements in samples from slightly younger populations is laterally restricted beneath a posterior process that is somewhat longer than in specimens from older populations; and, in the Pa elements of the youngest species of *Kockelella* (Fig. 5.42), lateral processes lengthen and also bifurcate, and the basal cavity beneath them is

Fig. 5.42. Elements and the apparatus typical of a species of *Kockelella*, typical genus of the Kockelellidae.

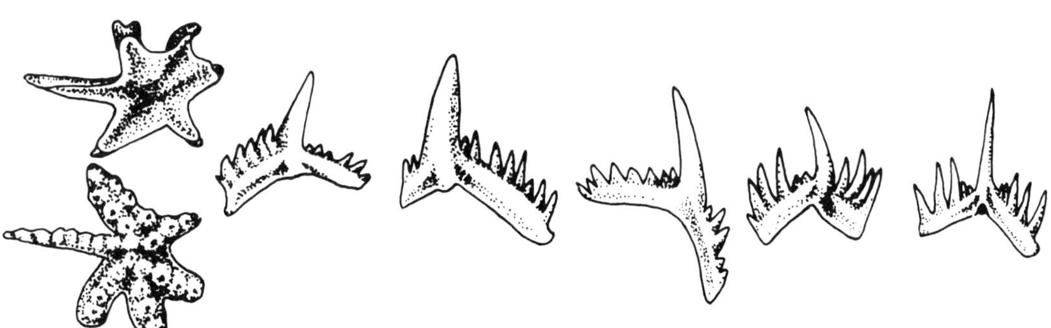

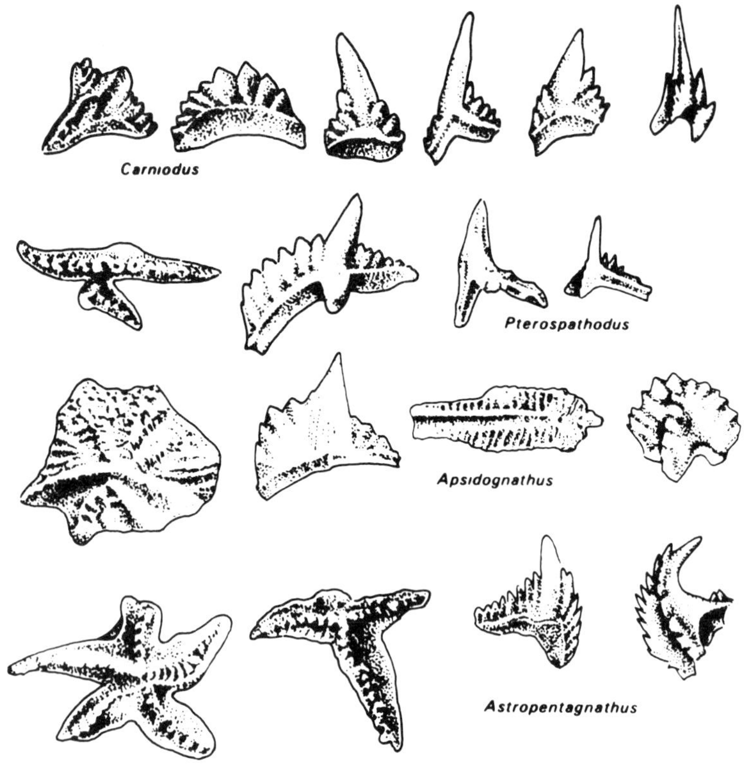

Fig. 5.43. Elements and apparatuses typical of species of genera of the Pterospathodontidae.

further restricted, so that the profile of the basal margin is no longer circular as viewed from above but is distinctly digitate. Such elements are formally stelliscaphate.

In outline, Pa elements of *Ancoradella ploeckensis* Walliser are closely similar to those of the stelliscaphate ones of *Kockelella variabilis* Walliser, but the attachment surface is more complex. That is, the basal cavity is reduced to a subcusp pit, which is surrounded by a broad zone of recessive basal margin. In short, these Pa elements are stelliplanate. Nothing more is known of the apparatus of *A. ploeckensis*, but the close similarity in Pa elements certainly suggests [as Barrick and Klapper (1976) have noted] that *Ancoradella* evolved from *Kockelella*, because transformation of the Pa elements characteristic of *Kockelella variabilis* into those typical of *Ancoradella ploeckensis* would involve only a minor shift in the controls of basal-margin formation in Pa elements—and that at a probably advanced stage in ontogeny.

5.9.3 Family Pterospathodontidae Cooper, 1977

On the strength of the similarity between Pa and Pb elements of *Pterospathodus amorphognathoides* Walliser and those of species of *Ancyrodelloides* (Fig. 5.40) described in a multielement sense by Murphy and Matti (1982), I regard the Pterospathodontidae as a stock of ozarkodinide conodonts, not a Silurian prioniodontide family, as advocated in the *Treatise.* To be sure, elements thought by Barrick and Klapper (1976) to have occupied M and S positions in the apparatuses of two of the three known species of *Pterospathodus* (Fig. 5.43) resemble those in the apparatuses of *Distomodus* species more than comparable structures in the apparatuses of contemporary *Ozarkodina* species. However, the association between diagnostic Pa and Pb elements and specimens supposed to be M and S elements of *Pterospathodus* is not unequivocally confirmed by published data on co-occurrence (e.g., Rexroad and Nicoll, 1972; Barrick and Klapper, 1976), and morphologic similarity in P elements becomes, for me, a more compelling reason for reclassification of the family.

The Pa position in the apparatus of *Pterospathodus* species was occupied by stelli- or pastiniscaphate pectiniform elements with simple or bifurcate lateral processes that may be adenticulate or may bear denticles on one or both sides. It seems obvious that such structures developed from *Ozarkodina*-like carminate elements through gradual lateral expansion of one or both sides at midlength (Fig. 5.39F), just as Pa elements of *Amydrotaxis* (Fig. 5.40) and *Ancyrodelloides* (Fig. 5.40) are thought to have developed in the Early Devonian. The Pb position in the apparatus of the type-species of *Pterospathodus* was occupied by a distinctive angulate element, which bears at midheight a stout lateral rib or ridge that projects out and downwardly on the outer side to form a chevron-shaped lateral lappet. An element of almost precisely the same conformation occupied the Pb position in the apparatuses of various species of Early Devonian *Ancyrodelloides.*

The stelliscaphate Pa elements of *Astropentagnathus* (Fig. 5.43), *Aulacognathus*, and *Apsidognathus* (Fig. 5.43) are only slightly different in plan from those of *Pterospathodus*, and all three genera were included with *Pterospathodus* in the *Treatise* concept of the Pterospathodontidae. However, if elements such as the ones Walliser (1964) referred to the form species *Ambalodus galerus* and *Pygodus lyra* were also components of the *Apsidognathus* apparatus, as has been suggested by several authors, then all bets are off, at least for *Apsidognathus.* Unfortunately, I have no other idea about relations or alternative classification of *Apsidognathus* and continue to show it in Fig. 5.36 as a member of the Pterospathodontidae. *Carniodus*, as reconstructed by Bischoff (1986) and others, is also assigned to the Pterospathodontidae and not to the prioniodontide family Rhipidognathidae as was done in the *Treatise.* The apparatus of the type species is illustrated in Fig. 5.43.

5.9.4 Family Polygnathidae Bassler, 1925

The few polygnathid ozarkodinides that have been fully reconstructed have a seximembrate skeletal apparatus characterized in all but the earliest species known by carminiplanate Pa elements, angulate Pb elements, dolabrate (or bipennate) M elements, and an S series with bipennate Sc, extensiform digyrate Sb, and alate Sa elements, the latter with a denticulate posterior process. Pa elements formed by early species of the stem genus, *Polygnathus* (Fig. 5.44) (e.g., *P. pireneae*, *P. dehiscens*, *P. gronbergi*), are carminiscaphate; however, in slightly younger species (e.g., *P. laticostatus*, *P. inversus*) the underside of Pa elements is planate, a condition that was attained by development of a wide zone of recessive basal margin at progressively earlier ontogenetic stages and consequent restriction of the initially broad basal cavity to a pit in forms representing later stages of either growth or phylogeny. The pit is commonly elongate, situated at element midlength, and enclosed within a keellike ridge that projects below the lower surface, is narrowly grooved on its under edge, extends the complete length of the element, and represents (as does the pit it encloses) the under edge of the element prior to the stage at which the basal margin began to recede.

In Pa elements of a few species (e.g., *Polygnathus spicatus*) there is no discernible keel in the immediate vicinity of the pit, which is surrounded instead by a flat area of varying width that mimics a structure termed a *pseudokeel.* Such structures are common to the undersides of Pa elements of *Siphonodella* species and may, in most cases, be used to distinguish those structures from comparable Pa elements of *Polygnathus* species.

Soon after the character of *Polygnathus* became established in populations of Early Devonian conodonts descended from *Ozarkodina selfi* Lane and Ormiston (1979), three main lineages were founded. These are identified in Fig. 5.45 as the *linguiformis*, *robusticostatus*, and *costatus* stocks. The nature of these stocks and their many branches and the subsequent Early and Middle Devonian history of their component species have been explored in some detail by Weddige (1977) and Weddige and Ziegler (1979), who drew Fig. 5.45 and made the interpretations that I summarize in the following paragraphs.

The three main stocks recognized within *Polygnathus* by Weddige and Ziegler are distinguished in the fossil record by the outline and

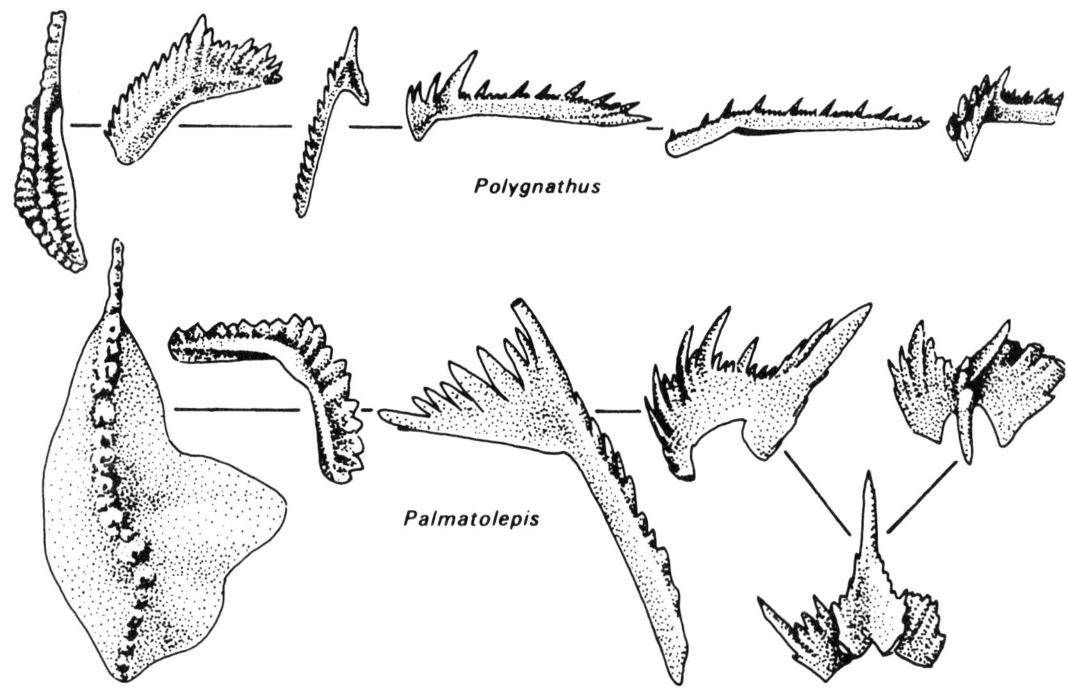

Fig. 5.44. Elements and apparatuses typical of species assigned to *Polygnathus* (Polygnathidae) and *Palmatolepis* (Palmatolepidae).

upper-surface features of Pa elements. Essential characters of the *linguiformis* stock appear first in Pa elements of *P. inversus* Klapper and Johnson (Fig. 5.45.13), which are conspicuously asymmetrical and develop a platform that is deflected laterally and downward posteriorly to form a "tongue" whose upper surface bears one or several incomplete transverse ridges. Pa elements of species representing later stages in development of the *linguiformis* stock are readily distinguished by the posterior platform "tongue", and by their distinct asymmetry. Those that make up the central, or *bultyncki*, branch of this stock (Fig. 5.45.20 and 5.45.24) have elongate Pa elements with deep troughs (known technically as *adcarinal troughs*) on either side of the central carina and a well-developed, transversely ribbed "tongue" at the posterior end of the platform. Species of the so-called *pinguis* branch, on the other hand (Fig. 5.45.23, 5.45.25, 5.45.26, and 5.45.27), have stout, compact, coarsely ornamented Pa elements that commonly reach great size and, as noted by Weddige and Ziegler (1979), "... appear aberrant" in collections. The delicate Pa elements of *Polygnathus linguiformis alveolus* and its younger derivatives (Fig. 5.45.19, 5.45.22, and 5.45.28), which define the *alveolus* branch, have a tiny "tongue" that bears only a few transverse ridges, whereas strong asymmetry characterizes Pa elements of *P. serotinus* and other species of the *serotinus* branch (Fig. 5.45.9 and 5.45.34).

Pa elements of species in the *robusticostatus* stock and its three branches (Fig. 5.45.1, 5.45.2, 5.45.3, 5.45.16, 5.45.18, 5.45.21, 5.45.33, 5.45.36, 5.45.37, and 5.45.41) have platforms that are about equally well developed on either side of the carina, which extends all the way to the posterior tip. Pa elements of *Polygnathus cooperi* Klapper, thought by Weddige and Ziegler (1979) to be the ancestral species in this stock, are slightly asymmetric, have a laterally and downwardly deflected "tongue" at the posterior end, and thus resemble comparable elements of species in various branches of the *linguiformis* stock. However, transverse ribs on the posterior "tongue" of *P. cooperi* Pa elements are broken at midwidth by nodes, which serve to continue the carina to the posterior end of the element. Thus the Pa elements of *P. cooperi* are transitional in morphology between those typical of the *linguiformis* and the *robusticostatus* stocks.

In the early Middle Devonian, *Polygnathus cooperi* apparently gave rise to forms with subsymmetrical Pa elements that lack a posterior "tongue" and are distinguished by a carina that extends to the posterior tip. In species of the *zieglerianus* branch (Fig. 5.45.1, 5.45.2, 5.45.3, and 5.45.41), platform segments of Pa elements become reduced in width and the carina juts out beyond them for slight to considerable distances posteriorly. Pa elements of the morphologically conservative *robusticostatus* branch (Fig. 5.45.33), on the other hand, are symmetrical, retain both platform segments, and differ from those of ancestral *P. cooperi* only in lacking any trace of a posterior "tongue." Symmetrical Pa elements of the *trigonicus* branch (Fig. 5.45.16, 5.45.18, 5.45.36, and 5.45.37) are blunt, robust, and have wide adcarinal troughs that extend far to the posterior.

Pa elements typical of the *Polygnathus cos-*

Fig. 5.45. Lower and Middle Devonian lineages in *Polygnathus*. Species represented by the elements figured are identified in the text. Modified, with additions, from Weddige (1977) and Weddige and Ziegler (1979).

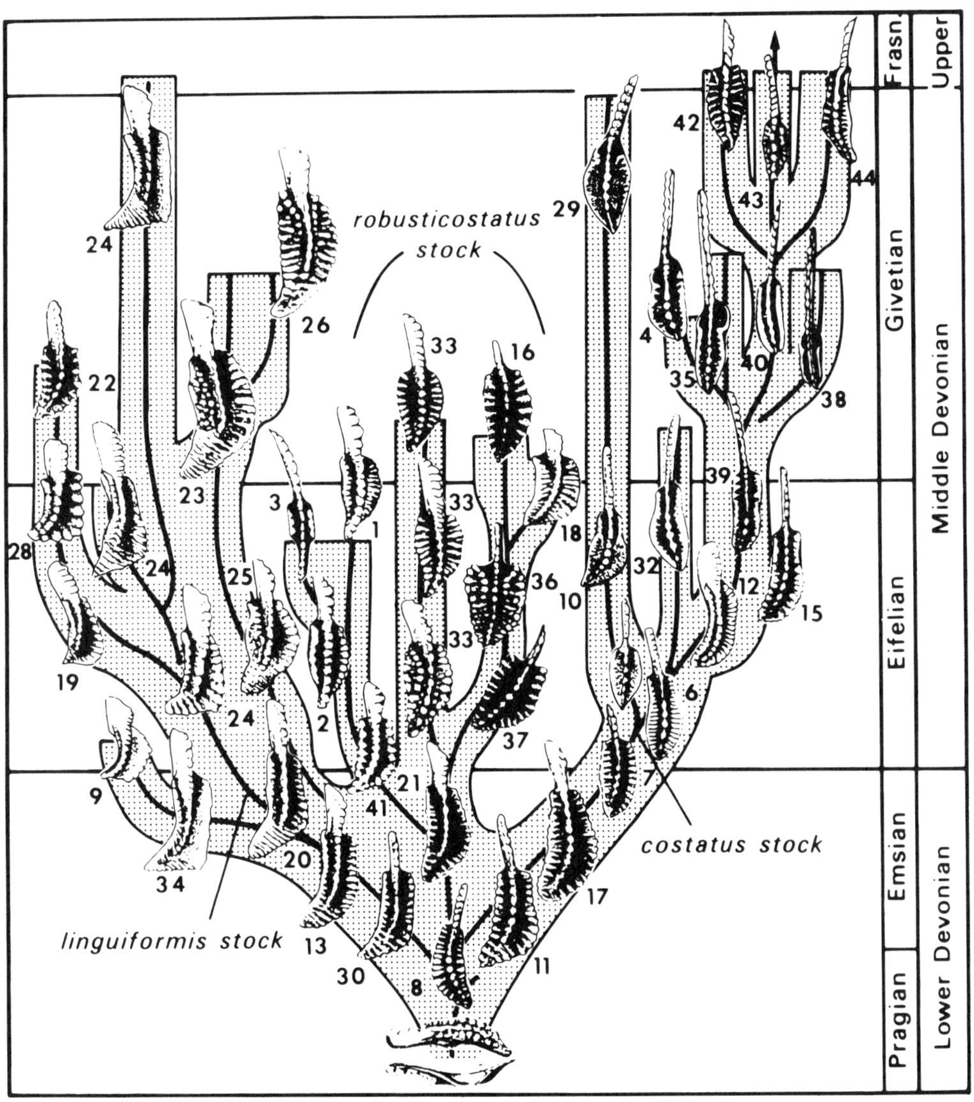

tatus stock (all the ones above species 7 along the right side of Fig. 5.45) are essentially symmetrical but, according to Weddige and Ziegler (1979), the carina "either disappears or becomes obscure near the posterior end." Later in the history of all branches of this stock, there is at least a tendency for margins of the anterior segments of Pa-element platforms to turn upward and for the transverse ridges on these upturned segments to curve toward the carina and ultimately to fuse to form ridges parallel to the carina. The complex but distinctive feature thus formed is termed a *rostrum*, and the ridges that border it *rostral ridges*. Also, as platform margins became upturned for their entire length, Pa elements of later Middle Devonian species in the *costatus* stock became narrower and anterior portions of their platforms tended to become constricted. Just above the level represented by the top of Fig. 5.45, highly variable populations of conservative species of the *costatus* stock developed explosively into lineages of *Polygnathus* and related genera characteristic of the Late Devonian.

More than 115 species of *Polygnathus* have been based on material from rocks younger than any represented in Fig. 5.45. Of these, approximately 75 are restricted to the Upper Devonian; 5 appear in the Upper Devonian and continue into the Lower Carboniferous (or Dinantian); and 35 are known only from the Lower Carboniferous. These species represent several developmental groups, distinguished by morphology of their Pa elements. However, the phylogeny of only one of these (the *Polygnathus nodocostatus* group) has been worked out in any detail (Helms, 1961), and it is only recently (Klapper and Lane, 1985) that we have been provided an insight into what the *Polygnathus* story may have involved in environments closer to shore than those represented by most collections of Upper Devonian conodonts described to date.

Ancyrognathus (Fig. 5.46), represented by a succession of distinctive Late Devonian species, developed in the latest Middle Devonian from some as yet unknown ancestor in the *Polygnathus costatus* stock (Fig. 5.45) or perhaps from a member of the Palmatolepidae such as *Mesotaxis asymmetricus*. There is currently no information available to me about the skeletal apparatus of any of the 9 species now assigned to *Ancyrognathus*, but Pa elements are pastiniplanate structures that consist of a laterally bowed anteroposterior axis with a carina on the upper surface and a keel on the lower one and a lateral lobe that bears a distinctive row of nodes on its upper surface and a well-developed keel on its lower surface; keels on the lower surface meet at a small, generally three-cornered basal pit. The pastiniplanate Pa elements of *Ancyrognathus* are similar in outline to those of the early species of *Palmatolepis* referred by some authors to the subgenus *Manticolepis* but lack the conspicuous central node and sigmoid carina of the latter and differ in other aspects of surface ornamentation. They are also reminiscent of the pastiniplanate elements of the Middle Ordovician prioniodontides *Eoplacognathus* and *Cahabagnathus*, which are surely not very closely related. Convergence in shape suggests, however, that the triramous pastinate condition was of considerable long-term significance in skeletal architecture, and it may someday be possible to relate development of such elements to episodically recurrent features in the life history or surroundings of conodonts.

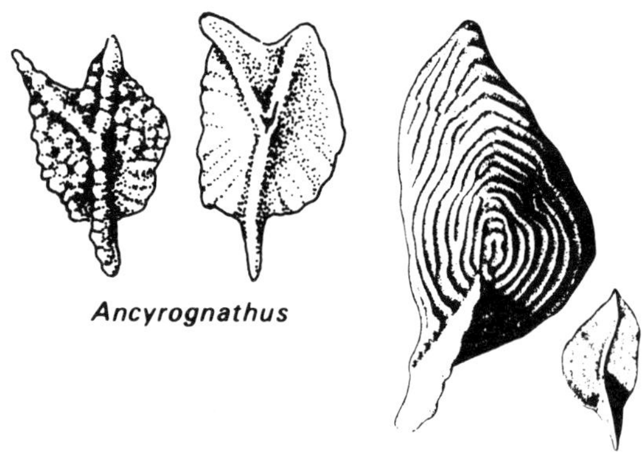

Fig. 5.46. Pa elements of species of *Ancyrognathus* and *Polylophodonta* (Polygnathidae). Figures of *Polylophodonta* from Helms (1961).

5.9.5 *Family Palmatolepidae, new*

Species currently assigned to *Schmidtognathus*, *Klapperina*, *Mesotaxis* (all illustrated in Fig. 5.47), and *Palmatolepis* (Fig. 5.44), together with several others that have commonly been

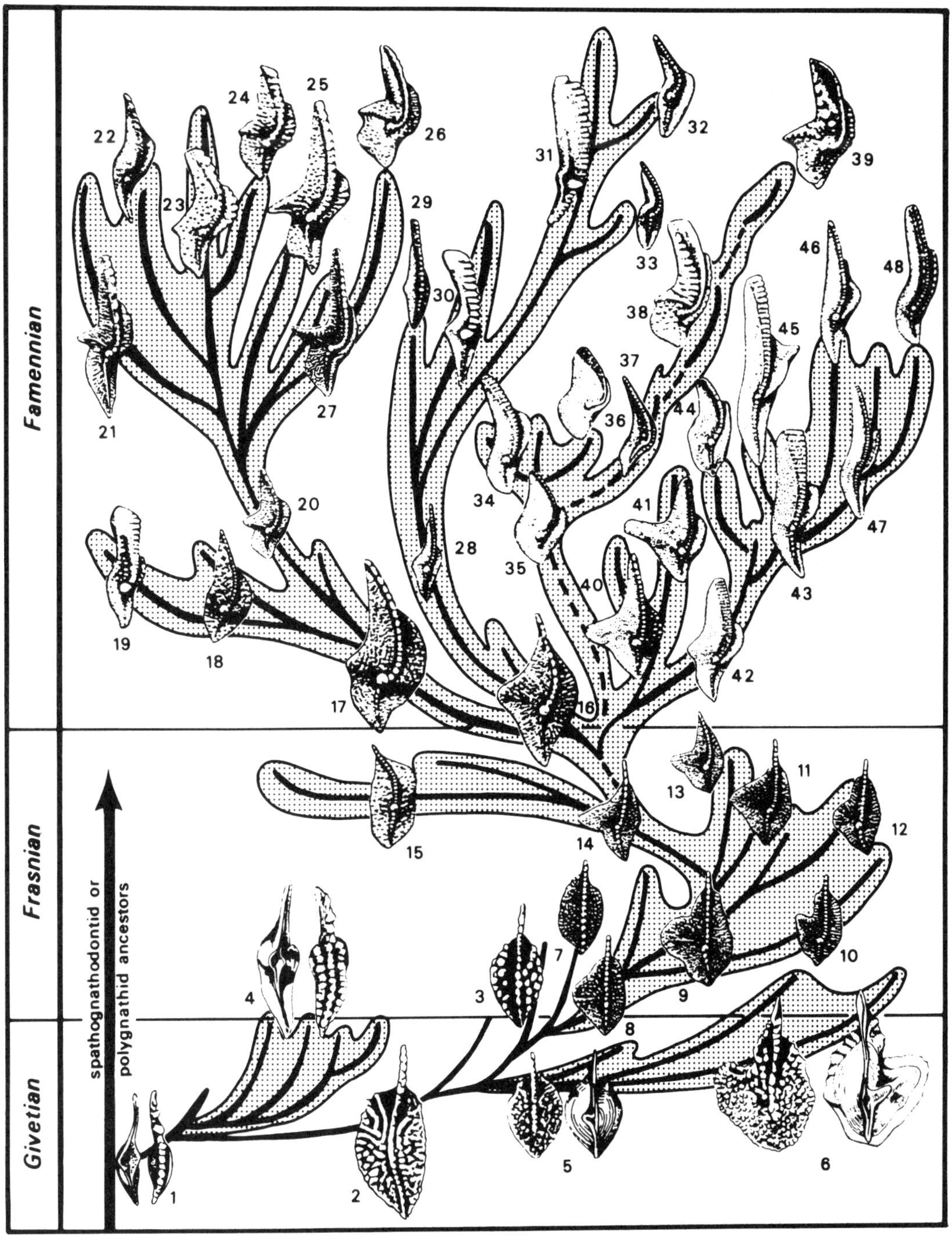

Fig. 5.47. Elements typical of major species of the Palmatolepidae. Modified, with additions and omissions, from Helms and Ziegler, in Clark et al. (1981). (1) "*Polygnathus*" *latifossatus* Wirth; (2) "*Polygnathus*" *limitaris*; (3) "*Polygnathus*" *cristatus*; (4) *Schmidtognathus wittekindti*; (5) *Klapperina disparilis*; (6) *K. disparalvea*; (7) *Mesotaxis asymmetricus*; (8) *Palmatolepis transitans*; (9) *P. punctata*; (10) *P. proversa*; (11) *P. hassi*; (12) *P. unicornis*; (13) *P. subrecta*; (14) *P. gigas*; (15) *P. linguiformis*; (16) *P. triangularis*; (17) *P. perlobata perlobata*; (18) *P. crepida*; (19) *P. termini*; (20, 23) *P. perlobata schindewolfi*; (21) *P. perlobata maxima*; (22) *P. perlobata helmsi*; (24) *P. perlobata postera*; (25) *P. perlobata sigmoidea*; (26) *P. rugosa ampla*; (27) *P. perlobata grossi*; (28, 29) *P. minuta minuta*; (30) *P. minuta schleizia*; (31) *P. gracilis gracilis*; (32) *P. gracilis gonioclymeniae*; (33) *P. gracilis manca*; (34) *P. quadrantinodosa inflexoidea*; (35) *P. quadrantinodosa inflexa*; (36, 37) *P. marginifera*; (38) *P. rugosa trachytera*; (39) *P. rugosa rugosa*; (40) *P. quadrantinodosalobata*; (41) *P. subperlobata*; (42) *P. tenuipunctata*; (43) *P. glabra prima*; (44) *P. klapperi*; (45) *P. glabra acuta*; (46) *P. glabra lepta*; (47) *P. glabra pectinata*; and (48) *P. glabra distorta*.

included in a group of "wide-plated" *Polygnathus* species (e.g., *P. limitaris*, *P. cristatus*), are here segregated as the Palmatolepidae. I have no information on composition of the skeletal apparatus of species of either *Schmidtognathus* or *Klapperina* or on that of the *Polygnathus* species with broadly plated Pa elements that were presumably their ancestors. However, Klapper and Philip (1971, 1972) have reconstructed and described the apparatus of *Mesotaxis asymmetricus*, and reconstructions of the apparatuses of several species of *Palmatolepis* have been based independently on coprolitic associations (Lange, 1968), natural assemblages (Puchkov, Klapper, and Mashkova, 1982) and cluster analyses (van den Boogaard and Kuhry, 1979). These are all in such close agreement that firm knowledge of the general features of the more important members of the family seems assured.

The best-known palmatolepids had a sexi- or septimembrate skeletal apparatus in which the Pa position was typically occupied by broad, scalelike carminiplanate elements, with profusely nodose or pustulose upper surfaces and a short free blade. In species of *Mesotaxis* (Fig. 5.47.7) and in most species of *Palmatolepis*, the Pb position was occupied by distinctively bowed anguliplanate elements, with a high, bladelike anterior process, a conspicuous cusp, and a posterior process with a surficially nodose platformlike brim on one or both sides. Pb elements like this have in the past been referred to some 20 different species of the form genus *Nothognathella*.

M elements of *Mesotaxis* and *Palmatolepis*, identified in form taxonomy as species of *Synprioniodina* and *Palmatodella*, respectively, are bipennate ramiform structures, in the latter case with a long, profusely denticulated and downwardly directed anterior process. The three or four types of elements that make up the symmetry-transition series include alate Sa, extensiform digyrate Sb, and bipennate Sc elements, which have been referred in form taxonomy to *Scutula*, *Falcodus*, and *Prioniodina*. Processes of S elements all have "hindeodelloid" denticulation, which consists of sets of short, needlelike denticles separated from one another at regular intervals by longer, broader denticles.

As noted, the skeletal apparatuses of species of *Schmidtognathus* and *Klapperina* have not been reconstructed or described, nor have those of the presumably ancestral species with broadly plated Pa elements. The relationship of these species to each other and to the better-known species of *Mesotaxis* and *Palmatolepis* must thus be determined from features of their Pa elements.

Pa elements of *Klapperina disparilis* (Fig. 5.47.5), *K. disparalvea* (Fig. 5.47.6), and *K. disparata* are broad, surficially nodose, and carminiplanate. They have short free blades and a carina that is indistinct posterior of midlength and apparently always fails to reach the posterior end of the platform. On the under side, near midlength of the median keel, is a distinct basal pit that expands very distinctly to one side. The curiously asymmetric basal pit seems to distinguish Pa elements of *Klapperina* from those, for example, of *Polygnathus cristatus* (Fig. 5.47.3), whose subcentral basal pit tends to be nearly symmetrical, or from morphologically similar Pa elements of *Palmatolepis* (Fig. 5.47.8 to 5.47.48), which commonly have no obvious basal pit at all.

The carminiplanate Pa elements built by species of *Schmidtognathus* (Fig. 5.47.4) have a large, asymmetric, *Klapperina*-like basal pit that is situated near midlength of the platformed portion of the element. On the other hand, the large basal pit of carminiplanate Pa elements of "*Polygnathus latifossatus*" Wirth (Fig. 5.47.1), which is supposed to be the ancestor of the Palmatolepidae, is located near the anterior end of the weakly developed platform.

Palmatolepis, which is abundantly represented in a biofacies that is known from many parts of the world, has been studied intensively and provides the principal basis for the high-resolution biostratigraphy of Upper Devonian rocks described by Ziegler (1962, 1971) and various co-workers in numerous publications. Major features in the phylogeny of *Palmatolepis* and its predecessors are shown schematically in Fig. 5.47, which is adapted largely from a much more extensive diagram prepared for the 1981 *Treatise* volume by Helms and Ziegler.

Within *Palmatolepis* Helms and Ziegler (in Clark et al., 1981) recognize a number of distinct branches, most (but not all) of which are shown schematically in Fig. 5.47. Initial Frasnian radiation is represented in Fig. 5.47 by Pa elements of the group of species numbered 8 to 15. This species group is identified by some authors as the subgenus *Manticolepis*. Note that Pa elements of these early species of *Palmatolepis* have a prominent lobe on the outer side; a straight or only slightly sinuous carina; an undistinguished central node; a downwardly deflected posterior tip; and no parapet along the inner margin.

The several Famennian lineages recognized within *Palmatolepis* diverge from *P. triangularis* (Fig. 5.47.16), whose Pa elements are related in many morphologic particulars to those of the Frasnian "manticolepids" but differ in having a more distinctly sinuous carina; a more conspicuous central node; and a platform that is flexed upward (instead of downward) posteriorly.

Sketches 17 and 20 to 27 in Fig. 5.47 are Pa elements of a group of species (or subspecies) that includes the type of *Palmatolepis*, *P. perlobata*. Pa elements of species in this group are large and conspicuously sinuous; have narrow outer lateral lobes that generally bear a distinct secondary carina; and develop a low, ridgelike parapet on the inner side anterior to the central node.

The *Palmatolepis* lineage that begins with species 42 in Fig. 5.47 is distinguished by Pa elements with long, mostly smooth platforms that lack an outer lateral lobe but bear a serrate or ridgelike parapet on the inner side, anterior of the central node. Species in this lineage are inluded by many authors in the subgenus *Panderolepis*.

van den Boogaard and Kuhry (1979) refer most of the *Palmatolepis* species in the stock that begins with sketch 35 (Fig. 5.47) to the subgenus *Conditolepis*. Pa elements of most species in this group lack an outer lateral lobe, have ovate to quadrate platforms with a carina that is weakly developed posterior of a large central node; and a sharp-edged parapet that rims the inner margin to a point well posterior of the central node.

Pa elements depicted in sketches 26, 38, and 39 in Fig. 5.47 have been taken to represent three chronologic subspecies of *Palmatolepis rugosa*. *P. rugosa ampla* (sketch 26) is thought to be a derivative of the *P.* (*Palmatolepis*) stock, which is where it is placed in Fig. 5.47. However, Sandberg and Ziegler (1973) postulate development of *P. rugosa trachytera* (sketch 38) and *P. rugosa rugosa* (sketch 39) from a species in the *Conditolepis* stock that is related to the one with Pa elements like those shown in sketches 36 and 37. If the phylogenetic arrangement of these three subspecies is as shown in Fig. 5.47 (and we may always have to guess about that), some nomenclatural adjustments will obviously have to be made. The more important lesson, however, is that gross morphology of single elements may not always be the best guide to relationship. All of these "subspecies" are said to be connected through morphologically intermediate Pa elements with other members of the lineages to which they are assigned in Fig. 5.47.

Species in the *Palmatolepis* branch represented by sketches 28 to 33 in Fig. 5.47 are characterized by small, narrow Pa elements in which the smooth-surfaced platform withdraws phylogenetically from the anterior end and becomes widest and best developed posterior of the central node. Anterior of the central node, the main axis of the element is developed as a conspicuous blade. A more important character of this stock, which was referred to the subgenus *Deflectolepis* by Müller (1956), may be the discovery by van den Boogaard and Kuhry (1979) that Pb elements are not "nothognathellan" but are arched pastinate forms of a type that has been referred in form taxonomy to several species of *Tripodellus* Sannemann. It is difficult to see how these structures can be homologized with the "nothognathellan" Pb elements of species in other *Palmatolepis* lineages, so it may turn out that this group will ultimately merit separate generic distinction. When (and if) that day comes, the venerable name *Tripodellus* Sannemann, 1955 (not *Deflectolepis* Müller, 1956) will be available.

Conventional wisdom has it (e.g., Ziegler and Lane, 1987) that *Polygnathus latifossatus* Wirth (Fig. 5.47.1), progenitor of the Palmatolepidae, developed from species of the *Polygnathus varcus* stock (the far righthand branch in Fig. 5.45). Species in that stock built carminiplanate Pa elements with long, subquadrate blades; relatively short, narrow, parallel-sided platforms, and passably large basal cavities situated at or near the junction of blade and platform. Wirth (1967), however, suggested when he established the species that *P. latifossatus* evolved directly from *Ozarkodina* (=?*Mehlina*) *semialternans* (Wirth), whose Pa elements are carminate and thus lack platform development anywhere along the blade. Ziegler, Klapper, and Johnson (1976) also regard such a sequence as a phylogenetic possibility, but they carefully point out that "the origin of *P. latifossatus* is not clearly understood."

Although I have neither the expertise nor the critical collections on which to base further speculation on the origin of *Polygnathus latifossatus* and the Palmatolepidae, it is surely not unreasonable to point out that origin of *P. latifossatus* in the *P. varcus* stock requires *P. latifossatus* to reinvent narrowly platformed Pa elements with very large basal cavities. Those characters figured prominently, of course, in Pa elements formed in the early history of the Polygathidae, but they were soon replaced by the smaller pits and broader platforms typical of the Pa elements of all later Early and Middle Devonian species. Such reinvention is not impossible, of course, but scenarios that involve reversals in long-term trends are commonly frowned on by phylogeneticists—particularly if reasonable alternatives are close at hand.

It may also be noted in this connection that I have already described several instances of obviously iterative development of closely similar types of Pa elements in ozarkodinide populations widely separated in time (e.g., in Silurian *Pterospathodus*, Early Devonian *Ancyrodelloides*, and Late Devonian *Ancyrodella*), and Ziegler and Lane (1987) regard such iterative bursts as important events in the overall evolution of Silurian, Devonian, and Early Carboniferous conodont faunas. Development of the Palmatolepidae may represent another such iteration, and I confess that the attractiveness of this possibility is the principal basis for recognizing the family Palmatolepidae.

Recognition of the Palmatolepidae as an independent development from the Spathognathodontidae will not change our understanding of the internal history of the group, but it will (if substantiated) require some minor nomenclatural adjustment. That is, *Polygnathus latifossatus* could no longer be kept in *Polygnathus*, for to do so would make that venerable genus polyphyletic. It would also be necessary to reassign *Polygnathus limitaris* (Fig. 5.47.2), *P. cristatus* (Fig. 5.47.3), and the other species with broadly plated Pa elements that are supposed to be descendants of *P. latifossatus*. As a temporary measure, of course, *P. latifossatus* might be included in *Schmidtognathus* and the broadly plated species in *Mesotaxis*, which has included one of their number (*M. asymmetricus*) for more than a decade. I have not followed either of these temporary routes here, or in Figs. 5.36 or 5.47, but have used the rubric "*Polygnathus*" to indicate informal assignment. Formal reassignment might well be delayed until the skeletal apparatuses of the species in question have been reconstructed and evaluated.

5.9.6 *Family Elictognathidae Austin and Rhodes, 1981*

Pa elements of species assigned to this family are carminiplanate, like those of *Polygnathus*, but lack a well-defined keel on the attachment surface posterior to the basal pit. A more or less well-developed keel, with a median groove, occurs anterior to the basal pit, but its position posterior to the pit is occupied by a broad flat area termed a *pseudokeel*, which may have a thin groove for some part or all of its length, or by a low, broad keel that is considerably less distinct than the one anterior of the pit.

Nomenclature of this family is complicated, as is that of *Siphonodella* (Fig. 5.48), the genus taken to be most typical of it. That is, *Siphonodella* was first named *Siphonognathus* by Branson and Mehl (1934), but that name could not be used for conodonts because it had been used previously for a fish. Hence, in 1944 Branson and Mehl proposed the substitute name *Siphonodella*. Meanwhile, Huddle (1934) had used the name *Falcodus* and Cooper (1939) the form-generic names *Elictognathus* and *Dinodus* for elements that were probably parts of the apparatus of species of *Siphonodella*. Although I believe the synonymy among *Falcodus*, *Elictognathus*, *Dinodus*, and *Siphonodella* to be probable, it will take additional work to prove

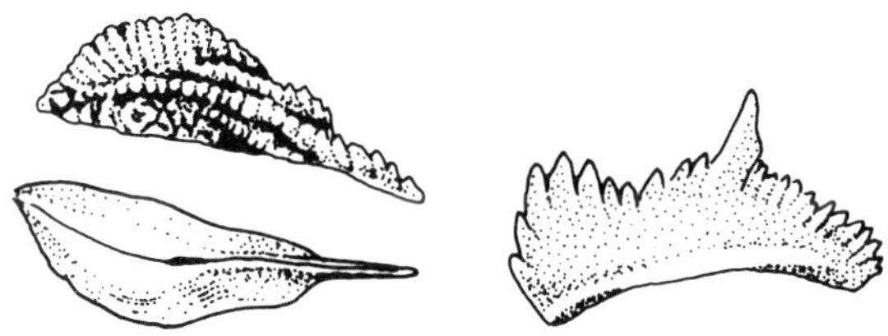

Fig. 5.48. Elements typical of multielement *Siphonodella* (Elictognathidae).

it. When (and if) that happens, *Falcodus* will clearly be the oldest valid name for the typical genus of the family Elictognathidae, and *Elictognathus*, *Dinodus*, and *Siphonodella* will be junior synonyms. The family name Elictognathidae will stand, however, because it is the oldest one proposed for any of the genera (synonyms or not) now included in it.

Although a great deal of attention has been paid to the morphology of elictognathid Pa elements, there is no published reconstruction of the complete skeletal apparatus for any of the 20 species now included in the family. No one, to my knowledge, has ventured even a guess as to composition of the apparatus of any species of *Alternognathus* (Fig. 5.49.1 to 5.49.3), but Sandberg et al. (1978) speculated that the Pb position was occupied in the apparatus of at least some species of *Siphonodella* by anguliplanate elements of the sort commonly referred to *Elictognathus*; that the M position included elements that have been described in form taxonomy as *Falcodus angulus*; and that various form species of *Dinodus* occupied positions in the S series. As Sandberg and his colleagues noted, however, no specimens of *Dinodus*, *Elictognathus*, or *Falcodus angulus* have been reported from rocks with elements of *Siphonodella praesulcata* (Fig. 5.49.4) or from collections with early representatives of *Siphonodella sulcata* (Fig. 5.49.5). These observations were taken to mean either that *S. praesulcata* and the earliest representatives of *S. sulcata* would have to be removed from *Siphonodella* on grounds of apparatus incompatibility, or that the occurrence of *Dinodus*, *Elictognathus*, and *Falcodus angulus* with the Pa elements typical of all other species of *Siphonodella* is a result of "similarity of niches." I suggest that there is also the possibility that the really distinctive features of *Siphonodella* developed gradually and that they might not yet have appeared in populations of *S. praesulcata*, the earliest and least *Siphonodella*-like species of the genus, or even in early populations of *S. sulcata*. Those early representatives may have evolved Pa elements with the essential characters of *Siphonodella*, but they might have retained the generalized Pb, M, and S-series elements of their ancestors. In short, evolution of *Siphonodella* may not have affected all components of the skeletal apparatus equally or at the same time, and there is no reason to assume *a priori* that non-Pa elements were morphologically the same in the apparatuses of every species.

With the exception of *Siphonodella praesulcata* Sandberg, all known Late Devonian representatives of the Elictognathidae belong in *Alternognathus*, which was created by Ziegler and Sandberg (1984) for species previously referred with question to *Polygnathus* or included in *Scaphignathus*, whose species have Pa elements that are closely similar in morphology but were probably derived from *Pandorinellina*, not from *Mehlina*.

The oldest known species of *Alternognathus*, *A. pseudostrigosus* (Dreesen and Dusar) (Fig. 5.49.1), formed carminiplanate Pa elements with a narrow, asymmetrically developed platform that bears only a few nodes marginal to the sigmoidally curved carina, and with a narrow, flat undersurface that has an elongate basal pit but no very distinct keel. Younger species of *Alternognathus* (Fig. 5.49.2 and 5.49.3) are characterized by Pa elements with more elaborately developed platforms, which have a median carina that is separated by a slight depression from the anterior blade and marginal rows of nodes or short transverse ridges.

Although there is currently a substantial stratigraphic gap between the youngest known specimens of *Alternognathus* and the oldest

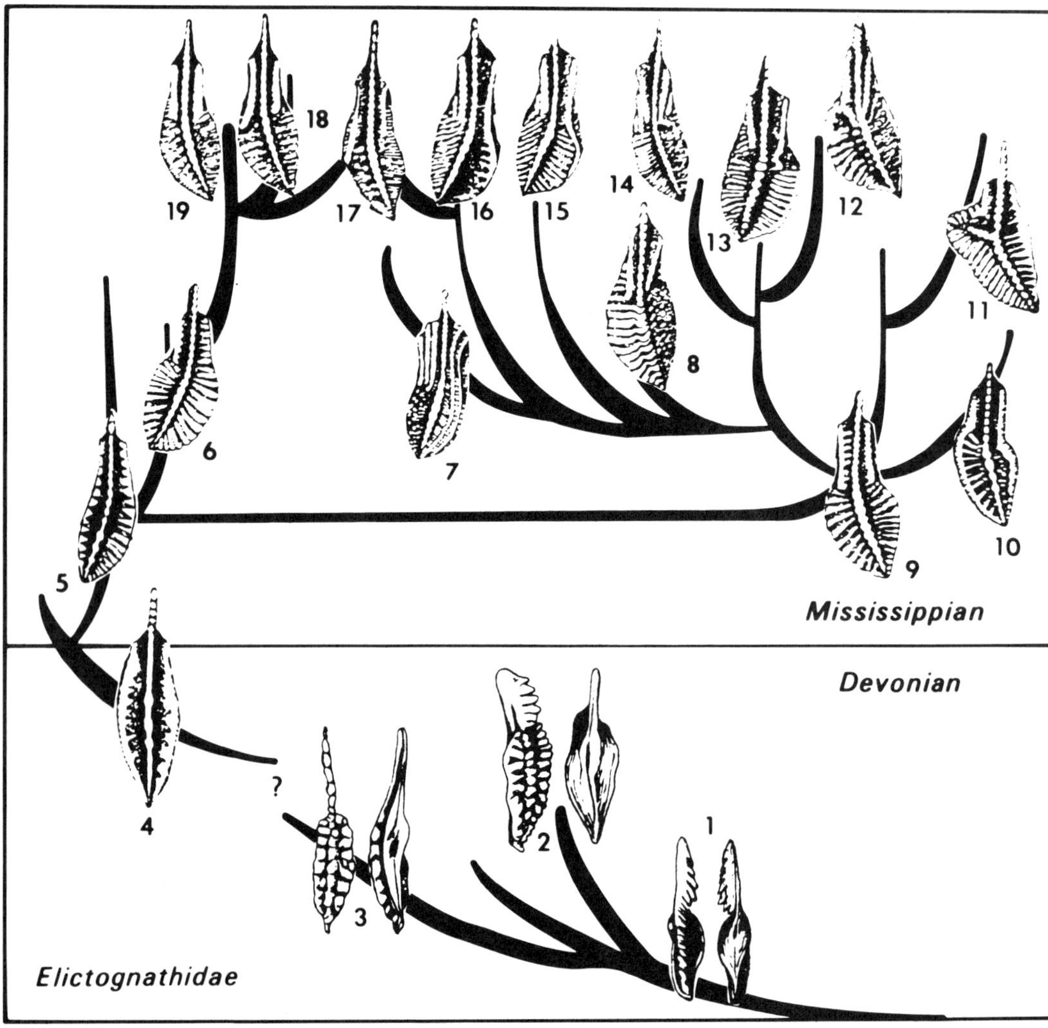

Fig. 5.49. Phylogeny of the Elictognathidae. Figures 4 to 19 and pattern shown for species of *Siphonodella* from Sandberg, Ziegler, Leuteritz, and Brill (1978). (1) *Alternognathus pseudostrigosus*; (2) *A. buelensis*; (3) *A. regularis*; (4) *Siphonodlella praesulcata*; (5) *S. sulcata*; (6) *S. duplicata*; (7) *S. sandbergi*; (8) *S. carinthiaca*; (9) *S. duplicata*, M2; (10) *S. cooperi*, M1; (11) *S. lobata*; (12) *S. crenulata*, M1; (13) *S. duplicata* sensu Hass; (14) *S. quadruplicata*; (15) *S. cooperi*, M2; (16) *S. obsoleta*; (17) *S. isosticha-obsoleta*; (18) *S. isosticha*; and (19) *S.* cf. *isosticha*.

representatives of *Siphonodella*, it has for some years been a part of the conventional wisdom that *S. praesulcata* (Fig. 5.49.4), the oldest *Siphonodella*, evolved from some part of the stock now included in *Alternognathus*. This seems eminently logical on morphologic grounds, although other avenues of development may be suggested when more complete information is made available as to other components of the *Alternognathus* and *Siphonodella* apparatuses.

Siphonodella praesulcata Sandberg, the oldest known species of the genus and the only one represented in Devonian rocks, developed narrow, symmetrical Pa elements that are carminiscaphate in early developmental stages but become carminiplanate in later ones, which exhibit a raised, flat pseudokeel posterior to the elongate basal pit. The platform of *S. praesulcata* Pa elements has a central carina that is separated from transversely nodose marginal ridges by wide adcarinal troughs. Ziegler and Sandberg (1984) suggest that *S. praesulcata* evolved from *Alternognathus pseudostrigosus*, the simplest and presumably most generalized member of that genus, and not from one of the probably more specialized, younger species.

As indicated in Fig. 5.49, *Siphonodella praesulcata* was succeeded early in the Carboniferous by *S. sulcata* (Fig. 5.49.5), which was evidently the progenitor of highly variable *S. duplicata* (Fig. 5.49.6, 5.49.9, and 5.49.13). Following an interval of substantial variation, the *duplicata* zones, features characteristic of Pa elements in various segments of the *S. duplicata* populations stabilized, and younger populations with those characters are recognized as independent species, most of which ranged to the end of the Kinderhookian Epoch.

5.9.7 *Family Gnathodontidae, new*

Conodonts assembled here (Fig. 5.36) developed late in the Devonian from *Bispathodus stabilis* (Spathognathodontidae). They are characterized by a basically seximembrate skeletal apparatus in which Pa elements are carminiscaphate, and Sc elements are alate, with a denticulated posterior process. The sides of gnathodontid Pa elements flare laterally posterior to midlength, and their upper surfaces may be smooth or ornamented by a few scattered nodes or by longitudinal or radial rows of nodes. Little attention has been paid to the complete skeletal apparatus, so taxonomy of gnathodontid conodonts is based almost entirely on features of Pa elements.

The oldest gnathodontids, from rocks of latest Famennian (Devonian) age, are referred to *Protognathodus* (Fig. 5.50.2 to 5.50.6), whose species formed carminiscaphate Pa elements with attachment surfaces in broadly expanded basal cavities (or "cups") that occupied the posterior half of the under surface. Pa elements of the three Late Devonian species of *Protognathodus* differ primarily in the manner in which the upper surface of the posterior cup is ornamented; one species (Fig. 5.50.2) formed Pa elements with smooth-surfaced cups; Pa elements of another (Fig. 5.50.4) bear short rows of nodes on either side of a subcentral carina; and homologous elements of a third species (Fig. 5.50.3) have a node or two on either side of the carina. In Pa elements of these Late Devonian species, and in those of a fourth one that appeared early in the Mississippian (Fig. 5.50.5), anterior ends of cup segments are directly opposed on opposite sides of the blade. In Pa elements of two somewhat younger Mississippian species (Fig. 5.50.6 and 5.50.7), however, anterior ends of cup segments join sides of the blade at slightly different points. This condition heralds one that is common to the Pa elements of all species of *Gnathodus*, but it is combined in Pa elements of *Protognathodus praedelicatus* (Fig. 5.50.6) and *P. cordiformis* (Fig. 5.50.7) with features of surface ornamentation that are more like the Pa patterns of older species of *Protognathodus* than those of somewhat younger species of *Gnathodus*.

Shortly after *Protognathodus praedelicatus* (Fig. 5.50.6) appeared in Early Mississippian seas, it was joined by geographically widespread populations of conodonts with similar but substantially more elaborate Pa elements. These conodonts, which represent the first species of *Gnathodus*, differ from *P. praedelicatus*, their presumed ancestor, in a number of ways. First, Pa elements are conspicuously asymmetric. The anterior end of the cup on the concave (or "inner") side of the element invariably joins the blade at a point well anterior of the point at which the anterior end of the cup on the convex (or "outer") side meets the blade. Furthermore, nodes on the upper surface of the inner cup segment commonly join to form a distinctive ridge- or comblike *parapet*, which may be short and weakly defined or long and prominent. In Pa elements of most species of *Gnathodus*, the outer cup is more broadly expanded than the inner one; it may be essentially smooth, may bear irregularly distributed nodes, or may have longitudinally, radially, or concentrically arranged rows of nodes or low ridges.

In the highly variable populations of *Gnathodus* that spread with the initial explosive burst, Lane, Sandberg, and Ziegler (1980) recognized three major groups. The group that centers on *G. delicatus* (Fig. 5.50.8), and also includes *G. cuneiformis* (Fig. 5.50.11), is characterized by Pa elements with a long, well-defined parapet. The groups typified by *G. typicus* (Fig. 5.50.9) and *G. punctatus* (Fig. 5.50.10) have Pa elements with short parapets. The latter groups are distinguished by the fact that

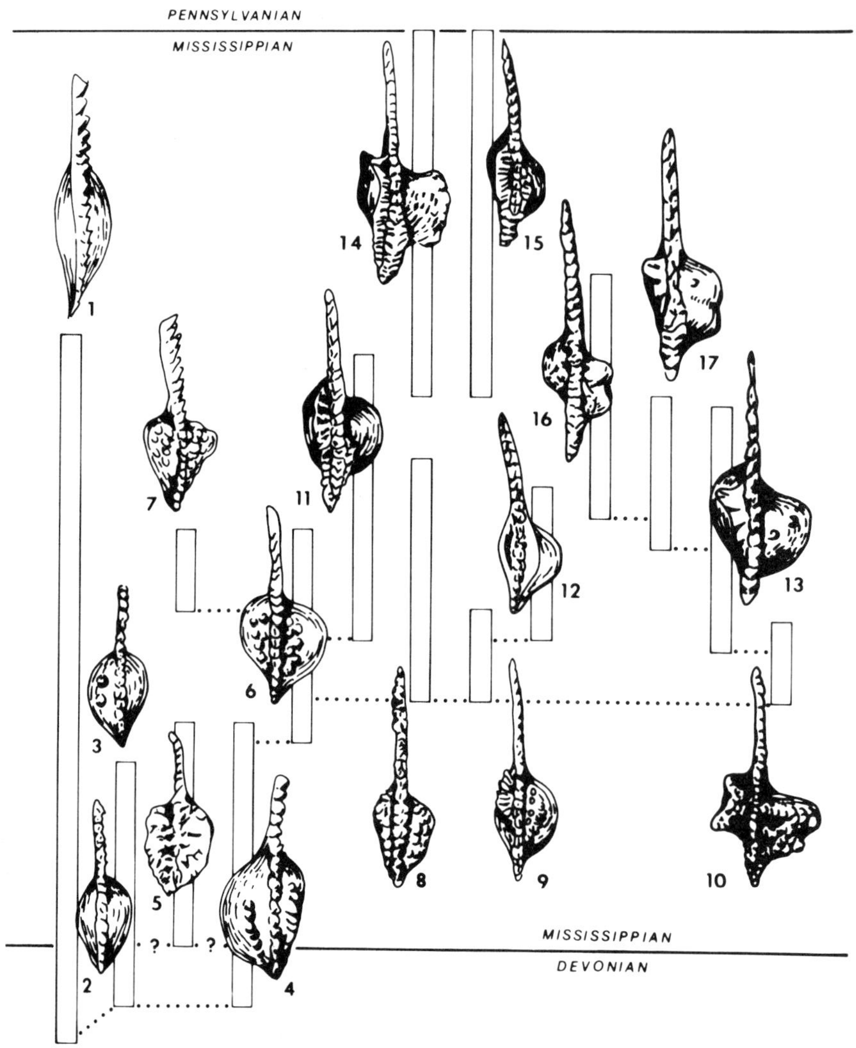

Fig. 5.50. Phylogeny of the Gnathodontidae. Relations determined by Lane, Sandberg, and Ziegler (1980). (1) *Bispathodus stabilis*; (2) *Protognathodus meischneri*; (3) *P. collinsoni*; (4) *P. kockeli*; (5) *P. kuehni*; (6) *P. praedelicatus*; (7) *P. cordiformis*; (8) *Gnathodus delicatus*; (9) *G. typicus*; (10) *G. punctatus*; (11) *G. cuneiformis*; (12) *G. typicus*, M2; (13) *G. semiglaber*; (14) *G. bilineatus*; (15) *G. girtyi*; (16) *G. texanus*; and (17) *G. pseudosemiglaber*.

denticles toward the posterior end of the carina tend to spread laterally and become transversely ridgelike in Pa elements of species in the *G. punctatus* group, but remain simple and unexpanded in Pa elements of the *G. typicus* group.

In somewhat later Mississippian rocks, Pa elements of the *delicatus*, *typicus*, and *punctatus* types can also be recognized, but they tend to be simpler—a pattern that is especially well illustrated by the several members of the *G. punctatus* group shown in Fig. 5.50.10, 5.50.13, 5.50.17, and 5.50.16, but is also clearly evident in features of Pa elements from the two stratigraphically superposed samples of *G. typicus* illustrated by Fig. 5.50.9 and 5.50.12.

Lane, Sandberg, and Ziegler (1980), who are largely responsible for the phylogenetic interpretation summarized in Fig. 5.50, recognized that *Gnathodus bilineatus* (Fig. 5.50.14) and related later Mississippian species formed Pa elements that combined features of the two early Mississippian groups: that is, a long, well-developed parapet and prominently widened denticles in the posterior segment of the carina. Lane and his colleagues did not speculate on relations between *G. bilineatus* and earlier species of *Gnathodus*. However, since simplification in the morphology of Pa elements appears to have been the hallmark of evolutionary development in the three major groups of early Mississippian species, it seems unlikely that *G. bilineatus*, with its highly complex Pa elements, is very closely related to either the *typicus* or *punctatus* groups, even though posteriorly widened denticles are characteristic of the latter. Thus, if there is a relationship between *G. bilineatus* and any group of early Mississippian gnathodontids, one might be postulated with the *delicatus* group, which specialized in building Pa elements with long parapets and might also have taken up denticle widening in mid-Mississippian times.

Belka (1985) derives *Gnathodus bilineatus* from his species, *G. praebilineatus*, which is said to be "... a perfect homeomorph of *Gnathodus delicatus*." Although this would seem to vindicate my "guess" that the *G. bilineatus* stock had its origins in the *delicatus* group, I should also note that Belka chooses *G. semiglaber* (Fig. 5.50.13), a member of the *punctatus* group, as the likely ancestor, and not a species of the *delicatus* group. This would require reversal of morphologic trends in the *G. punctatus* group recognized by Lane, Sandberg, and Ziegler (1980) by requiring that posterior denticles of the carina revert to simplicity in *G. praebilineatus* (only to become complex again in *G. bilineatus* itself), and that the parapet become longer, despite a tendency in the *punctatus* group for the parapet to begin short and become shorter!

Gnathodus girtyi (Fig. 5.50.15), which appeared at about the same time as *G. bilineatus*, might be a somewhat simplified member of the same group. However, Belka (1985) has recently described a new species, *G. austini*, which is said to have Pa elements transitional in morphology between those of *G. texanus*, an end member of the *punctatus* group, and *G. girtyi*. Derivation of *G. girtyi* from *G. austini*, however, would require that a short parapet, with a high anterior node, become longer and more uniform in height; and that the posteriorly widened denticles of *G. texanus*, *G. austini*'s presumed progenitor, be abandoned in favor of a simpler arrangement. I opt at this stage for a scenario that requires fewer reversals in trend—that is, for a relationship to the *delicatus* group (and, perhaps, *G. bilineatus*), which has a long parapet and posteriorly simple denticles.

5.9.8 Lochriea and Vogelgnathus

Two additional Mississippian lineages, referred tentatively to the Spathognathodontidae in Fig. 5.36, merit discussion here, separate from the Gnathodontidae. One lineage is formed by the several species of *Lochriea* (also known as *Paragnathodus*) (Fig. 5.51), the other by the two known species of *Vogelgnathus* Norby and Rexroad (Fig. 5.51). Species of both *Lochriea* and *Vogelgnathus* are represented by bedding-plane assemblages, hence there are few mysteries about skeletal anatomy. There are substantial questions, however, about their relationships to other conodonts.

Lochriea commutatus (Branson and Mehl) has a seximembrate skeletal apparatus in which the Pa position was occupied by carminiscaphate elements that are closely similar to those of *Protognathodus meischneri* (Fig. 5.50.2) in that the cup is smooth on its upper surface. Indeed, it would be easy to confuse Pa elements of the two species which, however, differ in relative height of carina and basal cavity and in the fact that denticles of the carina tend to widen laterally and develop a distinctive cancellate pattern in *L. commutatus*. *L. cracoviensis* Belka, the oldest known species of *Lochriea*, is distinguished by Pa elements with cups of more elliptical outline and laterally expanded, surficially cancellate denticles on both anterior blade and posterior carina.Other species of *Lochriea*, like other species of *Protognathodus*, developed Pa elements whose cups are variously ornamented by nodes and ridges.

Lochriea cracoviensis appears stratigraphically in the biozone just above the one in which *Protognathodus cordiformis* is last recorded and just below the one in which *L. commutatus* and *Gnathodus bilineatus* (Fig. 5.50.14) make their debut. It is easy to postulate a relationship between *L. cracoviensis* and *L. commutatus*, but difficult to relate either species to *Para-*

gnathodus or *Gnathodus*. That is, Pa elements of *P. cordiformis* (Fig. 5.50.7) have slightly asymmetric cups that are heart shaped as viewed from above and conspicuously ornamented by nodes that are randomly scattered or arranged in rows, whereas Pa elements of *L. cracoviensis* and *L. commutatus* have unornamented cups that are broadly ovate in superior view and a carina that includes laterally expanded denticles with a curious polygonal pattern on part of their surfaces. Pa elements of *Gnathodus bilineatus* have carinae with laterally expanded denticles, but they also have a long, well-organized parapet and a prominent outer cup that bears a complex pattern of surficial nodes and ridges.

Belka (1985) speculates that *Lochriea cracoviensis* developed independently from *Bispathodus stabilis* instead of from any of the early Mississippian gnathodontid stocks shown in Fig. 5.36. I find that speculation attractive, but it will need support from future studies. Pending such studies, I have removed *Lochriea* from the Gnathodontidae in Fig. 5.36 and have squeezed it in under the somewhat crowded umbrella of the Spathognathodontidae.

Vogelgnathus (Fig. 5.51), also included with the Spathognathodontidae in Fig. 5.36 (and Appendix A), is closely similar to *Lochriea* and may not merit separate generic status. The two best-known species have a seximembrate skeletal apparatus that includes an alate Sa element with a long, denticulated posterior process and carminiscaphate Pa elements whose laterally expanded, posteriorly situated cups are smooth on their upper surfaces. Species of *Vogelgnathus* were widespread members of late Mississippian conodont faunas, but their phylogenetic relationships are uncertain. Norby and Rexroad (1985), who provided both name and diagnosis for *Vogelgnathus*, note that the Pa elements of many species of *Gnathodus* pass through early developmental stages that are reminiscent of the mature Pa elements of *Vogelgnathus*. They suggest that this might be an example of paedomorphosis achieved through neoteny. Derivation from *Bispathodus stabilis*, independently of *Gnathodus* or other gnathodontids, does not seem to have been considered. However, in view of the preceding discussion of *Lochriea* this seems at least an attactive possibility.

If future studies demonstrate that *Lochriea* and *Vogelgnathus* represent a lineage separate from the Spathognathodontidae and Gnathodontidae, it will probably be desirable to create and name a new family for them. I recommend that this not be done, however, until the relationships of *Diplognathodus* and related genera (Sweetognathidae) have also been considered, because they may also have roots in the same spathognathodontid plexus. At present the history of both groups is punctuated by so many gaps that phylogenies cannot be made out very clearly.

Fig. 5.51. Elements and apparatuses typical of species of *Lochriea* and *Vogelgnathus* (Spathognathodontidae).

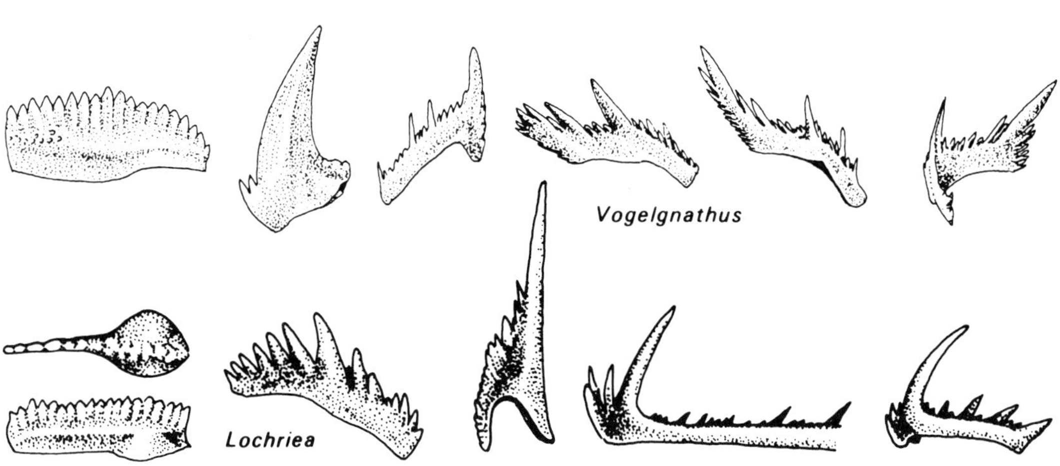

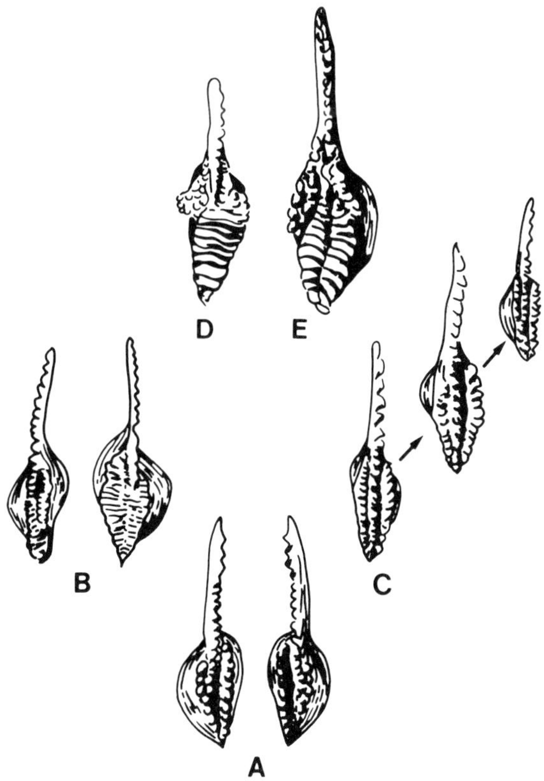

Fig. 5.52. Upper views of Pa elements typical of various genera of the Idiognathodontidae. (A) *Declinognathodus*; (B) *Idiognathoides*; (C) *Neognathodus*; (D) *Idiognathodus*; and (E) *Streptognathodus*.

5.9.9 Family Idiognathodontidae Harris and Hollingsworth, 1933

Idiognathodontid conodonts are distinguished by carminiscaphate Pa elements whose upper surfaces typically bear three longitudinal rows of nodes or denticles. One of these rows, the *carina*, is a posterior continuation of the *free blade*, which is assigned an anterior position and may account for more than half the total length of the element. Conceptually, at least, the other denticle rows are marginal to the carina; in fact, the carina curves laterally at varying distances posterior to the end of the blade to join one of the marginal rows or is replaced across much of the platform by a median groove or trough, so that many idiognathodontid Pa elements appear to have only two denticle rows. The complete skeletal apparatus is known for only a few of the species now included in the Idiognathodontidae. In those species, the apparatus is seximembrate, includes an alate Sa element with a denticulated posterior process, and is in other respects closely similar in composition to that of the presumably ancestral Gnathodontidae.

Declinognathodus noduliferus (Fig. 5.52A), the oldest known idiognathodontid, is distinguished by carminiscaphate Pa elements in which the carina curves laterally to join the outer marginal denticle row a short distance posterior of the end of the blade. In Pa elements of species of *Idiognathoides* (Fig. 5.52B), which appears shortly after *Declinognathodus*, the posterior end of the blade curves abruptly to one side and joins the outer marginal denticle row wihout forming even a short carina. Furthermore, Pa elements of some species of *Idiognathoides* exhibit Class III symmetry; that is, sinistral and dextral specimens differ in morphologic detail, but not in overall pattern.

Early species of *Neognathodus* (Fig. 5.52C) built subsymmetrical Pa elements in which a well-developed carina extends nearly to the posterior end of the cup between bordering marginal rows. In Pa elements of later species, however, the carina joins the outer marginal row at points closer and closer to the anterior end of the cup; in Pa elements of the youngest *Neognathodus* species known, only the anteriormost denticle of the marginal row is separate from the carina, and the overall morphology is thus quite similar to that of the Pa elements of *Declinognathodus*.

In the Pa elements of *Streptognathodus* (Fig. 5.52E) the carina is median in position but extends for only a short distance onto the anterior end of the platform. Its position over the remainder of the platform is taken up by a median trough or groove, and complex aggregates of secondary nodes or denticles commonly develop adjacent to the blade on the anterior shoulders of the cup. In Pa elements of closely related species of *Idiognathodus* (Fig. 5.52D), early ontogenetic stages resemble the mature elements of *Streptognathodus*. In later stages the median trough disappears from the platform, which is crossed instead by a series of transverse ribs or ridges.

Declinognathodus, *Idiognathoides*, and *Neognathodus* are confined to the Lower Pennsylvanian, but *Streptognathodus* and *Idiognathodus* range through the Pennsylvanian into the Permian. Named species of the latter two genera are numerous, and the Pa elements by which they are identified are common in samples from Pennsylvanian and lower Permian rocks. Because species of *Streptognathodus* and *Idiognathodus* were components of low-diversity faunas, one might expect that relations among these species would have been worked out in great detail, since detailed biostratigraphies of upper Carboniferous and lower Permian rocks will depend on critical analyses of this group of ozarkodinide conodonts. Unfortunately, there is as yet no meaningful taxonomy in place for the *Streptognathodus-Idiognathodus* plexus, although direction signs have recently been posted and several intriguing studies are in progress.

van den Boogaard (in van den Boogaard and Bless, 1985) has recently documented the ontogenetic development of Pa elements in two species of *Idiognathodus*, *I. delicatus* and *I. magnificus*. The sketches in Fig. 5.53 are adapted from his more detailed photographic record. Note that tiny specimens that represent the earliest growth stages have an elongate cup and a median carina that extends to the posterior end of the element between subparallel marginal ridges. Such specimens, of course, have the essential characters of *Streptognathodus* and would surely be assigned to that genus in form taxonomy. In successively larger specimens, the platform broadens, transverse ridges appear across its posterior tip, and lobate areas occupied by nodes develop on the platform's anterolateral shoulders. In the largest Pa elements of *Idiognathodus delicatus* illustrated by van den Boogaard and Bless (1985), the carina enters only the anteriormost part of the platform, which is otherwise corrugated by transverse ridges that are bordered anterolaterally and laterally by conspicuously nodose marginal lobes. These features of Pa elements are the ones on which concepts of *Idiognathodus* species are based.

In 1971 Merrill (in Lane, Merrill, Straka, and Webster, 1971) noted somewhat cryptically, that "... *Streptognathodus* appears to have been both ancestor and successor to *Idiognathodus* several times during the Pennsylvanian." That is, stratigraphic intervals dominated by *Streptognathodus* alternate with those dominated by *Idiognathodus* through the Pennsylvanian. This has surely complicated understanding of this plexus and represents a major unsolved problem in conodont taxonomy.

If early Pennsylvanian (Morrowan) *Streptognathodus* spawned *Idiognathodus*, as stratigraphic evidence seems to indicate, and if *Idiognathodus* species retained a *Streptognathodus* stage in their early ontogeny, as ontogenetic data summarized schematically in Fig. 5.53 suggest, later Pennsylvanian and Permian history of the *Streptognathodus-Idiognathodus* plexus may have involved neotenic shifts of juvenile *Streptognathodus* characters into adult stages. *Streptognathodus*, the initial ancestor, would thus seem to become more and more diverse in later Pennsylvanian and early Permian conodont faunas, or episodically through the Pennsylvanian as paedomorphosis affected various lineages of *Idiognathodus*. If such an evolutionary program was followed in the developmental history of this important group of

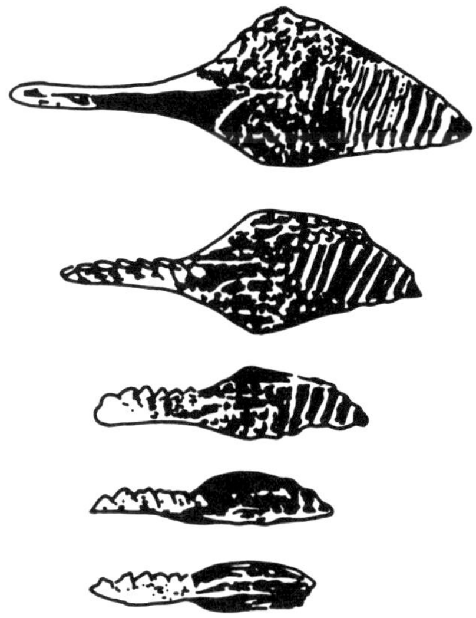

Fig. 5.53. Ontogenetic sequence in Pa elements of *Idiognathodus*. Smallest specimens have morphologic features of *Streptognathodus*. Drawn from photographs in van den Boogaard and Bless (1985).

conodonts, I suspect it involved several lineages, which may have been affected to different degrees and at different times. Sorting this out will require careful biometric studies of collections from rock sequences that represent long-continued stability of depositional conditions. I doubt that it will be worked out in the cyclic deposits by which the Carboniferous and Lower Permian are represented in much of the world, despite the fact that collections from such rocks are commonly rich in specimens that represent the *Strepognathodus-Idiognathodus* plexus.

5.9.10 Family Anchignathodontidae Clark, 1972

Ozarkodinide conodonts included in this family have a basically seximembrate skeletal apparatus that includes carminiscaphate Pa elements, angulate Pb elements with relatively short processes, and alate Sa elements that lack any trace of a posterior process. At present one anchignathodontid species is assigned to *Aethotaxis*, which is peculiar in that its apparatus appears to have lacked elements in the P positions, and all the others are included in *Hindeodus*, which is probably interpreted far too broadly. The oldest anchignathodontids known are of Early Mississippian (Kinderhookian) age; the youngest are of earliest Triassic age.

The rootstock of the Anchignathodontidae is formed by a succession of Carboniferous, Permian, and earliest Triassic species herein assigned collectively (and probably quite loosely) to *Hindeodus*. Only a few of these species have been diagnosed and described in fully modern, multielement terms, so future work may well establish that several different lineages are involved. In line with the practice in many other families, each of those lineages would merit recognition as a genus, and there are names already available for most of them!

The oldest anchignathodontid known to me, *Hindeodus crassidentatus* (Branson and Mehl) (Fig. 5.54), represents a group of Early Mississippian (Kinderhookian) species in which carminiscaphate Pa elements have a cup-shaped basal cavity beneath their posterior half and a finlike anterior blade that consists of three or four denticles that decline in length and decrease in width posteriorly. These elements were originally diagnosed in form taxonomy as various species of *Spathodus* or *Spathognathodus*, and it is not possible at present to determine if they represent one or dozens of species in a multielement sense. In any event, Pa elements of this type are regularly associated in collections from Kinderhookian strata with bowed angulate pectiniform elements that have relatively short subequal anterior and posterior processes and have been identified previously as species of *Subbryantodus* Branson and Mehl, 1934. Such elements qualify morphologically and fraternally as Pb components of species in the *H. crassidentatus* group. Maybe, after this group has been revised and its taxonomy brought up to date, *Subbryantodus* will turn out to be the oldest available generic name for it.

Cudotaxis priceslingi Chauff, based on material from somewhat younger, middle Mississippian rocks, also seems to be a member of the *crassidentatus* group, and I see no way to distinguish it generically from other species in this plexus.

Hindeodus scitulus (Hinde) (Fig. 5.54) represents a second component of the Mississippian species-group here referred loosely to *Hindeodus*. *H. scitulus* has a distinctive car-

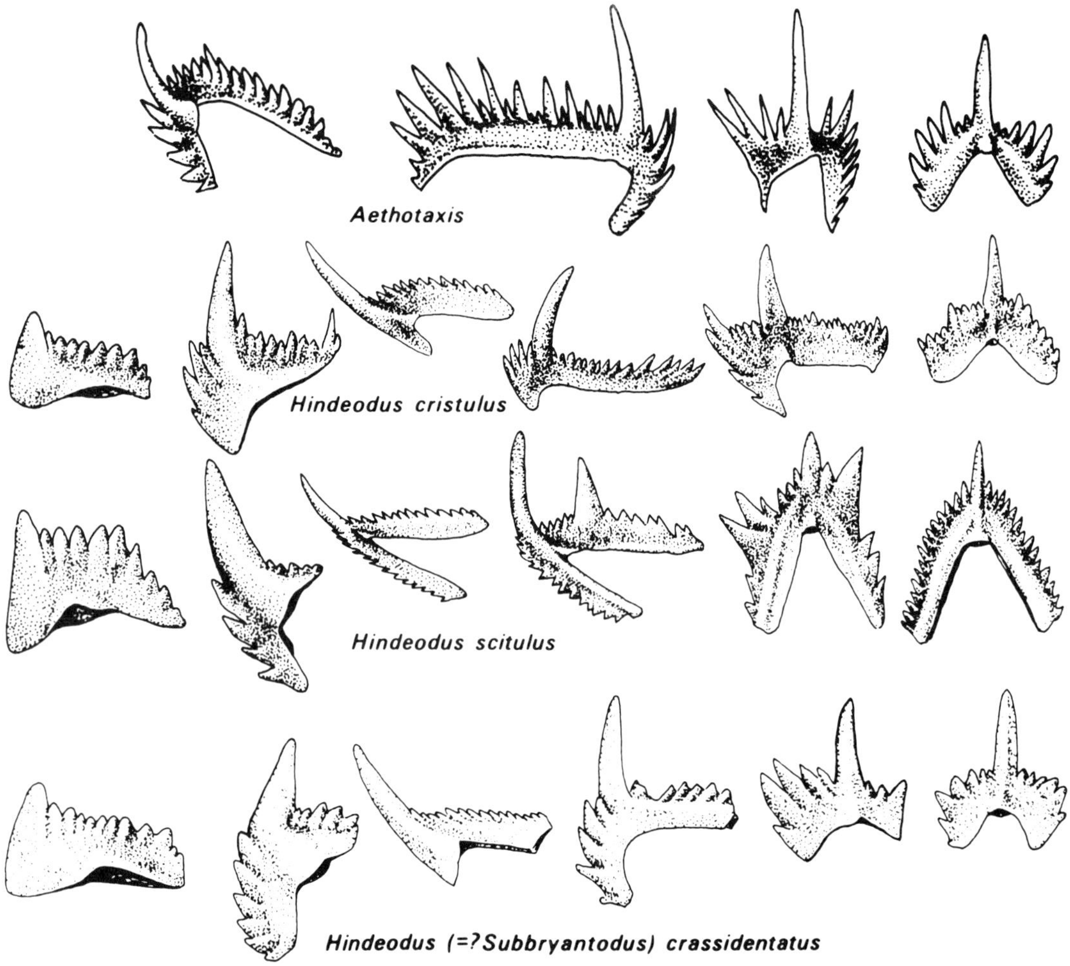

Fig. 5.54. Elements and apparatuses typical of species of *Hindeodus* (?*Subbryantodus*), *Hindeodus*, and *Aethotaxis* (Anchignathodontidae). Sa elements on right, followed to left by Sb, Sc, M, Pb, and Pa elements. P elements of *Aethotaxis* not known.

miniscaphate Pa element in which the relatively small cup-shaped basal cavity is restricted to the midportion of the underside and the laterally compressed anterior process consists primarily of a single tall, broad denticle of subtriangular lateral outline. The laterally bowed, *Subbryantodus*-like Pb element has a large cusp and short processes, and M and S-series elements are the distinctive forms commonly identified by students of Mississippian conodonts as form species of *Apatognathus*. The alate Sa element lacks a posterior process, and the M element is bipennate (or synprioniodiniform), which is not the case with M elements of other *Hindeodus* species.

Hindeodus cristulus (Youngquist and Miller) (Fig. 5.54), the type-species, represents the third group of species of *Hindeodus*. Pa elements are like those of the *crassidentatus* group in having a basal cavity that extends in cuplike fashion to the posterior end of the unit and like those of *H. scitulus* in conformation of the bladelike anterior process. But the most diagnostic feature of the apparatus of *Hindeodus* species, in my view, is the curious extensiform digyrate element that occupies the Sb position in the symmetry-transition series. One lateral process of that element tends to be bent sharply upward and posteriorly a short distance from the cusp, whereas the other commonly develops a distinctive radiate cluster of long, needle-like denticles at its distal end.

Species of the *Hindeodus cristulus* group appear first in rocks of middle Mississippian

(Meramecian) age, and the lineage clearly extends through the remainder of the Carboniferous, the Permian, and into the lowermost Triassic. Initially, in 1970, I interpreted Pa elements of the youngest species known, *H. typicalis* (Sweet), as components of the unimembrate apparatus of *Anchignathodus typicalis*, whereas occupants of the Pb, M, and S positions were regarded as components of the multimembrate apparatus of *Ellisonia teicherti* Sweet. Later (Sweet, 1973) it became obvious that *Anchignathodus typicalis* and *Ellisonia teicherti* were names for different parts of the apparatus of the same species, which ultimately (Sweet, 1976; Sweet, in Ziegler, ed., 1977) was shown to be closely similar in skeletal architecture to other species of *Hindeodus* as interpreted in a multielement sense. Before all of this had been sorted out, however, Clark (1972) had based a family-group name on *Anchignathodus* (superfamily Anchignathodontacea), and this must prevail for the family under consideration despite the fact that *Anchignathodus* has been considered a junior subjective synonym of *Hindeodus* since at least 1977!

The skeletal apparatus of *Hindeodus cristulus* (Youngquist and Miller), the Mississippian type-species, is now well known, as are those of Late Permian *H. julfensis* (Sweet) and Late Permian-Earliest Triassic *H. typicalis* (Sweet). For the most part, however, Pennsylvanian and Permian species have been neglected, and their skeletal apparatuses are largely unknown. In fact, a majority of authors who deal with late Paleozoic and early Triassic conodont faunas recognize only one species in this long interval, *H. minutus* (Ellison) which, despite substantial differences in morphology of its various skeletal elements, is even regarded as the senior synonym of *H. typicalis*, whose types are from the Lower Triassic. The limited material available to me suggests, however, that critical analysis of larger collections that span greater stratigraphic intervals will result in recognition of many species. But that is a job for the future.

5.9.11 Family Sweetognathidae Ritter, 1986

The group of Carboniferous and Permian species referred to *Diplognathodus*, *Sweetognathus*, and *Neostreptognathodus* makes up a distinctive and probably related stock that also includes ancestors of *Homoiranognathus*, the less well-known Permian *Rabeignathus* and *Iranognathus*, and the Early Triassic *Isarcicella*. In the *Treatise*, Clark and I included all these genera (except *Homoiranognathus*) with *Hindeodus* and *Aethotaxis* in the Anchignathodontidae. However, information published since the *Treatise* manuscript was prepared makes me uncomfortable with that assignment. Thus, I now treat *Diplognathodus* and its kin as an independent group with familial status. The family Sweetognathidae was established (as Sweetognathinae) for this group by Ritter (1986) who, however, included it within the family Anchignathodontidae.

In terms of the morphology of their Pa elements, species of *Diplognathodus* (Fig. 5.55) are in many respects homeomorphic replays of the simpler species of Late Devonian-Early Mississippian *Protognathodus* and mid- to late Mississippian *Lochriea*. The Pa elements in question (Fig. 5.55) are carminiscaphate structures with a cup that expands widely to the sides and extends to the posterior tip. Upper surfaces of the cup are characteristically smooth, but rare specimens in a large sample may have a node or denticle or two. Especially distinctive of the Pa elements of *Diplognathodus*, however, is division of the median denticle row into a high anterior *free blade* and, posterior of it and above the cup, a lower carina that may consist of a few denticles but is more typically fused to a smooth ridge or *spatula*, which has a subquadrate lateral profile and drops off steeply to the posterior end of the cup.*Diplognathodus* was said to have a multimembrate apparatus at the time it was established by Kozur and Merrill (in Kozur, 1975), but no illustrations were provided. It is significant to note, however, that a hibbardelliform (i.e., alate) element with a long posterior process was listed as a component.

In the same year that Kozur and Merrill (1975) established *Diplognathodus*, Perlmutter interpreted an aggregation of elements that form a recurrent group in several samples from the Lower Permian of Kansas as the skeletal apparatus of *D. expansus*. Perlmutter's reconstruction, which includes an alate Sa element with a long posterior process, follows the plan mentioned vaguely by Kozur and Merrill in their generic diagnosis. However, Perlmutter's

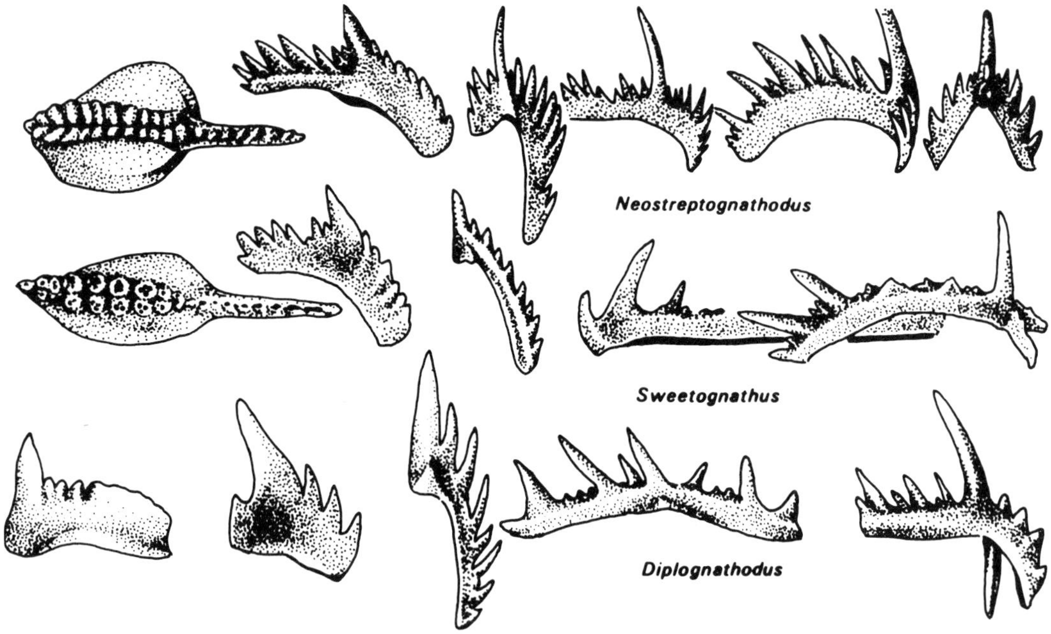

Fig. 5.55. Elements and apparatuses of species assigned to genera of the Sweetognathidae. Sa elements on right, followed by Sb, Sc, M, Pb, and Pa elements to left. Sb elements of *Diplognathodus* and *Sweetognathus* have not been described but are probably much like those of *Neostreptognathodus.*

decision to group these elements together has recently been questioned by von Bitter and Merrill (1985), who regard the ramiform elements Perlmutter interpreted as occupants of the M and S postions in the apparatus of *D. expansus* as components of the apparatus of a phantom species of *Cavusgnathus* (family Cavusgnathidae) that apparently lacked the Pa elements typical of that genus. In short, von Bitter and Merrill conclude that *D. expansus* and, by extension, other species of the genus, had a unimembrate apparatus composed only of Pa elements. In reaching this contrary opinion, von Bitter and Merrill seem to have been influenced at least partly by their conviction that *Diplognathodus* is a close relative of *Hindeodus* and hence should have an apparatus more like that of the latter than the one Perlmutter proposed for *D. expansus.*Specifically, von Bitter and Merrill seem to be most critical of Perlmutter's inclusion of an alate element with a long posterior process in the apparatus of a species assigned to *Diplognathodus*, since the Sa element of *Hindeodus* species lacks a posterior process. In 1977 I also recognized this as a problem, but suggested that differences in the ramiform complement of their skeletal apparatuses might indicate that *Diplognathodus* and *Hindeodus* were not as closely related as morphologic similarities in their Pa elements might suggest.

If the apparatus of *Diplognathodus expansus*, as reconstructed by Perlmutter (1975), is taken as representative of the apparatus of other species of *Diplognathodus*, it becomes profitable to explore a possibility I suggested in 1977—that is, that *Diplognathodus* is more closely related to *Lochriea commutatus* than to *Hindeodus.* The principal problem with this suggestion is that the first occurrence of the oldest named species of *Diplognathodus, D. orphanus*, is separated from the last occurrence of *L. commutatus* by an interval equivalent to the entire Morrowan Series—an interval interpreted to represent about the first 15 million years of Pennsylvanian time. The impact of this information is minimized, however, by noting that von Bitter and Plint-Geberl (1982) have illustrated poorly preserved Pa elements from Upper Mississippian rocks in eastern Canada that appear to represent *Diplognathodus* and are from a stratigraphic interval that is well within the range of *Lochriea* and *Vogelgnathus.*

Close similarity in morphology between Pa elements of *Diplognathodus expansus* Perlmutter (1975) and *Sweetognathus merrilli* Kozur (1975) suggests that the *Sweetognathus* stock developed from diplognathodontan ancestors in the Early Permian. Species of *Sweetognathus* (Fig. 5.55), like those of *Diplognathodus,* had a quinqui- to seximembrate skeletal apparatus in which alate elements in the Sa position have a long, denticulated posterior process and structures in the Pa position are carminiscaphate elements with a relatively short anterior blade that continues posteriorly across the broadly expanded basal cup as a carina that is basically an adenticulate ridge, but may be replaced by a single row of low broad nodes or supplemented laterally by a pair of subparallel rows of such nodes. A characteristic feature of the Pa elements formed by species of *Sweetognathus* is development of a pustulose pattern on the upper surface of posterior nodes. Such a pattern was also developed on Pa elements by species of *Lochriea,* which may suggest a relationship. Evolutionary development of the *Sweetognathus* lineage has recently been traced by Ritter (1986).

Ritter (1986) has demonstrated that species of *Rabeignathus* Kozur (1978) (Fig. 5.56) have a quinqui- to seximembrate skeletal apparatus, similar to that of *Sweetognathus.* The two known species of *Rabeignathus* are recognized, however, by their distinctive Pa elements, which are like the double-rowed carminiscaphate elements of some *Sweetognathus* species but differ in having one or two additional subparallel rows of somewhat more irregularly pustulose nodes laterally. *Rabeignathus,* which almost certainly developed from *Sweetognathus,* has a very short range in the Early Permian.

Homoiranognathus was established by Ritter (1986) for a single Early Permian species (*H. huecoensis*) known only from carminiscaphate Pa elements on whose upper surface an adenticulate, pustulose median ridge is flanked laterally by two to four subparallel rows of pustulose nodes. Elements of *H. huecoensis* are surely distinctive in appearance, but I am not

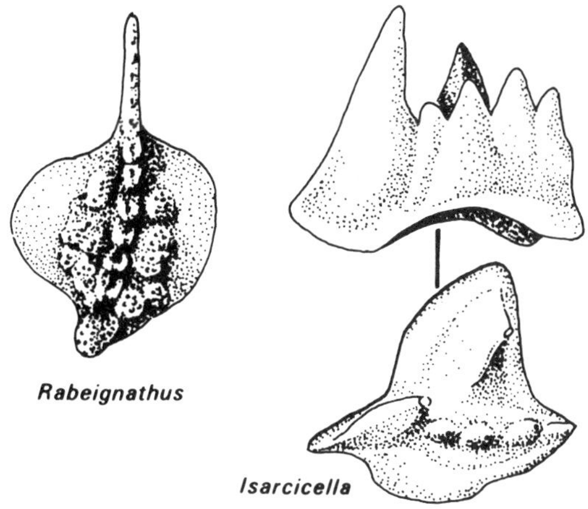

Fig. 5.56. Pa elements typical of species assigned to *Rabeignathus* and *Isarcicella* (Sweetognathidae).

yet completely convinced that this pattern (or that of Pa elements of *Rabeignathus*) merits recognition at the generic level.

Sweetognathus whitei (Rhodes, 1963) was also the apparent ancestor, in the Early Permian, of *Neostreptognathodus* Clark (1972) (Fig. 5.55), a stock of stratigraphically important sweetognathid species with Pa elements characterized by subparallel rows of nodes separated by a well-marked median groove or channel. It is also of interest to note that nodes in the posterior rows of Pa elements of *Neostreptognathodus* species seem not to have the distinctively pustulose surface that is characteristic of comparable nodes in the Pa elements formed by species of *Sweetognathus, Homoiranognathus, Rabeignathus,* and *Lochriea.* The significance of these features is difficult to assess with information presently available. It is possible, of course, that *Neostreptognathodus* represents a stock developed *de novo* from *Diplognathodus,* whose stratigraphically discontinuous record extends almost to the end of the Permian.

The Late Permian Pa elements for which Kozur, Mostler, and Rahimi-Yazd (1976) established *Iranognathus* are, in many respects, homeomorphic replays of those on which Ritter (1986) based *Homoiranognathus.* Evolutionary patterns in Permian stocks are currently difficult to reconstruct because collections from critical intervals and facies have yet to be made or described in detail. However, it seems unlikely that the two species of Late Permian *Iranognathus* represent phylogenetic continuations of the *Homoiranognathus* stock, which was spawned by *Sweetognathus whitei* in the late Wolfcampian. It is more likely (to me, at least) that *Iranognathus* represents a short-lived homeomorphic development from *Diplognathodus* in the latest Permian.

Problems associated with interpreting the origin and relationships of *Iranognathus* are duplicated with respect to *Isarcicella isarcica* (Huckriede) (Fig. 5.56) of the earliest Triassic. Carminiscaphate Pa elements of this stratigraphically important species are similar in lateral aspect to those of *Diplognathodus*, but the simple median carina is denticulated for its full length, and a single conical denticle may be developed on lateral flanks of the cusp on one or both sides of the carina. No *Diplognathodus* has been described from either the very latest Permian or the earliest Triassic; hence ancestry of *Isarcicella* has customarily been sited in *Hindeodus* of the Anchignathodontidae. It has been assumed (by me, among others) that ramiform elements of species of the two genera were vicarious. A more likely interpretation is that *Hindeodus typicalis* and *Isarcicella isarcica* represent the end members of two different major lineages of the Ozarkodinida, the *Ozarkodina-Hindeodus* and *Diplognathodus* lineages, each of which "did its thing" repeatedly through the Carboniferous and Permian into the earliest Triassic, when nearly contemporaneous extinction of both lines marked the end, also, of the Ozarkodinida.

5.9.12 Family Cavusgnathidae Austin and Rhodes, 1981

Ozarkodinide conodonts included here formed seximembrate skeletal apparatuses in which the alate Sa element has a long, denticulated posterior process and the Pa position is occupied by bowed carminiscaphate elements with a relatively long anterior free blade and a generally narrow, double-rowed posterior platform. At its posterior end, the blade may join the platform at a point intermediate between the anterior ends of the two posterior platform rows, or it may be continuous with one or the other of those rows. Apparatuses of many cavusgnathid species exhibit Class III symmetry; that is, right and left members of a pair may be bowed in opposite directions even though the blade always joins the anterior end of the same platform row, or right and left members of a pair may differ in size and style of denticulation or in other morphologic features.

Species of *Patrognathus*, *Cloghergnathus*, *Capricornignathus*, and *Taphrognathus* (all illustrated in Fig. 5.57) formed Pa elements in which the blade joins the anterior end of the platform at a point intermediate between its two marginal denticle rows. *Patrognathus* appeared in the Late Devonian and ranges into the Lower Mississippian Isosticha-Upper Crenulata Zone; the other genera of this group appear in the Texanus Zone, some two zones higher in the Mississippian, and are either confined to that zone or range only into the next one higher. If it were not for this disjunct stratigraphic distribution it would be very difficult to effect a reliable separation of the four genera. Pa elements of species of *Cloghergnathus* differ from those of *Patrognathus* only in the profile of the upper margin of the free blade, whereas one of the marginal rows of denticles in Pa elements of *Capricornignathus* may be replaced by two subparallel marginal rows of nodes that are laterally confluent. These features are certainly sufficient to distinguish the Pa elements of closely related species of the same genus, but I am skeptical of their significance on the generic level.

Clydagnathus, which made its debut in the Devonian only a bit later than *Patrognathus* and also ranges upward into the Lower Mississippian Isosticha-Upper Crenulata Zone, includes cavusgnathid species that built Pa elements in which the posterior end of the blade is continuous with the anterior end of the row that forms the right margin of the double-rowed posterior platform. This is also the case with *Cavusgnathus*, which appears in the late part of the range of *Taphrognathus* and continues upward to the top of the Mississippian, and of *Adetognathus*, which evolved in the latest Mississippian and ranges through the Pennsylvanian into the lower part of the Permian.

I find it difficult to point to any single feature or particularly significant combination of features that would distinguish the Pa elements of

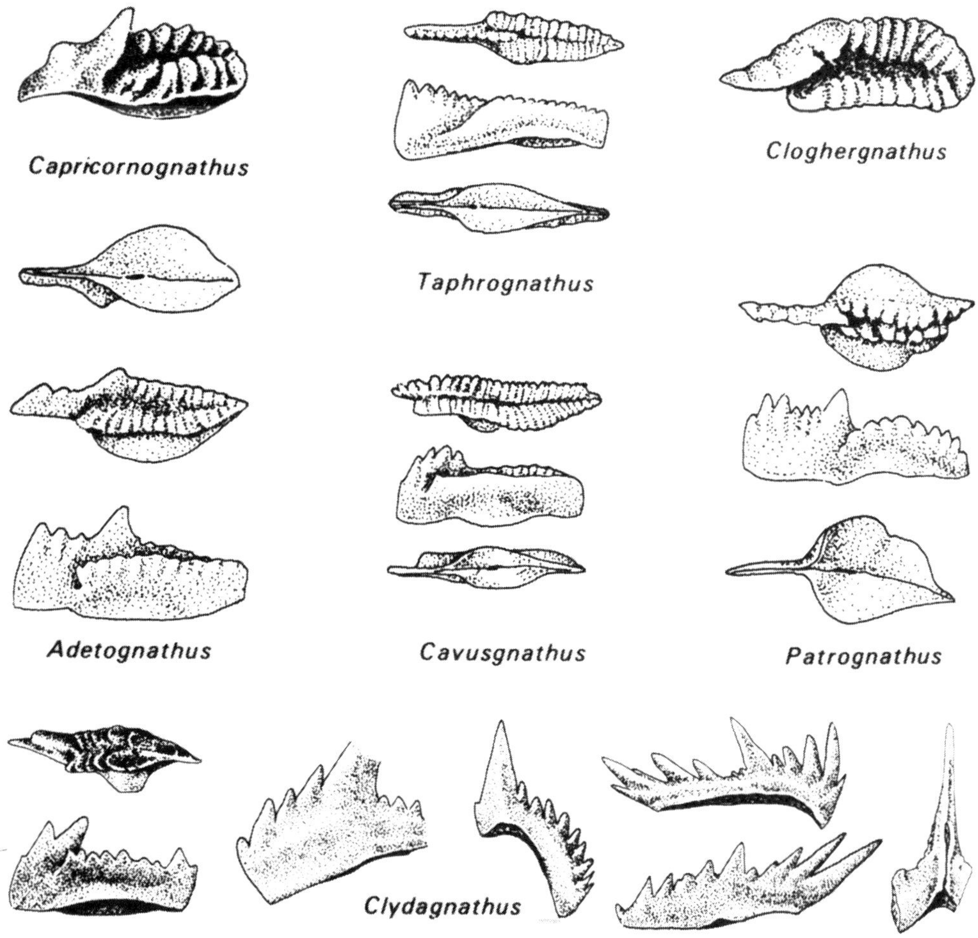

Fig. 5.57. Elements and apparatuses typical of species of the Cavusgnathidae.

Clydagnathus, *Cavusgnathus*, and *Adetognathus*. Apparatuses of *Cavusgnathus* species apparently exhibit Class IIIa symmetry in that right and left Pa elements are distinguished solely by being bowed in different directions. Right and left Pa elements of *Adetognathus*, on the other hand, are morphologically somewhat different and thus exhibit Class IIIb symmetry. In addition, Pa elements formed by species of *Adetognathus* have essentially no fixed blade, whereas in comparable Pa elements of *Cavusgnathus* species as much as half the length of the blade may be incorporated into the marginal platform row with which the blade is confluent.

I would ascribe my consistent inability to separate specimens representing species of *Cavusgnathus* and *Adetognathus* to my lack of experience with these genera were it not for the fact that specialists on Carboniferous conodonts appear to have problems that are at least as great as mine! In any event, elements of cavusgnathid conodonts are common in rocks that record marginal marine environments that were characterized by shallow water of variable salinity, and they are useful as guides to such environments.

Ancestors of cavusgnathid conodonts have been identified by Sandberg and Ziegler (1979) in Late Devonian populations of spathognathodontid ozarkodinides referred to *Pandorinellina insita* (Stauffer). In those populations, Pa elements (Fig. 5.38) are dominantly single-rowed, carminate pectiniform structures with a prominent finlike blade that is deflected to the right at its posterior end. However, minor segments of the same populations apparently developed double rows of nodelike denticles posterior to the blade. Variants of these are thought to have given rise either to *Patrognathus*, with a blade that joins the platform between its two marginal rows, or to *Clydagnathus*, in whose Pa elements the blade is continuous with the nodose row on the right side of the platform. One might guess that *Taphrognathus*, *Cloghergnathus*, and *Capricornignathus* developed in mid-Mississippian time from the *Patrognathus* branch of the family, and that *Cavusgnathus* and, later, *Adetognathus*, are continuations of the *Clydagnathus* branch. Gaps currently separate *Patrognathus* and *Clydagnathus* from the first representatives of the later cavusgnathid stocks, however, so it can not be ruled out that the later cavusgnathids (i.e., *Taphrognathus*, *Cavusgnathus-Adetognathus*, *Capricornignathus*, and *Cloghergnathus*) developed iteratively from as yet unidentified or undocumented pandorinellinid ancestors. This possibility would not be worth mentioning except for the fact that it was clearly the pattern of development widely adopted in various different lineages of the Ozarkodinida.

5.9.13 *Family Mestognathidae Austin and Rhodes, 1981*

The five species of *Mestognathus* Bischoff (1957) (Fig. 5.58), which constitute the only known members of this family, are the subjects of a very recent monographic study by von Bitter, Sandberg, and Orchard (1986). That study focuses on morphology, ontogeny, and distribution of the Pa elements of *Mestognathus* species, either because available collections were inadequate to the job of reconstructing complete apparatuses or because only the Pa elements of *Mestognathus* species were mineralized by the conodonts that formed them.

Fig. 5.58. Upper, side, and under views of a Pa element of *Mestognathus* (Mestognathidae).

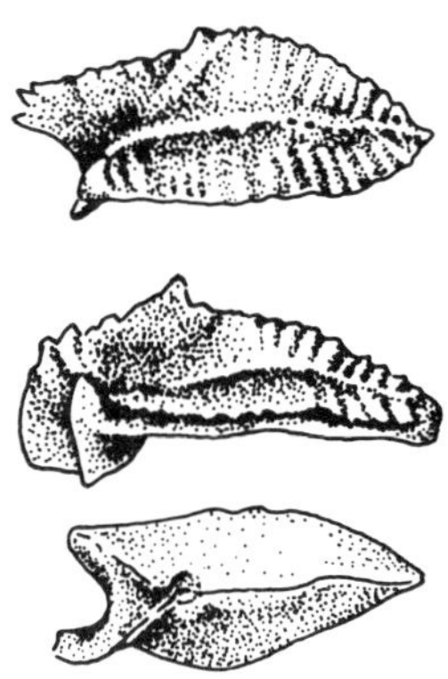

Pa elements of *Mestognathus* species are carminiplanate pectiniform structures with a high anterior blade and a posterior V-shaped platform with slightly upturned, transversely ridged margins that are separated by adcarinal troughs from a longitudinal, submedial carina that extends to the posterior tip of the platform. The blade, which is divided into free and fixed segments of approximately equal length, is continuous with the ridged right margin of the platform. The anterior end of the opposite margin is developed into a distinctively denticulated parapet to which von Bitter, Sandberg, and Orchard (1986) assign premier taxonomic importance.

The underside of *Mestognathus* Pa elements is marked by a grooved median keel, which encloses a small basal pit anterior of midlength and is bordered posterior of the pit by narrow zones of recessive basal margin (termed an "eversion strip" by von Bitter, Sandberg and Orchard). A secondary keel, which may include a branch of the median groove, is developed beneath the anterior end of the carina and parapet segment.

Phylogenetic development of *Mestognathus* involved gradual change in morphology of the parapet segment of the left platform margin; a gradual increase in development of the anterior notch; increasing definition of the secondary keel; and gradual development of broader and broader zones of recessive basal margin marginal to the primary keel and posterior to the basal pit. Pa elements that represent early stages in the ontogeny of *Mestognathus* species are similar morphologically to Pa elements of adult *Cavusgnathus* in having scaphate under surfaces and no zones of recessive basal margin. This, together with the fact that the latter character also becomes more pronounced in its development phylogenetically, suggests to von Bitter, Sandberg and Orchard (1986) that *Mestognathus* evolved from *Clydagnathus* (family

Cavusgnathidae), whose scaphate Pa elements have a double-rowed platform and a blade that joins the right marginal row.

Pa elements of *Mestognathus* are homeomorphic with those of *Scaphignathus* (Fig. 5.41), which is thought to have evolved in the Late Devonian from *Pandorinellina insita* (Fig. 5.38), as did *Clydagnathus*, the proximate ancestor of *Mestognathus*. Short of a thorough restudy and revision of *Pandorinellina*, the Cavusgnathidae (including *Clydagnathus*), and *Mestognathus*, however, I see no way of accommodating these remarkable homeomorphs (and their common ancestors) in the same suprageneric category. Thus, somewhat anomalously, I suppose, I retain *Scaphignathus* in the Spathognathodontidae, whereas its younger homeomorph, *Mestognathus*, is accommodated in a separate family, the Mestognathidae. Possibly the best way to insure taxonomic symmetry would be to establish a separate family also for *Scaphignathus*, which would raise the possibilility, of course, of a separate family for each conodont genus!

Mestognathus is interpreted as a denizen of harsh, nearshore marine environments characterized by high salinity and probably landward of the shallow, inshore environment inhabited by its progenitor, *Clydagnathus*. von Bitter, Sandberg, and Orchard (1986) speculate that restriction to such an extreme environment may help explain the rarity of *Mestognathus* and the fact that it may have mineralized only the Pa elements of its skeletal apparatus.

5.10 Order Unknown

5.10.1 Family Coleodontidae Branson and Mehl, 1944

This family is retained for *Coleodus* Branson and Mehl, 1933, *Archeognathus* Cullison, 1938, and *Neocoleodus* Branson and Mehl, 1933, and possibly related forms such as *Stereoconus* Branson and Mehl, 1933, and *Mixoconus* Sweet, 1955. Elements of species assigned to these genera are shown in Fig. 5.59.

In connection with their thorough recent study of *Archeognathus*, Klapper and Bergström (1984) point out that the Coleodontidae

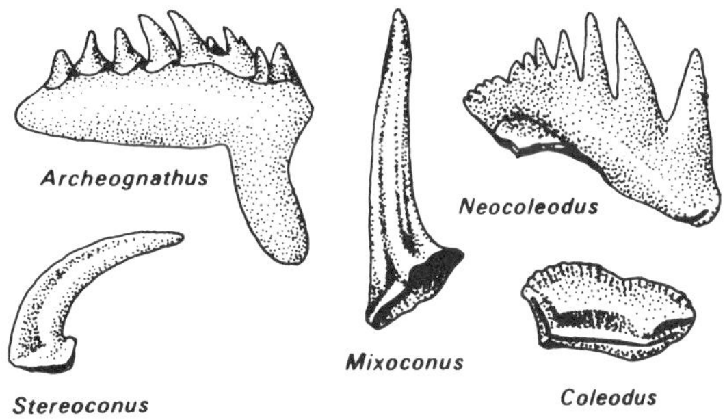

Fig. 5.59. Elements typical of species included in the Coleodontidae.

is difficult, if not impossible, to diagnose properly because the type-species of *Coleodus*, its type genus, is based on fragmentary specimens that make comprehensive morphologic interpretation impossible. In general, however, species in this taxonomically isolated group formed relatively large elements that have hyaline, fibrous crowns, which (in the case of *Stereoconus* and *Mixoconus*) consist of robust coniform elements or chains of such elements weakly (in the case of *Archeognathus*) or stoutly connected (in *Coleodus* and *Neocoleodus*) with their neighbors at the base. Undersides of crowns are flat, broadly convex, or longitudinally grooved, and in numerous specimens of *Archeognathus* and *Coleodus* crowns surmount a prominent basal structure that consists of an elongate bar that exhibits a conspicuous downward projection and apparently lacks a cavity or excavation. No basal structures have been observed with elements of *Stereoconus*, *Mixoconus* or *Neocoleodus*.

Barskov, Moskalenko, and Starostina (1982) illustrate numerous bonelike features in the basal structure of Siberian specimens referred to *Coleodus*, but Klapper and Bergström (1984) were unable to identify comparable features in the basal structure of *Archeognathus*. Although the crowns of elements assignable to species of the Coleodontidae clearly exhibit the internal structure of conodonts, the basal structures have no counterparts elsewhere in the Conodonta. No relationship to other groups is apparent. The Coleodontidae, if a natural unit, might represent a separate class of the Conodonta.

References

Al-Rawi, D. (1977). Biostratigraphische Gliederung der Tentaculiten-Schichten des Frankenwaldes mit Conodonten und Tentaculiten (Unter- und Mittel-Devon; Bayern, Deutschland). *Senckenbergiana Lethaea* **58**, 25–80.

Aldridge, R. J. (1972). Llandovery conodonts from the Welsh Borderland. *Bull. British Mus. Nat. Hist. Geol.* **22**(2), 127–226.

Amsden, T. W., and Sweet, W. C. (1983). Upper Bromide Formation and Viola Group (Middle and Upper Ordovician) in eastern Oklahoma. *Oklahoma Geol. Survey Bull.* **132**, 1–76.

Barnes, C. R., and Poplawski, M. L. S. (1973). Lower and Middle Ordovician conodonts from the Mystic Formation, Quebec, Canada. *J. Paleont.* **47**(4), 760–790.

Barrick, J. E., and Klapper, G. (1976). Multielement Silurian (late Llandoverian-Wenlockian) conodonts of the Clarita Formation, Arbuckle Mountains, Oklahoma, and phylogeny of *Kockelella. Geol. et Palaeont.* **10**, 59–98.

Barskov, I. S., Moskalenko, T. A., and Starostina, L. P. (1982). Novye dokazatel'stva prinadlezhnosti konodontoforid k pozvonochnym. *Paleont. Zhurnal* **1982**(1), 80–86. [Translated to English in *Paleontological Journal,* 1 1982, **16**(1), 82–90].

Bassler, R. S. (1925). Classification and stratigraphic use of the conodonts. *Geol. Soc. Am. Bull.* **36**, 218–220.

Belka, Z. (1985). Lower Carboniferous conodont biostratigraphy in the northeastern part of the Moravia-Silesia Basin. *Acta Geol. Polonica* **35**(1–2), 33–60.

Bergström, S. M. (1979). Whiterockian (Ordovician) conodonts from the Hølonda Limestone of the Trondheim Region, Norwegian Caledonides. *Norsk Geol. Tidsskr.* **59**, 295–307.

——— (1983). Biogeography, evolutionary relationships, and biostratigraphic significance of Ordovician platform conodonts. *Fossils and Strata* **15**, 35–58.

Bergström, S. M., and Sweet, W. C. (1966). Conodonts from the Lexington Limestone (Middle Ordovician) of Kentucky, and its lateral equivalents in Ohio and Indiana. *Bull. Am. Paleont.* **50**(229), 271–441.

Bischoff, G. (1957). Die Conodonten-Stratigraphie des rheno-herzynischen Unterkarbons mit Berücksichtigung der Wocklumeria-Stufe und der Devon/Karbon-Grenze. *Abh. Hess. Landesamtes Bodenforsch.* **19**, 1–64.

——— (1986). Early and Middle Silurian conodonts from midwestern New South Wales. *Courier Forschungsinst. Senckenberg* **89**, 1–337.

Boogaard, M. van den, and Bless, M. J. M. (1985). Some conodont faunas from the Aegiranum marine band. *Proc. K. Nederlandse Acad. Wetenschap, ser. B* **88**(2), 133–154.

Boogaard, M. van den, and Khury, B. (1979). Statistical reconstruction of the *Palmatolepis* apparatus (Late Devonian conodontophorids) at the generic, subgeneric, and specific level. Scripta Geologica **49**, 1–57.

Bradshaw, L. E. (1969). Conodonts from the Fort Peña Formation (Middle Ordovician), Marathon Basin, Texas. *J. Paleont.* **43**(5), 1137–1168.

Branson, E. B., (1938). Stratigraphy and paleontology of the Lower Mississippian of Missouri. Part I. *Univ. Missouri Studies* **13**(3), 1–208.

Branson, E. B. and Branson, C. C. (1947). Lower Silurian conodonts from Kentucky. *J. Paleont.* **21**, 549–556.

Branson, E. B., and Mehl, M. G. (1933). Conodont studies no. 1: Conodonts from the Harding Sandstone of Colorado; Bainbridge (Silurian) of Missouri; Jefferson City (Lower Ordovician) of Missouri. *Univ. Missouri Studies* **8**(1), 5–72.

——— (1934). Conodont studies no. 3: Condonts from the Grassy Creek Shale of Missouri. *Univ. Missouri Studies* **8**(3), 171–259.

——— (1938). The conodont genus *Icriodus* and its stratigraphic distribution. *J. Paleont.* **12**, 156–166.

——— (1944). Conodonts. Pp. 235–246 in *Index fossils of North America* (ed. H. W. Shimer and R. R. Shrock). Wiley, New York.

Broadhead, T. W., and McComb, R. (1983). Paedomorphosis in the conodont family Icriodontidae and the evolution of *Icriodus. Fossils and Strata* **15**, 149–154.

Bultynck, P. (1972). Middle Devonian *Icriodus* assemblages (Conodonta). *Geol. et Palaeont.* **6**, 71–86.

Carls, P., and Gandl, J. (1969). Stratigraphie und Conodonten des Unter–Devons der östlichen Iberischen Ketten (NE-Spanien). *N. Jb. Geol. Pal. Abh.* **132**, 155–218.

Chatterton, B. D. E., and Perry, D. G. (1977). Lochkovian trilobites and conodonts from northwestern Canada. *J. Paleont.* **51**, 772–796.

Chauff, K. (1985). Phylogeny of the multielement conodont genera *Bactrognathus*, *Doliognathus* and *Staurognathus. J. Paleont.* **59**(2), 299–309.

Clark, D. L. (1972). Early Permian crisis and its bearing on Permo-Triassic conodont taxonomy. *Geol. Palaeont.* **SB1**, 147–158. Clark, D. L., and Ethington, R. L. (1962). Survey of Permian conodonts in western North America. *Brigham Young Univ. Geol. Studies* **9**(2), 102–114.

Clark, D. L., Sweet, W. C., Bergström, S. M., Klapper, G., Austin, R. L., Rhodes, F. H. T., Müller, K. J., Ziegler, W., Lindström, M., Miller, J. F., and Harris, A. G. (1981). Conodonta. In *Treatise on invertebrate paleontology* Pt. W, Suppl. 2 (ed. R. A. Robison). Geol. Soc. America and Univ. Kansas, 202 pp.

Cooper, B. J. (1975). Multielement conodonts from the Brassfield Limestone (Silurian) of southern Ohio. *J. Paleont.* **49**, 984–1008.

——— (1977). Toward a familial classification of Silurian conodonts. *J. Paleont.* **51**, 1057–1071.

Cooper, C. L. (1939). Conodonts from a Bushberg-Hannibal horizon in Oklahoma. *J. Paleont.* **13**, 379–422.

Croft, J. S. (1978). Upper Permian conodonts and other microfossils from the Pinery and Lamar Limestone members of the Bell Canyon Formation and from the Rustler Formation, west Texas. Unpubl. M. Sc. Thesis, The Ohio State University, Columbus, 176 pp.

Drygant, D. M. (1974). Prostye Konodonty Silura i nizov Devona Volyno-Podolya (Simple conodonts of the Silurian and lowermost Devonian of the Volyn-Podolian area). *Paleont. Sb. Lvov Univ.* **G10**, 64–70.

Dzik, J. (1976). Remarks on the evolution of Ordovician conodonts. *Acta Palaeont. Polonica* **21**, 395–455.

——— (1983). Relationships between Ordovician Baltic and North American Midcontinent conodont faunas. *Fossils and Strata* **15**, 59–85.

Dzik, J., and Drygant, D. M. (1986). The apparatus of panderodontid conodonts. *Lethaia* **19**, 133–141.

Ethington, R. L., and Brand, U. (1981). *Oneotodus simplex* (Furnish) and the genus *Oneotodus* (Conodonta). *J. Paleont.* **55**, 239–247.

Ethington, R. L., and Clark, D. L. (1982). Lower and Middle Ordovician conodonts from the Ibex area, western Millard County, Utah. *Brigham Young Univ. Geol. Studies* **28**(2), 1–160.

Fåhraeus, L. E. (1984). A critical look at the Treatise family-group classification of Conodonta: An exercise in eclecticism. *Lethaia* **17**, 293–305.

Fåhraeus, L. E., and Nowlan, G. S. (1978). Franconian (Late Cambrian) to early Champlainian (Middle Ordovician) conodonts from the Cow Head Group, western Newfoundland. *J. Paleont.* **72**, 444–471.

Fortey, R. A., Landing, E., and Skevington, D. (1982). Cambrian-Ordovician boundary sections in the Cow Head Group, western Newfoundland. Pp. 95–129 in *The Cambrian-Ordovician boundary: Sections, fossil distributions, and correlations* (ed. M. G. Bassett and W. T. Dean). Nat. Mus. Wales, Geol. Ser. 3, 27 pp.

Furnish, W. M. (1938). Conodonts from the Prairie du Chien (Lower Ordovician) beds of the Upper Mississippi Valley. *J. Paleont.* **12**, 318–340.

Gagiev, M. H. (1979). [Conodonts from the Devonian/Carboniferous boundary deposits of the Omolon Massif]. Guidebook, Tour 9, Biostratigraphy and fauna of Devonian-Carboniferous boundary deposits. 14th Pacific Science Congress, Khabarovsk, USSR, August 1979, **Suppl. 2**, 104 pp. (In Russian, with English diagnoses of new genera and species.)

Harris, R. W. (1964). Subgenera of the conodont genus *Multioistodus* in Simpson-Burgen (Ordovician Arbuckle) conodonts from Oklahoma. *Okla. Geol. Notes* **24**, 108–118.

Harris, R. W., and Harris, B. (1965). Some West Spring Creek (Ordovician Arbuckle) conodonts from Oklahoma. *Okla. Geol. Notes* **25**, 34–47.

Harris, R. W., and Hollingsworth, R. V. (1933). New Pennsylvanian conodonts from Oklahoma. *Am. J. Sci.*, ser. 5, **25**(147), 193–204.

Hass, W. H. (1959). Conodonts from the Chappel Limestone of Texas. *U. S. Geol. Surv. Prof. Paper 294J*, 365–400.

Helms, J. (1961). Die "*nodocostata*-Gruppe" der Gattung *Polygnathus*. *Geologie* **10**, 674–711.

Huckriede, R. (1958). Die Conodonten der Mediterranen Trias und ihr stratigraphischer Wert. *Paläont. Z.* **32**, 141–175.

Huddle, J. W. (1934). Conodonts from the New Albany shale of Indiana. *Bull. Am. Paleont.* **21**(72), 136 pp.

Jeppsson, L. (1974) [1975]. Aspects of Late Silurian conodonts. *Fossils and Strata* **6**, 79 pp.

——— (1983). Silurian conodont faunas from Gotland. *Fossils and Strata* **15**, 121–144.

Kennedy, D. J. (1980). A restudy of conodonts described by BRANSON & MEHL, 1933, from the Jefferson City Formation, Lower Ordovician, Missouri. *Geol. et Palaeont.* **14**, 45–76.

Klapper, G., and Barrick, J. E. (1983). Middle Devonian (Eifelian) conodonts from the Spillville Formation in northern Iowa and southern Minnesota. *J. Paleont.* **57**(6), 1212–1243.

Klapper, G., and Bergström, S. M. (1984). The enigmatic Middle Ordovician fossil *Archaeognathus* and its relations to conodonts and vertebrates. *J. Paleont.* **58**, 949–976.

Klapper, G., and Johnson, D. B. (1975). Sequence in conodont genus *Polygnathus* in Lower Devonian at Lone Mountain, Nevada. *Geol. et Palaeont.* **9**, 65–83.

Klapper, G., and Lane, H. R. (1985). Upper Devonian (Frasnian) conodonts of the *Polygnathus* biofacies, N.W.T., Canada. *J. Paleont.* **59**(4), 904–951.

Klapper, G., and Murphy, M. A. (1975). Silurian-Lower Devonian conodont sequence in the Roberts Mountains Formation of central Nevada. *Univ. Calif. Publ. Geol. Sci.* **111**, 1–62.

Klapper, G., and Philip, G. M. (1971). Devonian conodont apparatuses and their vicarious skeletal elements. *Lethaia* **4**, 429–452.

——— (1972). Familial classification of reconstructed Devonian conodont apparatuses. *Geol. et alaeont.* **SB1**, 97–105.

Klapper, G., and Ziegler, W. (1967). Evolutionary development of the *Icriodus latericrescens* group (Conodonta) in the Devonian of Europe and North America. *Palaeontographica Abt. A* **127**, 68–83.

Kozur, H. (1975). Beiträge zur Conodontenfauna des Perm. *Geol. Paläont. Mitt. Innsbruck* **5**(4), 1–44.

Kozur, H., Mostler, H. and Rahimi-Yazd, A. (1976). Beiträge zur Mikrofauna permotriadischer Schichtfolgen. Teil II. Neue Conodonten aus dem Oberperm und der basalen Trias von Nord- und Zentraliran. *Geol. Paläont. Mitt. Innsbruck* **5**(3), 1–23.

Landing, E. (1979). Conodonts and biostratigraphy of the Hoyt Limestone (Late Cambrian, Trempealeauan), eastern New York. *J. Paleont.* **53**, 1023–1029.

Landing, E., Barnes, C. R., and Stevens, R. K. (1986). Tempo of earliest Ordovician graptolite faunal succession: Conodont based correlations from the Tremadocian of Quebec. *Can. J. Earth Sci.* **23**(12), 1928–1949.

Lane, H. R., Merrill, G. K., Straka, J. J., II, and Webster, G. D. (1971). North American Pennsylvanian conodont biostratigraphy. Pp. 395–414. In *Symposium on Conodont Biostratigraphy* (ed. W. C. Sweet, and S. M. Bergström), *Geol. Soc. America Mem.* **127**.

Lane, H. R., and Ormiston, A. R. (1979). Siluro-Devonian biostratigraphy of the Salmontrout River area, east-central Alaska. *Geol. et Palaeont.* **13**, 39–96.

Lane, H. R., Sandberg, C. A., and Ziegler, W. (1980). Taxonomy and phylogeny of some Lower Carboniferous conodonts and preliminary standard post-*Siphonodella* zonation. *Geol. et Palaeont.* **14**, 117–164.

Lange, F. G. (1968). Conodonten-Gruppenfunde aus Kalken des tieferen Oberdevon. *Geol. et Palaeont.* **2**, 37–57.

Lindström, M. (1955). Conodonts from the lowermost Ordovician strata of south-central Sweden. *Geol. Fören. Stockholm Förh.* **76**(4), 517–603.

——— (1959). Conodonts from the Crug Limestone (Ordovician, Wales). *Micropaleont.* **5**, 427–452.

——— (1970). A suprageneric taxonomy of the conodonts. *Lethaia* **3**, 427–445.

Lindström, M., Racheboeuf, P. R., and Henry, J.-L. (1974). Ordovician conodonts from the Postolonnec Formation (Crozon Peninsula, Massif Armoricain) and their stratigraphic significance. *Geol. et Palaeont.* **8**, 15–28.

Lipniagov, O. M. (1979). The conodonts of Ct_1 a and Ct_1 b of the Donetz Basin. *Serv. geol. belg., Prof. Paper* **1979 5**(161), 42–49.

Löfgren, A. (1978). Arenigian and Llanvirnian conodonts from Jämtland, northern Sweden. *Fossils and Strata* **13**, 1–129.

McCracken, A. D., and Barnes, C. R. (1982). Restudy of conodonts (Late Ordovician-Early Silurian) from the Edgewood Group, Clarksville, Missouri. *Can. J. Earth Sci.* **19**, 1474–1485.

McCracken, A. D., Nowlan, G. S., and Barnes, C. R. (1980). *Gamachignathus*, a new multielement conodont genus from the latest Ordovician, Anticosti Island, Quebec. *Geol. Surv. Canada Paper* **80–1C**, 103–112.

McHargue, T. R. (1982). Ontogeny, phylogeny, and apparatus reconstruction of the conodont genus *Histiodella*, Joins Fm., Arbuckle Mountains, Oklahoma. *J. Paleont.* **56**, 1410–1433.

McTavish, R. A. (1973). Prioniodontacean conodonts from the Emanuel Formation (Lower Ordovician) of western Australia. *Geol. et Palaeont.* **7**, 27–58.

Merrill, G. K., and Merrill, S. M. (1974). Pennsylvanian nonplatform conodonts, IIa: The dimorphic apparatus of *Idioprioniodus*. *Geol. et Palaeont.* **8**, 119–130.

Miller, J. F. (1969). Conodont fauna of the Notch Peak Limestone (Cambro-Ordovician), House Range, Utah. *J. Paleont.* **43**, 413–439.

——— (1980). Taxonomic revisions of some Upper Cambrian and Lower Ordovician conodonts, with comments on their evolution. *Univ. Kansas Paleont. Contrib.* **99**, 44 pp.

Müller, K. J. (1956). Zur Kenntnis der Conodonten-Fauna des europaischen Devons. l. Die Gattung *Palmatolepis*. *Abh. Senckenb. Naturforsch. Ges.* **494**, 70 pp.

——— (1962). Zur systematischen Einteilung der Conodontophorida. *Paläont. Z.* **36**, 109–117.

Müller, K. J. and Müller, E. M. (1957). Early Upper Devonian (Independence) conodonts from Iowa, Part l. *J. Paleont.* **31**, 1069–1108.

Murphy, M. A., and Matti, J. C. (1982 Lower Devonian Conodonts (*hesperius-kindlei* Zones), central Nevada. *Univ. California Publications, Geol. Sci.* **123**, 83 pp.

Nicoll, R. S. (1980), The multielement genus *Apatognathus* from the Late Devonian of The Canning Basin, Western Australia. Alcheringa **4**, 133–152.

——— (1982). Multielement composition of the conodont *Icriodus expansus* Branson & Mehl from the Upper Devonian of the Canning Basin, Western Australia. *BMR J. Austral. Geol. Geophys.* **7**, 197–213.

Nicoll, R. S., and Rexroad, C. B. (1968) [1969]. Stratigraphy and conodont paleontology of the Salamonie Dolomite and Lee Creek Member of the Brassfield Limestone (Silurian) in southeastern Indiana and adjacent Kentucky. *Indiana Geol. Surv. Bull.* **40**, 73 p.

Nieper, C. M. (1969). Conodonts of Nora Fm.,

Toko Range, Glenormiston, pp. 12–15. In Hill, D., Playford, G. and Woods, J. T., eds., *Ordovician and Silurian Fossils of Queensland.* Queensland Palaeont. Soc. (Brisbane).

Norby, R. D., and Rexroad, C. B. (1985). *Vogelgnathus*, a new Mississippian conodont genus. *Indiana Geol. Surv. Occasional Paper* **50**, 1–14.

Norris, A. W., Uyeno, T. T., and McCabe, H. R. (1982). Devonian rocks of the Lake Winnipegosis-Lake Manitoba outcrop belt, Manitoba. *Geol. Surv. Canada Mem.* 392, 280 pp. (Also issued as *Manitoba Mineral Resources Div., Dept. Energy and Mines, Publ.* 771.]

Orchard, M. J. (1980). Upper Ordovician conodonts from England and Wales. *Geol. et Palaeont.* **14**, 9–44.

Pander, C. H. (1856). Monographie der fossilen Fische des silurischen Systems der russisch-baltischen Gouvernements. *Akad. Wiss. St. Petersburg*, 91 pp.

Paull, R. K. (1983). Definition and stratigraphic significance of the Lower Triassic (Smithian) conodont *Gladigondolella meeki* n. sp. in the western United States. *J. Paleont.* **59**(1), 188–192.

Perlmutter, B. (1975). Conodonts from the uppermost Wabaunsee Group (Pennsylvanian) and the Admire and Council Grove groups (Permian) in Kansas. *Geol. et Palaeont.* **9**, 95–115.

Puchkov, V. N., Klapper, G., and Mashkova, T. V. (1981). Natural assemblages of *Palmatolepis* from the Upper Devonian of the northern Urals. *Acta Palaeont. Polonica* **26**(3–4), 281–298.

Repetski, J. E. (1982). Conodonts from El Paso Group (Lower Ordovician) of westernmost Texas and southern New Mexico. *New Mexico Bur. Mines Min. Res. Mem.* **40**, 121 pp.

Repetski, J. E., and Ethington, R. L. (1983) *Rossodus manitouensis* (Conodonta), a new Early Ordovician index fossil. *J. Paleont.* **57**(2), 289–301.

Rexroad, C. B. (1981). Conodonts from the Vienna Limestone Member of the Branchville Formation (Chesterian) in southern Indiana. *Indiana Geol. Surv. Occasional Paper* **34**, 1–16.

Rexroad, C. B., and Nicoll, R. S. (1972). Conodonts from the Estill Shale (Silurian, Kentucky and Ohio) and their bearing on multielement taxonomy. *Geol. et Palaeont.* **SB1**, 57–74.

Rhodes, F. H. T. (1953). Some British Lower Palaeozoic conodont faunas. *Philos. Trans. Roy. Soc. London Ser. B.* **237**, 261–334.

——— (1963). Conodonts from the topmost Tensleep Sandstone of the eastern Big Horn Mountains, Wyoming. *J. Paleont.* **37**, 401–408.

Rieber, H. (1980). Ein Conodonten-cluster aus der Grenzbitumenzone (Mittlere Trias) des Monte San Giorgio (Kt. Tessin/Schweiz). *Ann. Naturhist. Mus. Wien* **83**, 265–274.

Ritter, S. M. (1986). Taxonomic revision and phylogeny of post-Early Permian crisis *bisselliwhitei* Zone conodonts with comments on Late Paleozoic diversity. *Geol. et Palaeont.* **20**, 139–165.

Sandberg, C. A., and Dreesen, R. (1984). Late Devonian icriodontid biofacies models and alternate shallow-water conodont zonation. *Geol. Soc. Am. Spec. Paper* 196, 143–178.

Sandberg, C. A., and Gutschick, R. C. (1984). Distribution, microfauna, and source-rock potential of Mississippian Delle Phosphatic Member of Woodman Formation and equivalents, Utah and adjacent states. Pp. 135–178. In *Hydrocarbon source rocks of the greater Rocky Mountain region* (ed. J. Woodward, F. F. Meissner, and J. L. Clayton), Rocky Mountain Assoc. Geologists, Denver.

Sandberg, C. A., and Ziegler, W. (1973). Refinement of standard Upper Devonian conodont zonation based on sections in Nevada and West Germany. *Geol. et Palaeont.* **7**, 97–122.

——— (1979). Taxonomy and biofacies of important conodonts of Late Devonian *styriacus*-Zone, United States and Germany. *Geol. et Palaeont.* **13**, 173–212.

Sandberg, C. A., Ziegler, W., Leuteritz, K., and Brill, S. M. (1978). Phylogeny, speciation and zonation of *Siphonodella* (Conodonta, Upper Devonian and Lower Carboniferous). *Newsl. Stratigr.* **7**, 102–120.

Sannemann, D. (1955). Oberdevonische Conodonten ($toII_\alpha$). *Senckenbergiana Lethaea* **36**, 123–156.

Satterfield, I. R. (1971). Conodonts and stratigraphy of the Girardeau Limestone (Ordovician) of southeast Missouri and southwest Illinois. *J. Paleont.* **45**, 265–273.

Savage, N. M., and Bassett, M. G. (1985). Caradoc-Ashgill conodont faunas from Wales and the Welsh Borderland. *Palaeont.* 28(4), 679–713.

Serpagli, E. (1974). Lower Ordovician conodonts from Precordilleran Argentina (Province of San Juan). *Boll. Soc. Paleont. Ital.* **6**, 17–98.

——— (1983). The conodont apparatus of *Icriodus woschmidti* Ziegler. *Fossils and Strata* **15**, 155–161.

Sparling, D. R. (1981). Middle Devonian conodont apparatuses with seven types of elements. *J. Paleont.* **55**, 295–306.

Stouge, S. S. (1984). Conodonts of the Middle Ordovician Table Head Formation, western Newfoundland. *Fossils and Strata* **16**, 145 pp.

Sweet, W. C. (1970). Uppermost Permian and Lower Triassic conodonts of the Salt Range and Trans-Indus ranges, West Pakistan, pp. 207–275. In *Stratigraphic Boundary problems: Permian and Triassic of West Pakistan* (ed. B. Kummel and C. Teichert), *Univ. Kansas Dept. Geol. Spec. Publ.* 4.

——— (1976). Skeletal anatomy of the Late Paleozoic and Early Triassic conodont genus *Hindeodus* REXROAD & FURNISH, 1964. *Geol. Soc. Am. Abs. Prog.* **8**, 512–513.

——— (1979). Late Ordovician conodonts and biostratigraphy of the western Midcontinent Province. *Brigham Young Univ. Geol. Studies* **26**(3), 45–74.

——— (1982). Conodonts from the Winnipeg Formation (Middle Ordovician) of the northern Black Hills. *J. Paleont.* **56**, 1029–1049.

——— (1984). Graphic correlation of upper Middle and Upper Ordovician rocks, North American Midcontinent Province, U.S.A. Pp. 23–35. In *Aspects of the Ordovician System* (ed. D. L. Bruton) Paleont. Contr. Univ. Oslo, No. 295.

Sweet, W. C., and Bergström, S. M. (1972). Multielement taxonomy and Ordovician conodonts. *Geol. et Palaeont.* **SBl**, 29–42.

Sweet, W. C. and Schönlaub, H. P. (1975). Conodonts of the genus *Oulodus* BRANSON & MEHL, 1933. *Geol. et Palaeont.* **9**, 41–59.

Teichert, C., Kummel, B., and Sweet, W. C. (1973). Permian-Triassic strata, Kuh-e-Ali Bashi, northwestern Iran. *Bull. Mus. Comp. Zool. Harvard Univ.* 145(8), 359–472.

Uyeno, T. T., and Klapper, G. (1980). Summary of conodont biostratigraphy of the Blue Fiord and Bird Fiord formations (Lower-Middle Devonian) at the type and adjacent areas, southwestern Ellesmere Island, Canadian Arctic Archipelago. *Geol. Surv. Canada Paper* 80-lC, 81–93.

van Wamel, W. A. (1974). Conodont biostratigraphy of the Upper Cambrian and Lower Ordovician of northwestern Öland, southeastern Sweden. *Utrecht Micropaleont. Bull.* **10**, 1–126.

von Bitter, P. H. and Merrill, G. K. (1983). Late Palaeozoic species of *Ellisonia* (Conodontophorida)—Evolutionary and palaeoecological significance. *Roy. Ontario Mus. Life Sci. Contr.* **136**, 1–56.

——— (1985). *Hindeodus, Diplognathodus* and *Ellisonia* revisited—an identity crisis in Permian conodonts. *Geol. et Palaeont.* **19**, 81–96.

von Bitter, P. H., and Plint-Geberl, H. A. (1982). Conodont biostratigraphy of the Codroy Group (Lower Carboniferous), southwestern Newfoundland, Canada. *Can. J. Earth Sci.* **19**, 193–221.

von Bitter, P. H., Sandberg, C. A., and Orchard, M. J. (1986). Phylogeny, speciation, and palaeoecology of the Early Carboniferous (Mississippian) conodont genus *Mestognathus. Roy. Ontario Mus. Life Sci. Contr.* **143**, 115 pp.

Walliser, O. H. (1964). Conodonten des Silurs. *Abh. Hess. Landesamtes Bodenforsch.* **41**, 1–106.

Wardlaw, B. R. and Collinson, J. W. (1986). Paleontology and deposition of the Phosphoria Formation. *Univ. Wyoming Contr. Geol.* 24(2), 107–142.

Weddige, K. (1977). Die Conodonten der Eifel-Stufe im Typusgebiet und in benachbarten Faziesgebieten. *Senckenbergiana Lethaea* **58**, 271–419.

Weddige, K., and Ziegler, W. (1979). Evolutionary patterns in Middle Devonian conodont genera *Polygnathus* and *Icriodus. Geol. et Palaeont.* **13**, 157–164.

Wirth, M. (1967). Zur Gliederung des höheren Paläozoikums (Givet-Namur) im Gebiet des Quinto Real (Westpyrenaen) mit Hilfe von Conodonten. *N. Jb. Geol. Paläont. Abh.* **127**, 179–244.

Youngquist, W. (1945). Upper Devonian conodonts from the Independence shale (?) of Iowa. *J. Paleont.* **19**, 335–367.

Ziegler, W. (1962). Taxionomie und Phylgenie oberdevonischer Conodonten und ihre stratigraphische Bedeutung. *Abh. Hess. Landesamtes Bodenforsch.* **38**, 1–166.

——— (1971). Conodont stratigraphy of the European Devonian. Pp. 227–284. In *Symposium on Conodont Biostratigraphy* (ed. W. C. Sweet, and S. M Bergström,) *Geol. Soc. America Mem.* **127**.

——— ed. (1977). Catalogue of conodonts. *E. Schweizerbart'sche Verlagsbuchhandlung 3*, 1–574.

Ziegler, W., Klapper, G., and Johnson, J. G. (1976). Redefinition and subdivision of the *varcus*-Zone (Conodonts, Middle-?Upper Devonian) in Europe and North America. *Geol. et Palaeont.* **10**, 109–132.

Ziegler, W., and Lane, H. R. (1987). Cycles in conodont evolution from Devonian to mid-Carboniferous. Pp. 148–163. In *Palaeobiology of Conodonts* (ed. R. J. Aldridge), Ellis Horwood, Chichester, 80 pp.

Ziegler, W., and Sandberg, C. A. (1984). *Palmatolepis*-based revision of upper part of standard Late Devonian conodont zonation. *Geol. Soc. Am. Spec. Paper* 196, 179–194.

Ziegler, W., Sandberg, C. A. and Austin, R. L. (1974) Revision of *Bispathodus* group (Conodonta) in the Upper Devonian and Lower Carboniferous. *Geol. et Palaeont.* **8**, 97–112.

6. EVOLUTIONARY PATTERNS

6.1 Introduction

As noted in Chapters 1 and 2, conodonts occur abundantly in most types of marine sedimentary rock; they are widely distributed geographically; and, as a group, they have a long stratigraphic range. Consequently, conodonts are exceptionally useful as providers of biostratigraphic information to geologists interested in a broad spectrum of problems, and those of us who study conodonts have expended most of our effort in supplying biostratigraphic service to the geologic community. We have had little time left over in which to address the more esoteric aspects of what might be considered truly paleobiologic topics.

On the plus side, however, our service orientation has resulted in the assembly of an objective, if widely dispersed, data base, which is several orders of magnitude larger than that available for any other group of Paleozoic or Triassic "macrofossils" and is also one that may be viewed within a biostratigraphic framework currently capable of resolving the 300-my history of the Conodonta into 152 divisions. In addition, of course, we now have a fairly clear idea that recurrent groups of morphologically different element types are more suitable than individual elements as the basis for species-level taxonomy; and, as I have tried to show in Chapter 5, species may now be assembled into supraspecific categories that provide considerable insight into the patterns of conodont evolution. At present, our organized collections provide little information as to the modes or processes by which evolutionary development was effected, but patterns may ultimately suggest processes that may also be tested against the objective evidence.

Micropaleontologists, as Lipps (1981) has emphasized, are not noted for generating, testing, or evaluating fundamental paleobiologic hypotheses, despite their access to finely tuned data bases that are the envy of their peers who study larger fossils. This is due partly, I believe, to the caution automatically provided by very large population samples from stratigraphically well-controlled sequences. In some parts of the geologic column, dense spacing of large samples provides an almost continuous record of the development of populations that can be seen to have had and to have maintained very great internal complexity for appreciable intervals of time. Not uncommonly, it is difficult or impossible to express what we know about such populations within a generally accepted systematic framework that was designed to handle a rigidly typological taxonomy. And this, of course, has resulted in hiding a good deal of very useful information in categories named and treated as species or genera in accord with principles established and rigorously maintained by the *International Rules of Zoological Nomenclature*.

But students of conodonts have had additional excuses for caution. That is, the subjects of their study represent a group of animals that evidently does not occur in the living biosphere. This means, of course, that there is little unequivocal anatomic information available for conodonts and that biologically meaningful systematic and ecologic concepts must be generated quite indirectly.

Now, however, there are in place at least the broad outlines of a defensible taxonomy, and it is probably time to take up the biologically (and philosophically) challenging task of evolutionary interpretation. To this end, I assemble in this chapter some observations on the evolutionary patterns I recognize in the 300-my history of the conodonts, and comments on a few examples that seem to suggest process as well. Perhaps my observations and comments will prove sufficiently outrageous to stimulate more penetrating studies of evolutionary mode and process. I hope so.

6.2 Diversity Patterns

In a very recent study, Ziegler and Lane (1987) outlined a repetitive, or cyclic, pattern in the evolutionary development of Devonian and early Carboniferous conodonts. In each of the six cycles they described, an interval of low diversity is followed by one of higher diversity, which is introduced by a short burst of innovation and ultimately truncated by an episode of extinction. Although Ziegler and Lane related these cycles largely to stages in the iterative development of various ozarkodinide stocks, an attempt was made also to fit in events in the history of the prioniodontide family Icriodontidae.

Ziegler and Lane (1987) did not consider events in the evolutionary history of conodonts younger than Mississippian or older than late Silurian, but they did note that patterns similar to the cyclic ones they recognized in the Devonian and Mississippian might also have characterized the pre-Devonian and post-Mississippian history of the Conodonta. Logs constructed by Clark (1972) from a series-by-series tabulation of conodont form genera also show a crudely controlled succession of diversity cycles, as does a log that I compiled recently (Sweet, 1985) to show the diversity of Ordovician and Silurian conodont species on a zone-by-zone basis.

As the basis for examining in this chapter some of the major events in the evolutionary history of the Conodonta, I present three logs that show fluctuations in species-level diversity on a zone-by-zone basis for the entire late Cambrian through Triassic range of the Conodonta. These logs, depicted in Figs. 6.1, 6.2, and 6.3, also include a log that shows fluctuations in the origination-extinction ratio, which Clark (1972) used effectively as an *index of evolution.* These logs permit recognition of 4 first-order

Fig. 6.1. Species-diversity log (upper solid line) and log of origination-extinction ratios (lower dashed line) for 40 Late Cambrian through earliest Devonian zones. Stippled segments of origination-extinction ratio log indicate times when more species evolved than became extinct; asterisks indicate times of significant extinction separating diversity cycles identified with Roman numerals I to VI.

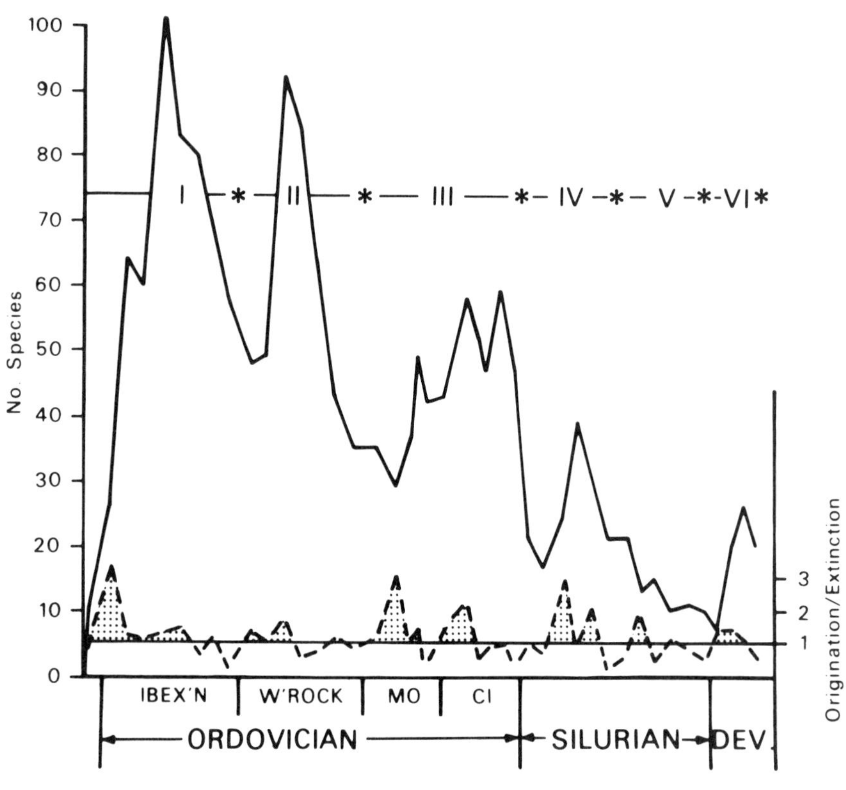

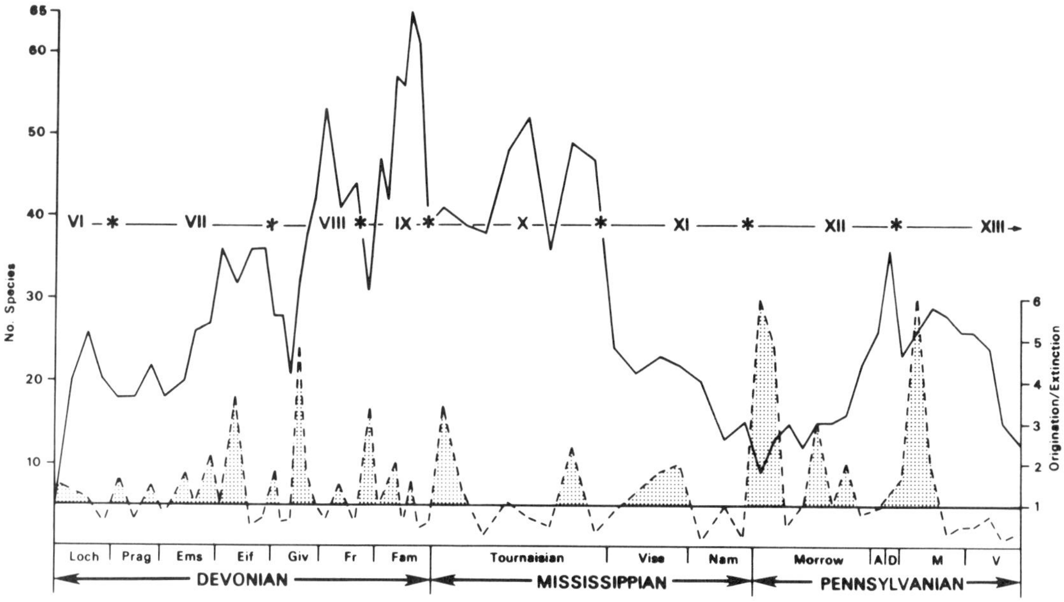

Fig. 6.2. Species-diversity log (upper solid line) and log of origination-extinction ratios (lower dashed line) for 64 Devonian and Carboniferous zones. Stippled segments of origination-extinction log indicate times when more species evolved than became extinct; asterisks indicate times of significant extinction separating diversity cycles identified by Roman numerals VI to XIII.

cycles in conodont diversity and also delineate about 20 second-order cycles that support the concept advanced by Ziegler and Lane (1987). The logs, of course, extend that concept to both earlier and later intervals of time than Ziegler and Lane considered.

In the tabulations depicted in Figs. 6.1, 6.2, and 6.3 I have included only the 1446 species (of 246 genera) that I regard as reasonably distinct in a multielement context, even though complete apparatuses have been established for only a few of them. Clearly, as additional data become available, certain peaks in the diversity logs may be emphasized and certain depressions filled in. But I suspect that the general pattern will remain much the same.

Fig. 6.3. Species-diversity log (upper solid line) and log of origination-extinction ratios (lower dashed line) for 48 Permian and Triassic zones. Stippled segments of origination-extinction logs indicate times when more species evolved than became extinct; asterisks indicate times of significant extinction separating diversity cycles identified with Roman numerals XIV to XX.

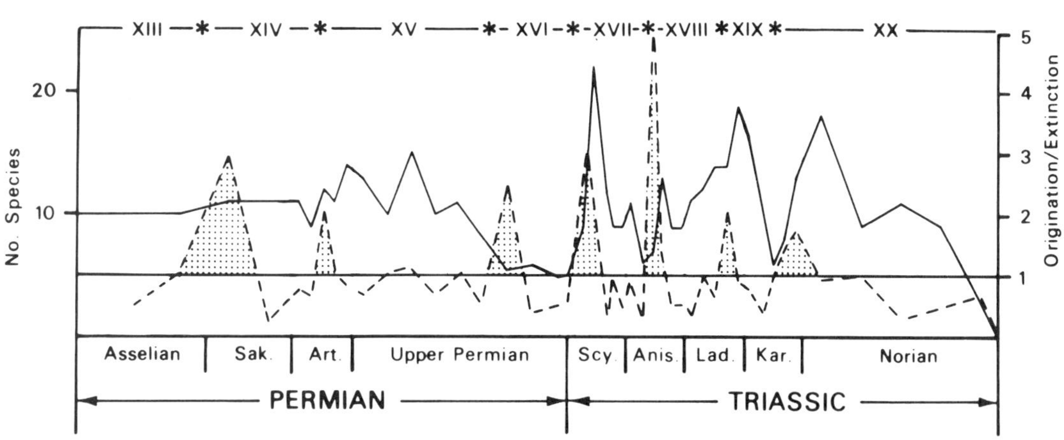

I have accepted at face value the zone-by-zone tabulations of species distributions published in recent years by various specialists, but I have also used my own judgment in many cases to determine the ranges of various species within the biostratigraphic scheme on which Figs. 6.1, 6.2, and 6.3 are assembled. In a number of cases, I have autocratically omitted records that are based on poorly preserved or indifferently illustrated material. For the most part, inclusion of such records would make little difference in the context of the larger data base, in which most occurrences and ranges are quite well documented.

6.2.1 *Overall pattern*

Data summarized in Figs. 6.1, 6.2, and 6.3 show what might be interpreted as a classical pattern of adaptive radiation in the very early Ordovician, with a very rapid rise to an all-time peak in species-level diversity (Fig. 6.1). This was followed by a cyclically interrupted decline in diversity through the later Ordovician and Silurian and an interval of renewed adaptive vigor in the Devonian and early Carboniferous (Fig. 6.2). Following a major extinction event near the Mississippian-Pennsylvanian boundary, diversity built up to a respectable level by mid-Pennsylvanian time, but then began a general decline that was continued in the Permian (Fig. 6.3) by a long interval during which diversity hovered around 10 and the "evolutionary index" was only rarely above the maintenance level. A brief episode of intense speciation built species diversity up above 20 for the last time in the early Triassic, but this innovative spurt was followed by a fitful decline through the remainder of the Triassic, to extinction at the end of the period.

Superimposed on this general pattern, however, and of considerably greater interest, are the records of 4 long-term cycles and at least 20 minor ones of much shorter duration, each of which includes an interval of relatively low diversity followed by one of higher diversity and terminated by a more or less abrupt drop in diversity that is anticipated in most cases by a sag in the index of evolution below 1 for a zone or two.

6.2.2 *Long-term cycles*

Conodonts reached their greatest diversity in the early part of the Ordovician, in the intervals identified in Fig. 6.1 as the Ibexian and Whiterockian epochs, but in late Whiterockian time declined precipitously in diversity to a low of fewer than 20 species in the early Mohawkian epoch. This episode corresponds in period to the one in which the most widespread part of the Sauk Sequence was deposited in the cratonic interior of North America (Sloss, 1963).

Note in Fig. 6.1 that in the late Middle Ordovician (Mohawkian) the diversity of conodont species again increased rapidly to a Late Ordovician high, which was followed (with a major second-order interruption at the Ordovician-Silurian boundary) by a general decline in abundance through the Silurian into the earliest Devonian. This major Middle Ordovician to earliest Devonian episode corresponds in duration, of course, to the time during which the transgressive-regressive Tippecanoe Sequence was formed in cratonic North America (Sloss, 1963).

In the cratonic interior of North America, an interval from the Early Devonian through the latest Mississippian was marked by formation of the transgressive-regressive Kaskaskia Sequence. In the diversity log of Fig. 6.2 that interval is clearly marked by an almost symmetrical "hump" between the Early Devonian and latest Mississippian and a diversity maximum of 65 species in latest Devonian time.

Above rocks of the Kaskaskia Sequence in much of the cratonic interior of North America are widespread Pennsylvanian and less widespread Permian and Triassic strata, which collectively record a curious transgressive-regressive cycle that Sloss (1963) named Absaroka. Note in Figs. 6.2 and 6.3 that outlines of the temporally asymmetric Absaroka cycle are clearly indicated by logs of post-Mississippian conodont species diversity, which show a major recovery in the Pennsylvanian and a prolonged, if somewhat fitful, decline in the Permian and Triassic.

There are a number of ways to interpret the close correspondence between the cratonic cycles of Sloss (1963), which surely record major eustatic episodes, and first-order segments of the species-diversity logs in Figs. 6.1, 6.2, and 6.3. The simplest, I believe, is to postulate a direct relationship between diversity of conodont species and the expansion (and subsequent reduction) of suitable habitats recorded by rocks of the transgressive-regressive sequences.

6.3 Second-Order Cycles

Superimposed on the first-order cycles recorded in the logs of Figs. 6.1, 6.2, and 6.3 is the record of 20 second-order cycles, each with the general characters of the ones recognized by Ziegler and Lane (1987) in the Devonian and Early Carboniferous. By and large, each of these cycles begins with an interval of low species diversity in which the ratio of species originations to species extinctions (the "evolutionary index") is below l.0, followed by an interval of higher diversity that is immediately preceded by an innovative episode signaled by a spurt in the evolutionary index. Cycles are terminated by more or less abrupt drops in species diversity which, for the most part, seem to be the expectable consequences of a decline in the evolutionary index in immediately preceding zones.

As is clear from the logs in Figs. 6.1, 6.2, and 6.3, whose horizontal scale is approximately proportional to the length of time represented by the intervals shown, second-order cycles were of greatly different durations and hence seem unlikely to have been the results of any regularly cyclic mechanism. As the following remarks will show, these cycles are complex internally, and it may be difficult to find a general explanation for them.

6.3.1 Cycles I and II

In the latest Cambrian, Ordovician, and Silurian history of the Conodonta, I recognize five second-order evolutionary cycles. These are identified by Roman numerals I to V in Fig. 6.1.

Cycle I, which began in the Late Cambrian with the appearance of the first conodonts in the shallow low-latitude seas of North America, Australia, and China, is largely a record of the rapid diversification of the *Teridontus* lineage, which founded the Paraconodontida and, in the very early Ordovician, launched the Prioniodontida. Early diversification of the Cavidonti was also accomplished early in Cycle I. The Proconodontida surely reached the acme of their evolutionary development then, and only *Polonodus* and *Pygodus* are known from rocks deposited during later Ordovician cycles. The Belodellida, which I interpret as descendants of the *Proconodontus* stock, appeared at the climax of diversity in Cycle I but were not conspicuous components of Ordovician conodont faunas until the interval of Cycle II.

The 40 conodont genera and 227 species that are largely confined in their known range to the interval of Cycle I represent two groups. One is made up of mostly short-lived experimental stocks such as the Clavohamulidae and Cordylodontidae, which appeared briefly during the initial radiation of cavidont and conodont lineages but were apparently not ancestral to anything else; whereas the other, typified, for example, by *Rossodus*, *Tripodus*, and *Prioniodus*, includes genera that founded or are intermediate parts of evolutionary lineages that, in modified form, assume greater importance in later cycles. These latter lineages were evidently subjected to more or less continual pruning during the prolonged regressive phase that characterized the waning stages in North America of the Sauk cratonic cycle, so there is no evidence of their catastrophic or mass extinction at the end of Cycle I.

Note that by early in the Ordovician conodonts were common not only in the tropical to subtropical waters of North America, Australia, Siberia, and probably parts of China, but also in seas at much higher latitudes, such as those that overspread Europe, western Australia, and western South America. Although there were remarkable parallels in the evolutionary development of low- and high-latitude Ordovician faunas, already evident in the interval of Cycle I, there was limited exchange between their characteristic faunas. Very possibly the existence through the Ordovician of well-developed conodont faunas from the equator almost to the pole contributed to the exceptionally high diversity of those faunas.

Cycle II is related, at least in North America where it is best known, to a short-lived transgressive interrruption of the major regressive phase that completed the Sauk cratonic cycle (Sloss, 1963). Conodont faunas characteristic of this cycle are a mixture of the tag ends of primitive lineages founded during Cycle I with advanced protopanderodontide and prioniodontide species with fully differentiated apparatuses (e.g., *Drepanoistodus*, *Amorphognathus*, *Paraprioniodus*, *Plectodina*, and *Phragmodus*). The Multioistodontidae are especially characteristic of this cycle in North America, and all of them but the parent stock, *Pteracontiodus*, are apparently restricted to it. In the later part of Cycle II 10 well-established lineages of high-latitude conodonts became extinct, as did a majority of the multioistodontids, which characterized the high-diversity episode of this cycle in low-latitude seas. Again, there is no evidence that these extinctions represent a mass-extinction event. They are, however, associated with an interval of the Ordovician in which shallow shelf seas came to be very restricted in extent.

6.3.2 *Middle and Late Ordovician Cycle III*

Secondary Cycle III is associated in North America with the transgressive phase of Sloss's (1963) Tippecanoe eustatic cycle. The low-diversity episode, in the early half of the Mohawkian epoch, was characterized in shallow, marginal environments by primitive species of the prioniodinide family Chirognathidae, especially of *Erismodus* and *Curtognathus*; by species of *Phragmodus* and *Plectodina*, which appeared in the high-diversity episode of Cycle II; and, in higher-latitude faunas, by the first species of the specialized prioniodontides *Icriodella* and *Rhodesognathus*. As Tippecanoe transgression continued, species that thrived in shallow, marginal environments gave rise to species adapted to more offshore, probably somewhat deeper-water, more normal marine environments. For example, the shallow-water *Phragmodus flexuosus* complex spawned *P. inflexus* and *P. undatus*, which came to be typical in later Ordovician time of deeper waters in more open-marine low-latitude environments. Of greater significance was development from the shallow-water *Plectodina aculeata* complex of a spate of forms such as "*P.*" *tenuis* and various species of *Aphelognathus*. As indicated in Chapter 5, I consider these species to be the founders of the order Ozarkodinida, which diversified greatly in the Silurian and dominates later Paleozoic faunas.

The event that terminated Cycle III is generally regarded as a mass extinction and is commonly attributed to the combined effects of eustatic drawdown and cooling of the world ocean consequent on Gondwana glaciation. Of the 35 genera of conodonts represented by species in latest Ordovician faunas, only 7 survived. In the Ordovician, these stocks were all characteristic of open-marine faunas in low and intermediate latitudes. None of the lineages typical of higher-latitude (or colder-water) Ordovician faunas continued into the Silurian.

6.3.3 *Silurian cycles*

In Fig. 6.1 the Silurian history of the Conodonta is divided into two second-order cycles, IV and V. Cycle IV includes a prolonged interval of low diversity in the Llandovery, followed by a high-diversity episode that was introduced by a significant spike in the evolutionary-index log and reached a peak in the Wenlock Epoch. Cycle IV is associated largely with diversification of the prioniodontide families Icriodellidae and Distomodontidae and the ozarkodinide families Spathognathodontidae, Plectospathodontidae, and Kockelellidae. The pattern established here was to continue in later Silurian and Devonian cycles, which are characterized by replays of the evolutionary bursts characteristic of Cycle IV and produced iteratively by development from the icriodellid, spathognathodontid, and possibly from successors of the distomodontid stocks.

Late Silurian conodont faunas are sparse and poorly known, and second-order Cycle V is rather weakly characterized. It is introduced by an interval of reduced evolutionary vigor, which was followed by an episode of innovation during which *Pedavis* arose from the icriodellid stock, *Polygnathoides* from the spathognathodontid stock, and *Coryssognathus* from the distomodontids. Recovery in diversity was slight, and conodonts declined almost to extinction in latest Silurian time.

6.3.4 *Devonian and Carboniferous cycles*

As noted in Section 6.2, Ziegler and Lane (1987) have recently recognized six second-order cycles in the evolutionary development of Devonian and Early Carboniferous conodonts. These are identified in Fig. 6.2 as Cycles VI to XI.

Cycle VI began with a short interval in the late Silurian and earliest Devonian during which the diversity of conodont species dropped to a low of 6. This critical interval in the history of the Conodonta was terminated in mid-Lochkovian time by an innovative burst, effected by vigorous speciation within *Ozarkodina* and iterative development from the spathognathodontid, icriodellid, and/or distomodontid stocks of a flood of species assigned to *Amydrotaxis*, *Ancyrodella*, *Pedavis*, *Pelekysgnathus*, and *Icriodus* (in a very broad sense). That burst introduced a short interval of high diversity in the late Lochkovian during which the number of conodont species built back up to about 26. Although temporarily vigorous, the *Amydrotaxis* and *Ancyrodelloides* stocks were fairly short-lived dead ends, and their disappearance, along with that of several similarly short-lived species of *Icriodus* s. l., produced the "extinction event" that brought Cycle VI to a close.

Cycle VII, like its immediate predecessor, represents yet another episode in iterative development of the Ozarkodinida, reinforced by bursts of vigorous speciation in shallow-water environments by the *Icriodus*, *Pedavis*, and *Pelekysgnathus* stocks. Particularly significant developments from *Ozarkodina* (Spathognathodontidae) were the *Eognathodus* lineage, which is confined primarily to Cycle V and, a bit later, the *Polygnathus* stock, which rose rapidly to importance in the offshore environments that became more widespread with the advance of Kaskaskia transgression. In the high-diversity interval of Cycle V, which peaked in the Emsian and Eifelian, at least 36 conodont species prowled Devonian seas—a sixfold increase from the low-diversity interval of Cycle VI.

The high-diversity interval of Cycle VII was terminated in early Givetian time by the disappearance of all but two of the lineages of *Polygnathus* established in earlier phases of the cycle, and diversity dropped to 21 in the mid-Givetian low-diversity interval that initiated Cycle VIII. A bit later in the Givetian (late Middle Devonian), iterative development of "*Polygnathus*" *latifossatus* from the spathognathodontid stock led rapidly to establishment of *Mesotaxis*, *Schmidtognathus*, *Klapperina* and, ultimately, *Palmatolepis*, which I have set aside in Chapter 5 as the new family Palmatolepidae. Species of these genera, which were apparently adapted to the increasingly widespread open marine environments of the late Middle Devonian, together with new forms developed from surviving members of the *Polygnathus* stock and short-lived species of *Ancyrodella* that also evolved iteratively from spathognathodontid ancestors, accounted collectively for a Frasnian high-diversity episode that numbered at its peak some 53 conodont species.

Ziegler and Lane (1987) recognize an abrupt decline in species diversity at the end of Cycle VIII and attribute it to a sharp extinction event just before the end of the Frasnian which, like the one that terminated Ordovician Cycle III, is commonly included in the short list of major Paleozoic mass extinctions. They note that "at this position almost all previous species of palmatolepids, ancyrodellids, ancyrognathids, and polygnathids became extinct." The tabulation I present in Fig. 6.2 also shows a rapid decline in species diversity in late phases of Cycle VIII, but not one that is as abrupt as suggested by Ziegler and Lane's comments. Although this disparity may result from the fact that I have assembled ranges on a biostratigraphic scale that is cruder than the one used by Ziegler and Lane, it may also suggest that the species to which they refer disappeared more gradually in the late Frasnian than has been assumed. If this were the case, conodonts might provide some support for the idea expressed by Johnson, Klapper, and Sandberg (1985) that at least part of the end-Frasnian mass extinction of benthic animals might have been the cumulative result of three earlier Frasnian deepening events and not the result of a single sudden shallowing event. That is, the drowning of reefs and reduction in extent of marginal, shallow-water environments that followed such deepening events might also have affected pelagic and nektobenthic elements of marine communities by eliminating or greatly reducing the marginal environments to which some were adapted and, perhaps, by removing previous barriers to migration. Both effects might well have operated gradually to lower endemism, increase cosmopolitanism, and thus reduce species diversity, which is just what my tabulations suggest happened in the late Frasnian history of the Conodonta.

The severe reduction in species diversity that marked the latest Frasnian-earliest Famennian phase of Cycle VIII continued for a brief interval into initial phases of the later Famennian Cycle IX. That low-diversity interval was rapidly supplanted in slightly later Famennian time by an intense innovative phase marked by the radiation of four new lineages of *Palmatolepis* and flowering of the *nodocostatus* group of *Polygnathus* species in offshore environments; by a final burst of speciation in the shallow-water *Icriodus-Pelekysgnathus* complex; by development from the spathognathodontid *Pandorinellina* of the first cavusgnathids; and by initial radiation of the *Bispathodus* lineage (Spathognathodontidae), to which we trace a majority of the stocks of conodonts characteristic of Carboniferous and Permian faunas. All this resulted in a mid-Famennian peak of 65 in species diversity, which was maintained only briefly. By a point in time near the end of the Devonian diversity had diminished to 39, and Ziegler and Lane (1987) note that only *Polygnathus communis*, three species of *Protognathodus*, one of *Siphonodella*, and the *Bispathodus*, *Patrognathus*, and *Pseudopolygnathus* stocks continued into the Early Carboniferous. The reduction in diversity of conodont species that marks the end of Cycle IX occurred at about the same time that Walliser (1983, 1984) has noted extinction of numerous other invertebrate groups, including the clymenoids, some ammonoids, and certain trilobites.

Cycle X began with a short interval of limited diversity in latest Devonian time, but appears in Fig. 6.2 to have been initiated in the early Tournaisian with a major innovative event that is attributable primarily to rapid radiation of *Siphonodella* (of the Elictognathidae), *Protognathodus* (of the Gnathodontidae), and *Pseudopolygnathus* (of the Spathognathodontidae). These evolutionary events, coupled with renewed speciation in the *Polygnathus* stock led to a mid-Tournaisian high-diversity interval characterized by more than 50 species. Diversity was maintained at a high level into the latest Tournaisian by development (from *Protognathodus*) of early species of *Gnathodus* and (from unknown prioniodinide ancestors) of the distinctive but short-lived bactrognathids (i.e., *Eotaphrus*, *Bactrognathus*, *Doliognathus*, *Dollymae*, and *Staurognathus*). The latter event is responsible for the conspicuous late-Tournaisian spike in the evolutionary-index log of Fig. 6.2.

Cycle XI, which occupied the Visean and early Namurian (or late Mississippian) segments of Carboniferous time, corresponds in duration to the regressive phase of Sloss's (1963) Kaskaskia eustatic cycle. The end of Cycle X was marked by a precipitous decline in species diversity, which I attribute primarily to the extinction or phyletic termination in late Tournaisian time of the *Protognathodus*, *Pseudopolygnathus*, and bactrognathid stocks as the offshore environments in which they were most at home diminished in area. However, shrinkage of those environments and increase in area of marginal ones characterized by much shallower water and more variable salinity and substrate probably offered an opportunity to the cavusgnathids, mestognathids, and prioniodinids, which all proliferated and thereby contributed to the slight recovery in species-level diversity that marks the high-diversity interval of Cycle XI. *Lochriea*, *Vogelgnathus*, and *Rhachistognathus* also evolved from ancestors in *Bispathodus* in the innovative phase of Cycle XI. The *Lochriea* and *Vogelgnathus* stocks are confined to this cycle, but the *Rhachistognathus* lineage survived through Cycle XII into the mid-Pennsylvanian.

The conspicuous decline in species-level diversity that marks the termination of Cycle XI resulted primarily from replacement of *Cavusgnathus* by *Adetognathus* and extinction in the late Mississippian of the *Gnathodus*, *Lochriea*, and *Vogelgnathus* lineages. However, during the early Morrowan (Pennsylvanian) interval of low diversity that initiated Cycle XII, populations of idiognathodontid conodonts descended from highly variable *Gnathodus girtyi* gradually segregated into discrete and somewhat less variable stocks identified as *Declinognathodus*, *Idiognathoides*, *Neognathodus*, *Idiognathodus*, and *Streptognathodus*. Radiation of these stocks kept the evolutionary index above the maintenance level of 1.0 through most of the Morrowan and promoted a slow recovery in species-level diversity, which peaked in the succeeding Atokan epoch with addition of the first representatives of *Diplognathodus* and the Gondolellidae and gradual phyletic evolution of the *Neognathodus* lineage.

Extinction of *Neognathodus* and probably also of *Rhachistognathus* in the late Desmoinesian (Pennsylvanian) marked the end of Cycle XII, which is also shown in Fig. 6.2 to be terminated by a rapid decline in species-level diversity. Continued gradual phyletic speciation in the *Idiognathodus-Streptognathodus* lineage, together with the appearance of *Aethotaxis*, *Ellisonia*, and new species of *Gondolella*, *Diplognathodus*, and *Hindeodus*, signaled a brief interval of innovation in the early part of the mid-Pennsylvanian Missourian Epoch and the somewhat later peak in species-level diversity that marked culmination of Cycle XIII. Note in Figs. 6.2 and 6.3 that in the latter half of Cycle XIII the evolutionary index was continuously below 1.0 and that species-level diversity declined gradually to 15 in the Virgilian and to 10 in the succeeding Asselian Stage of the Early Permian. Although species of *Hindeodus*, *Diplognathodus*, *Streptognathodus*, *Idiognathodus*, *Adetognathus*, *Gondolella*, and *Ellisonia* survived through much or all of this long interval of sorely restricted evolutionary vigor, only the *Hindeodus*, *Diplognathodus*, gondolellid, and *Ellisonia* lineages survived Clark's (1972) Early Permian crisis to provide the basis for secondary cycles in later Permian and Triassic seas.

6.3.5 Permian and Triassic cycles

Fig. 6.3 is a zone-by-zone summary of conodont species diversity through the Permian and Triassic periods, which together represent 78 million years of geologic time. As in summary logs for earlier parts of the Paleozoic, I also provide in Fig. 6.3 a zone-by-zone log of the ratio of origination to extinction of conodont species. That log may be used (with the customary caveats) as a crude index to temporal oscillations in evolutionary vigor.

Although at least 185 conodont species originated in the Permian-Triassic interval, species diversity hovered around 10 at any one time, dropping to a low of 5 in the earliest Triassic but rebounding rapidly in the same interval to a post-Mississippian high of 22.

Fig. 6.3 also outlines four second-order cycles in the Permian and another four in the Triassic .

The earliest of the Permian cycles, Cycle XIV in Fig. 6.3, followed a long interval in the late Pennsylvanian and Asselian when the evolutionary index was consistently below 1.0 and was provoked by the appearance in the Sakmarian of an early species of *Neogondolella* (*N. bisselli*), the first species of *Sweetognathus* (*S. merrilli*), and a distinctive species of *Diplognathodus* (*D. sakmariensis*). *N. bisselli* is thought to have evolved from *N. praebisselli*, which precedes it stratigraphically. However, these two species are separated by much of the Pennsylvanian from *N. clarki*, the earliest species with elements morphologically assignable to *Neogondolella*. For this reason, I suspect that the *N. praebisselli-N. bisselli* lineage is an independent Permian development from the long-ranging *Xaniognathus-Cypridodella* stock which, as I speculated in Section 5.8.5, was probably the progenitor of several *Gondolella* lineages in the Pennsylvanian and earliest Permian and the ancestor, as well, of numerous additional lineages of *Neogondolella*, *Neospathodus*, and *Epigondolella* later on in the Triassic.

Secondary Cycle XV follows the early Artinskian extinction of the typically Carboniferous genera *Streptognathodus*, *Idiognathodus*, and *Adetognathus*. Following a brief interval of slightly lowered diversity, a short innovative burst resulted in the appearance of the earliest species of *Neostreptognathodus*, *Sweetina*, and *Merrillina*. *Neostreptognathodus* apparently represents a modification of the *Diplognathodus-Sweetognathus* stock, whereas I have interpreted *Sweetina* and *Merrillina* in Chapter 5 as successive stages in a prioniodinide lineage that may have been the first of a series of such things to be derived iteratively from *Ellisonia*. The rather long late Permian segment of Cycle XV includes a high-diversity interval populated largely by species that seem mostly to represent subsequent chronospecific (i.e., phyletic) subdivision of the *Sweetina-Merrillina* and *Neostreptognathodus* lineages.

The very short late Permian (or early Dzhulfian) Cycle XVI shown schematically in Fig. 6.3 followed directly after extinction of the *Sweetognathus* and *Sweetina-Merrillina* lineages. Its introductory innovative burst is attributable to appearance in the latest Permian of five new species and extinction of just two. Three of these are assignable to *Neogondolella* and continue a second Permian lineage that began in early Late Permian time and originated either in Artinskian populations of *N. bisselli* or *de novo* from the stem stock of the Gondolellidae. A species of *Hindeodus* (*H. typicalis*) and one of *Diplognathodus* (*D. movshovitschi*) round out the incoming contingent. While these new species filled a void left by extinction of the last species of the *Sweetognathus* and *Merrillina* lineages, they do not appear to represent the same sort of iterative burst signaled by earlier Permian ones, and recovery in diversity in the succeeding Dzhulfian was slight.

Cycle XVII followed a prolonged interval of depressed evolutionary index in the latest Permian and is positively associated with initiation in the Early Triassic of *Neospathodus*, which represents yet another burst of speciation by the long-ranging *Xaniognathus-Cypridodella* complex, the stemstock of the Gondolellidae and the same one that produced the "naked" and "platformed" *Gondolella*s of the Pennsylvanian and earliest Permian and also, iteratively, at least two lineages of *Neogondolella* in the Permian. This early Triassic (or Scythian) burst was amplified by essentially concurrent development from the long-lived, conservative *Ellisonia* stock of species of *Pachycladina*, *Hadrodontina*, *Furnishius* and, ultimately, *Gladigondolella*. These events in the history of the two surviving branches of the Prioniodinida combined to produce a Scythian conodont fauna that is highly distinctive but obviously limited genetically and almost incestuously interrelated!

The high-diversity episode of secondary Cycle XVIII is related primarily to Anisian development in the Germanic Basin and perhaps in similarly restricted environments elsewhere of a group of species referred to *Chirodella* and related forms, all represented by very tiny elements and probably derived from (or part of) the long-ranging, generalized *Xaniognathus-Cypridodella* stock. At the same time, several species of *Neogondolella* (including the type, *N. mombergensis*) also made their debut. It is questionable if these species evolved from Scythian neogondolellas or developed iteratively from the *Xaniognathus-Cypridodella* stock. I favor the latter possibility primarily because no one has yet been successful in assembling a succession of samples that demonstrates morphologic intergradation between Scythian and Anisian neogondolellas.

Ladinian Cycle XIX includes a high-diversity episode that is clearly related to development from *Neogondolella* or the *Xaniognathus-Cypridodella* stock of a lineage commonly included in *Epigondolella*, but perhaps better separated under the name *Carinella*, which was proposed for it some years ago by Kiril Budurov. The innovative burst that preceded the high-diversity episode of Cycle XIX was reinforced by essentially concurrent development from *Cypridodella* of the *Pseudofurnishius-Mosherella* lineage.

Cycle XX, the final Triassic cycle and the last one in the history of conodonts, began in Karnian time following extinction of *Pseudofurnishius*, *Mosherella*, and the *Carinella* group of gondolellid conodonts. Its high-diversity episode, in the earliest Norian, was signaled by iterative development in the latest Karnian (probably from *Neogondolella*) of *Epigondolella sensu stricto* and *Misikella*, a replay of early Triassic *Neospathodus* and the only stock to survive to the very end of the Triassic.

6.4 Extinction

As just noted, both terminal and phyletic extinctions punctuated the entire evolutionary history of the Conodonta. However, Clark

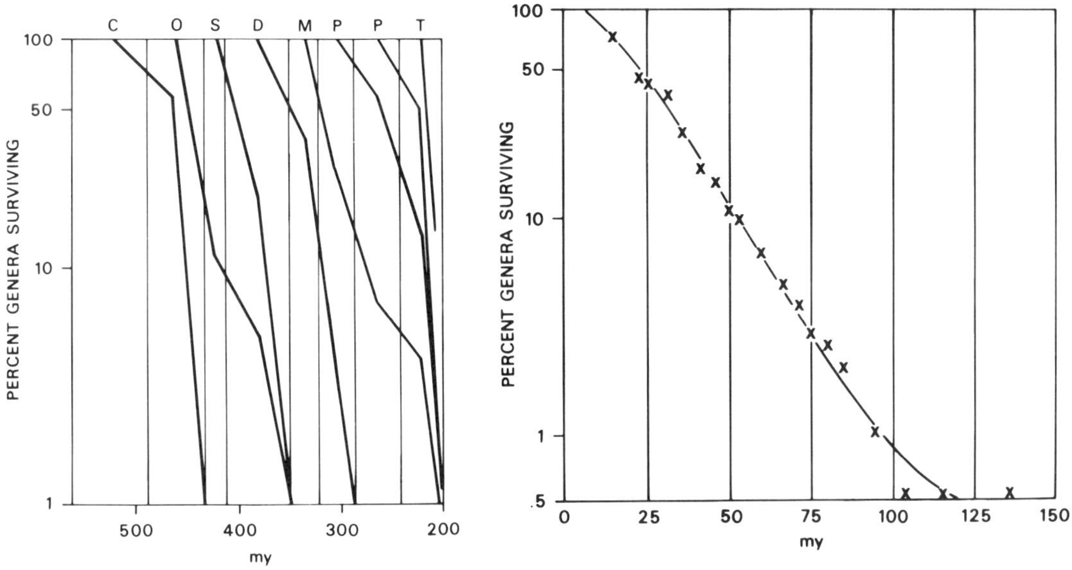

Fig. 6.4. Survivorship curves (left figure) for cohorts of conodont genera (e.g., all genera originating in Ordovician) and curve of longevity of conodont genera of all ages (on right). Redrawn from Clark (1983).

(1983, 1987) has recently focused particular attention on events in the late Paleozoic and early Mesozoic history of the phylum that preceded its extinction at the end of the Triassic.

In his 1983 study Clark presented a series of survivorship curves (Fig. 6.4) for Cambrian through Triassic cohorts of conodont genera and a summary curve for all conodont genera listed in the *Treatise* (Clark et al., 1981). Clark noted that these curves indicate that the average survival of systemic cohorts was 109 my and that mean survival of conodont genera was about 30 my. Thus, as Clark pointed out, an observer in the late Paleozoic with access to the same records would not have foreseen the end-Triassic extinction of the phylum but would probably have predicted survival into at least the Jurassic and possibly into the Early Cretaceous.

The record indicates that the Conodonta did not survive into the Jurassic or Cretaceous, however, but that the phylum became extinct in the latest Triassic. Conodonts were not the only group to become extinct or to suffer major declines in diversity at this time. Indeed, Sepkoski and Raup (1986), summarizing the opinions of others, note that "the Late Triassic contains one of the five largest mass extinctions of the Phanerozoic marine record." The data they present, however, like those for conodonts shown graphically in Fig. 6.5, show generally high extinction intensities throughout the Late Triassic, with a maximum in the latest Norian (or Rhaetic). This suggests to me, as it apparently did to Clark (1983), that the end-Triassic extinction of conodonts may have been the cumulative result of several factors, not of a single catastrophic event that terminated the phylum "well before its time."

Early in the Permian, at the end of secondary Cycle XIII, a majority of the typically Carboniferous stocks disappeared, following a long late Pennsylvanian and early Permian interval during which species-level diversity plummeted and the evolutionary index was consistently below the maintenance level. Permian recovery was surely not spectacular. It primarily involved a couple of bursts of speciation by the gondolellid stem group, one each by the *Ellisonia* and the *Diplognathodus* stocks, but very little action on the part of *Hindeodus*, which merely "showed the flag" through most of the Permian. Note in Fig. 6.5, however, that, despite these fitful attempts to restock Permian environments with new and vigorously successful conodont stocks, extinction intensity

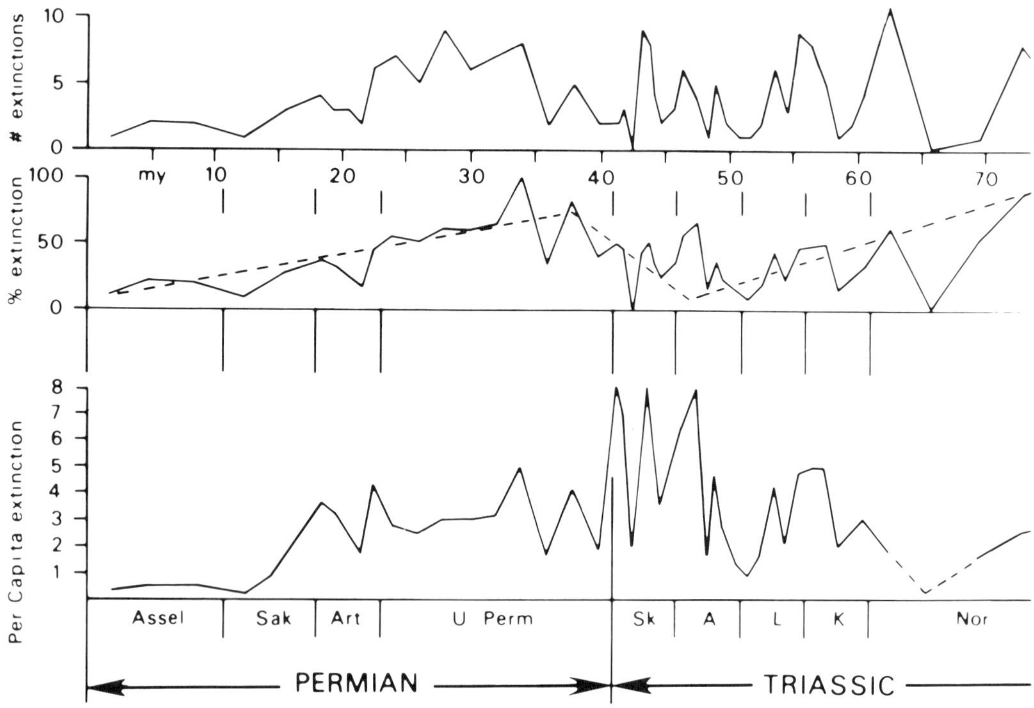

Fig. 6.5. Logs of various extinction metrics for Permian and Triassic conodonts. Top log plots simple number of extinctions for each zone. Central log plots extinctions as a percentage of standing crop of species in each zone, and dashed line merely suggests that extinction intensity increased through the Permian, decreased through the Early Triassic, and then increased again to extinction of the phylum in the latest Triassic. Lower log plots per capita extinction and is adjusted for time. All logs involve the same raw data.

gradually increased through the Permian to a high (or highs) near the end of the period, when overall diversity of the Conodonta dropped to an all-time low of 4 species.

With loss early in the Triassic of *Hindeodus*, the last gasp of the Ozarkodinida, the genome of the Conodonta was further diminished. During four intervals of the Triassic, the latest in the Karnian, brief bursts of iterative innovation by the two remaining prioniodinide stocks seem to represent responses to widening areas of suitable marine environment, but from early in the Middle Triassic (Anisian) the intensity of extinction again began a gradual increase, and this culminated in total extinction at the very end of the Norian.

From his own petrologic information and from eustatic data summarized by Vail, Mitchum, and Thompson (1977), Clark (1983) speculated that the Conodonta, weakened genetically by episodic pruning and increasingly modest recoveries through the Permian and Triassic, included among its three or four latest Triassic species none that was able to adapt to the shallower, higher-energy conditions probably produced at the end of the Triassic by a worldwide drop in sea level. In short, the phylum may have been caught with its adapative pants down. The shallow-water ellisoniids were gone and only the gondolellids remained—and this stock of prioniodinides had been more at home in deeper-water, more pelagic environments since its inception in the Pennsylvanian. Ultimate demise of the Conodonta may thus have been more the result of bad luck (being in the wrong place at the right time) than of cosmic accident or basic genetic incapacity.

6.5 Iterative Evolutionary Patterns

In the preceding discussion of diversity-related evolutionary patterns, I have several times tied the repetition of second-order diversity cycles to iterative development of various major groups of conodonts, particularly the Ozarkodinida and Prioniodinida. A similar pattern is apparent, however, in at least segments of the Panderodontida and Prioniodontida, and it may be general in the Conodonta as a whole.

As shown diagrammatically in Fig. 5.36, which summarizes my views on the phylogeny of the Ozarkodinida, elaboration of the sides and surfaces of Pa-element crowns was carried out repetitively in spathognathodontid populations at several widely spaced intervals in the Silurian and Devonian. As a consequence, apparatuses of the species that developed in each of these iterative bursts are closely similar. For example, *Ancyrodelloides* of the early Devonian has Pa and Pb elements that are very similar to those of early Silurian *Pterospathodus* and also essentially identical in morphology to those of later Devonian *Ancyrodella.* The species of early Silurian *Kockelella* experimented with elaboration of Pa elements in much the same manner as did early Devonian *Eognathodus*, primitive representatives of somewhat younger *Polygnathus*, and a number of later Devonian and early Carboniferous species assigned to *Scaphignathus* (Spathognathodontidae) and the Cavusgnathidae. And iterative development of the *Protognathodus-Gnathodus*, *Diplognathodus*, *Lochriea*, and *Rhachistognathus* lineages from *Bispathodus* (Spathognathodontidae) produced such a welter of species with morphologically similar Pa elements that supraspecific phylogenies are notoriously difficult to reconstruct or defend.

In Section 5.6 I have also noted my suspicion that *Parabelodina denticulata*, a panderodontide species with well-developed rastrate elements in its skeletal apparatus (and thus similar to species of *Belodina* and *Pseudobelodina*), evolved directly from a contemporary species of *Panderodus* (*P.bergstroemi*), not from some earlier species of *Belodina* or *Pseudobelodina*. Other evidence, mostly unpublished, suggests that *Panderodus* itself includes a number of lineages, each probably deserving of independent generic (or even familial) designation and representing chronologically distinct iterations of stocks characterized by skeletal elements of closely similar morphology.

Within the Prioniodontida there are quite a number of "cryptogenic" stocks for whose origin iterative development from basic lineages is also an attractive explanation. *Plectodina* and *Phragmodus*, for example, appear *de novo* in the early Middle Ordovician, and their species formed skeletal apparatuses that are, in a number of ways, closely similar to those built by several species in the Multioistodontidae (e.g., *Paraprioniodus costatus*). Because the latter precede *Phragmodus* and *Plectodina* stratigraphically, they have been considered (by me, among others) as attractive candidates for ancestors. However, multioistodontid elements are mostly hyaline, and there is little to recommend them in other morphologic particulars as hallmarks of the ancestors of either *Phragmodus* or *Plectodina.* I now suspect that those two lineages developed independently, and at slightly different times, from generalized prioniodontide ancestors in *Tripodus* (Oistodontidae). Although one may distinguish readily between many of the elements in apparatuses of *Phragmodus* and *Plectodina*, others (including certain Pa elements) are closely similar. Thus homeomorphy may not be as great as between *Ancyrodelloides* and *Ancyrodella*, but it is nevertheless substantial.

I suspect that iteration also characterized Silurian development of the prioniodontide family Distomodontidae, but this may be very difficult to establish objectively, and there are surely other possibilities, as indicated in Section 5.7.5. If the basic apparatus pattern of the distomodonts included unplatformed pastinate elements in the P positions, then iterative development may have been responsible for initiation of stocks such as those represented by *Icriodina*, *Hadrognathus* and, ultimately, even some or all of the icriodonts. Chatterton and Perry (1977) seem also to have suggested that iteration might have been the active pattern in icriodont evolution when they wrote that

> species that could be or have been assigned to the genus *Pelekysgnathus*, as it is broadly conceived, may have been derived from other genera during four different evolutionary events (origin of "*P.*" *index* in the Late Silurian, origin of *I. hadnagyi* in the Lochkovian, origin of [*Steptotaxis*] *furnishi* group in the Early Devonian and origin of "*P.*" *communis* in the late Devonian.

The recent suggestion by Sandberg and Dreesen (1984) that a group of Late Devonian species assigned currently to *Icriodus* may instead represent the iterative development from *Pelekysgnathus* of species with *Icriodus*-like Pa elements heightens my suspicion that iteration may have been the major operative pattern in icriodont evolution and that the Icriodontidae may thus be wildly polyphyletic!

Finally, I should repeat the conviction expressed again and again in Section 6.3.5 that late Paleozoic evolution of the prioniodinide families Gondolellidae and Ellisoniidae was iterative. A good deal of systematic taxonomic work badly needs doing in both these groups, hence some of the evidence for iterative development may currently be obscured by disparate classification of various late Paleozoic species. For example, early species of *Xaniognathus* s. l. may currently be hiding in *Gondolella*, some of whose species formed "naked" Pa elements with short posterior and essentially platformless anterior processes that are morphologic homeomorphs of the Pa elements on which I based the form genus *Xaniognathus* in 1970. The observation that the segminiplanate elements typical of *Gondolella* are commonly more closely related to the naked, *Xaniognathus*-like elements with which they sometimes occur stratigraphically than they are to somewhat younger gondolelliform elements, and the fact that ramiform elements are similar or identical in apparatuses of the two groups of species also suggest that *Gondolella* may have developed several times from *Xaniognathus*.

Debate as to whether species of the Permo-Triassic gondolellids *Neogondolella* and *Neospathodus* had unimembrate apparatuses or multimembrate ones represented by the recurrent groups I have assigned to various species of *Xaniognathus* and *Cypridodella* may also be part of this problem. That is, if *Neogondolella* and *Neospathodus* species developed iteratively from a *Xaniognathus* stem-stock, the ramiform elements that complete their skeletal apparatuses may currently be hidden taxonomically in *Xaniognathus* (or *Cypridodella*), despite evidence from a Triassic coprolite that *Neogondolella* had a multimembrate apparatus. Later iteration may well have been involved in development from *Cypridodella* of the *Carinella* and *Epigondolella* lineages. Should these suspicions be supported by further study, both taxonomy and phylogeny of the Gondolellidae will require basic overhauls.

Evidence of the sort just summarized compels me to suggest that central to each of the conodont orders was a rather slowly evolving stock characterized by an apparatus of conservative, generalized morphology from which shorter-lived, phylogenetically dead-ended lineages branched repetitively. The close relationship that seems to exist between iterative bursts and eustatic (or transgressive-regressive) events may suggest that the timing of major innovative bursts may have been provided by that of the eustatic events. Deepening cycles may have offered opportunities for iterative elaboration by the *Ozarkodina* lineage of the spathognathodontid stock, whereas shallowing associated with the regressive phases of those cycles may have offered opportunities for elaboration by the *Pelekysgnathus-Icriodus* and *Pandorinellina* stocks. In the Late Paleozoic and Triassic, episodic development of gondolellids during deepening phases and of ellisoniids during regressive ones may indicate a similar relationship between eustatic events and iterative episodes in the history of the Prioniodinida. In fact, this sharing of responsibility may well have contributed to final extinction of the phylum.

6.6 Evolutionary Trends

6.6.1 *Apparatus elaboration*

Founders of the Cavidonti and Conodonti (*Proconodontus* and *Teridontus*, respectively) built unimembrate apparatuses in which elements at all locations were morphologically similar or identical (Fig. 6.6). Very soon in the evolutionary history of both classes, however, different positions in the apparatus came to be identified by elements of at first slightly, then, a bit later, greatly different morphology. Fully elaborated apparatuses with six or seven morphologically different element types were present in several stocks by early in the Ordovician and characterized all but a few conodont species from the Middle Ordovician on.

6.6.2 *Apparatus reduction*

In species of a few Ordovician genera (e.g., *Eoplacognathus*, *Polyplacognathus*, *Scyphiodus*) and in many post-Ordovician ones, there appears to have been a concerted, if only superficial, effort to undo the hard work of morphologic differentiation carried out by their predecessors. That is, species in these stocks apparently formed apparatuses that were reduced just to P elements, or apparatuses in which elements in the P positions were the only ones mineralized. This conclusion is suggested by the fact that there is a stratigraphic reduction in the ratio between ramiform and pectiniform elements in several major groups (Fig. 6.7).

Merrill and Powell (1980) suggest that elements in the S and M positions may have been mineralized early in the ontogeny of certain species, that P elements may not have been mineralized until later, and that the early-formed S and M elements may have been partly or wholly resorbed by maturity. This intriguing hypothesis is supported by evidence from the Drum Limestone of Kansas, in which succes-

Fig. 6.6. Early diversification of the two basic conodont lineages, the Proconodontida on the right, the Protopanderodontida on the left. Numbers in circles indicate number of elements in apparatus. Note that earliest members of both lineages had only a single element type, but that diversification produced apparatuses with two to four element types by early in the Ordovician. Redrawn from Sweet (1985).

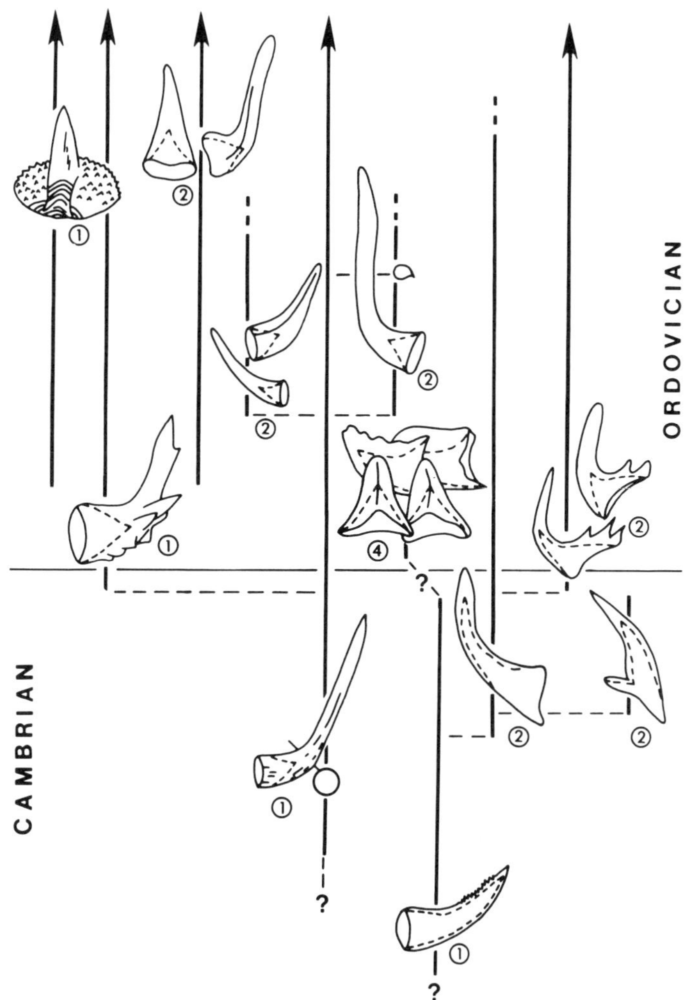

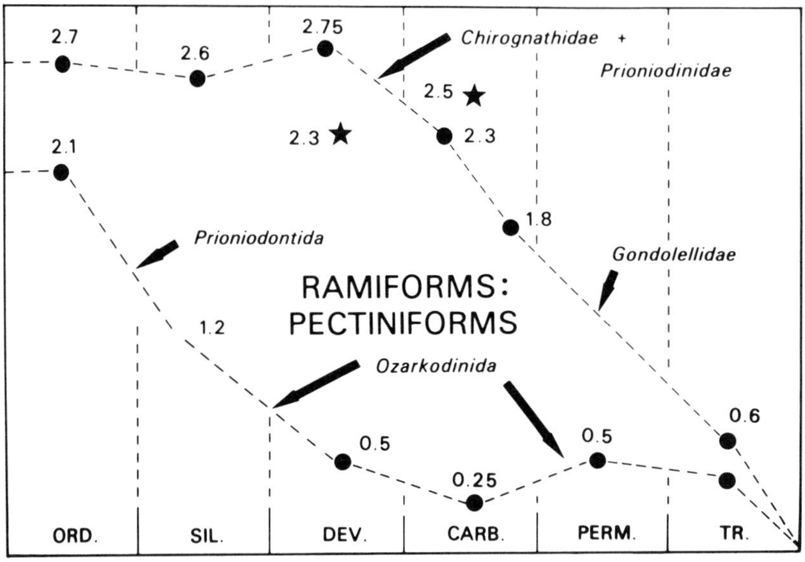

Fig. 6.7. Trends in ratio between ramiform and pectiniform elements in apparatuses of Ordovician through Triassic conodonts. It is not clear why pectiniform elements outnumber ramiform elements in many post-Silurian collections and the reduction in ratios did not take place at the same time in all taxonomic groups. Also, ratios derived from reports of discrete-element frequencies (on dashed lines) are commonly different from those derived from natural assemblages (stars) of similar age. Redrawn from Sweet (1985), with emendations.

sive beds seem to retain nearly pure records of successive stages in the ontogeny of single species.

In Fig. 6.7 I indicate that there is a substantial disparity in the ramiform-pectiniform ratios calculated from natural assemblages, which presumably represent complete skeletal apparatuses, and those determined from tabulated collections of discrete elements. It has been suggested that the natural assemblages, with mineralized elements in all positions, may represent juveniles of species in which mineralization of S and M elements ceased before maturity—or in which elements in those positions were actually resorbed in later ontogenetic stages. But there are other possible explanations. The one to which I subscribe postulates that mineralization was of critical importance to the function of P elements but perhaps of lesser importance (or of no importance) to the function of elements in other apparatus positions. Thus, under conditions of limited phosphate availability, elements in the S and M positions were simply not mineralized. Most Ordovician and Silurian species, and younger species represented by natural assemblages, may have inhabited environments in which there was phosphate enough to go around. Those represented largely, or entirely, by P elements may have adapted to phosphate-poor environments by giving up mineralization of elements in S and M positions or by resorbing in adult stages the phosphate stored in those elements in earlier ontogeny.

6.6.3 *Elaboration of elements in P positions*

In evolutionary differentiation and development of the apparatus in various lineages a major tendency was toward elaboration of pectiniform elements in the P positions. In a majority of Ordovician forms, P positions in even fully differentiated apparatuses were occupied by unplatformed pectiniform elements; in only a few lineages (e.g., Balognathidae, Polyplacognathidae) were platformed elements formed in those positions. In the Silurian, platformed elements were developed repeatedly in P positions by species with roots in the Spathognathodontidae, Distomodontidae, and Icriodellidae. In the Devonian and later in the Paleozoic, most ozarkodinide stocks included species with platformed pectiniform elements in the Pa position. The Prioniodinida delayed development of platformed P elements until the Mississippian, when the curious, short-lived Bactrognathidae invented them. Later, however, in the Pennsylvanian (and increasingly in the Permian and Triassic), prioniodinide stocks came to be characterized by platformed pectiniform elements in the Pa position.

In the absence of firm information as to the function of elements in any position in the conodont apparatus it is difficult to relate evolutionary elaboration of P elements to specific environmental or behavioral factors. However, from the fact that such elaboration characterized some or all of the lineages in nearly every major stock at one time or another in their history, it may be concluded that platform elaboration was surely of positive adaptive significance.

Nicoll (1987), who suggests that the skeletal elements of conodonts may have been internal supports for ciliated tissue in the food-gathering system of generally microphagous animals, concludes that P elements were opposed and operated in such a fashion as to roll, bruise, and crush particulate matter caught between them. Phylogenetic elaboration of P elements might thus have been related to changes with time in size, shape, and bruise- or crushability of food particles. If P elements functioned as teeth, as Jeppsson (1979, 1980) and others conclude, a similar relationship might also be postulated.

6.7 Developmental Strategies

Gould (1977) has provided signal service to the biologic community by emphasizing not only that "parallels between ontogeny and phylogeny are produced by heterochrony" but also that there may be a continuum of relationships between ecologic factors and the life-history patterns that may result from alterations in the timing by which various somatic and behavioral features develop. In short, Gould expanded on Van Valen's (1973) somewhat earlier observation that evolution might well be the control of development by ecology. Paleobiologists are thus encouraged to seek information on developmental patterns and potential ecologic controls in the materials they study.

The large, stratigraphically well-controlled collections available to students of conodonts should also be well suited to studies of developmental, or evolutionary, strategy. Because well-made collections commonly include elements that represent various stages in the ontogenetic history of the species that contributed to those collections, it should be possible to relate ontogenetic timing of the morphologic features recorded to developmental strategies in the lineages that formed them. And, if collections are carefully related to the petrologic and other biologic properties of the rocks from which they were obtained, it should also be possible, in time, to determine the nature and extent of any relationship between developmental strategy and environmental parameters.

Although information on the ontogenetic development of various skeletal elements has been provided by various authors, I am not aware that anyone has yet documented stages in the ontogeny of a conodont species from studies that involve all elements of the skeletal apparatus. Consequently, in the following sections I summarize pertinent information from a few recent studies of Pa elements that suggest that developmental strategies may have differed greatly from one group of conodonts to the next. We can only guess at the extent to which additional strategies may be suggested by future studies of complete apparatuses.

6.7.1 Recapitulation

Heterochrony involves "changes in the relative time of appearance and rate of development for characters already present in ancestors" (Gould, 1977) and may be effected either by acceleration or retardation. If a character that appears in the adult stage of an ancestor appears at a younger stage in development of a descendant, acceleration is the process involved and classical recapitulation the result. If, on the other hand, a character that appears in the young stage of an ancestor appears in, or is retained into, the adult stages of a descendant, retardation is the process and paedomorphosis the result.

Most studies of conodonts that consider evolutionary matters have focused on morphologic features that seem to have been "added onto" elements typical of stratigraphically previous populations. For example, it is now well known that evolution of *Pygodus anserinus* from *P. serra* involved addition by the former of a fourth denticle row to platformed P elements otherwise similar to the three-rowed ones of the latter. Additional examples are plainly in evidence in the phylogeny of *Polygnathus* outlined by Weddige (1977) and Weddige and Ziegler (1979) and in the one reconstructed for *Palmatolepis* by Helms and Ziegler (in Clark et al., 1981). Those phylogenies, which involved study of Pa elements only, are summarized in Figs. 5.45 and 5.47.

6.7.2 *Paedomorphosis*

Paedomorphosis characterizes developmental patterns of organisms that retain juvenile features of morphology, simplicity, or proportions into adult stages. In some paedomorphic stocks, said to have developed by *progenesis*, small individuals may be sexually mature, but in other respects are distinguished by juvenile characteristics that do not develop greatly following achievement of sexual maturity. *Neoteny*, on the other hand, is a paedomorphic process in which the development of morphologic features may be delayed, or retarded, without alteration in size or the timing of sexual maturity. Neotenic individuals, as a result, are generally of normal size at sexual maturity, but exhibit juvenile morphologic features.

In Section 5.9.9 it was noted that the Pa elements of *Idiognathodus* pass through a stage in early ontogeny that is reminiscent in morphology of the Pa elements of adult *Streptognathodus* and that *Streptognathodus* appears to have evolved from *Idiognathodus*, perhaps several times, late in the history of this complex group of ozarkodinides. This seems to be a reasonably well documented, if still fairly gross, example of neoteny—the retention of juvenile characters into later and later stages of ontogeny.

Broadhead and McComb (1983) have also outlined a scenario in which the evolution of early representatives of *Icriodus* (Prioniodontida, Icriodontidae) from *Pedavis* (Prioniodontida, Icriodellidae) may be indicated by paedomorphic loss, through neoteny, of the anterior lateral spur on P elements. While steps in the evolutionary sequence they postulate are logical and supported by objective evidence, Broadhead and McComb gave primary attention to characters of the Pa elements of *Pedavis* and *Icriodus* and neglected the substantial differences that distinguish elements in other apparatus positions. As indicated in Section 5.7.6, other considerations suggest that *Icriodus* may have evolved several times from *Pelekysgnathus*, or *de novo* from *Rotundacodina dubius*, that is, from lineages that are only distantly related to *Pedavis*. If those possibilities are considered, neoteny may not have been involved.

Mosher (1970) recognized that the Pa elements of certain Late Triassic species of *Epigondolella* (Gondolellidae) retain juvenile characters into later and later ontogenetic stages and may have achieved this paedomorphic condition through neoteny. In a more recent study, based on somewhat more extensive collections, Orchard (1983) has substantiated Mosher's interpretation for at least two *Epigondolella* species. And, in a related vein, I note that neoteny may also have been involved in the development of *Mosherella* from *Pseudofurnishius* and in the origin of *Misikella* either in populations of *Neogondolella* or those of more generalized gondolellid ancestors in the Late Triassic.

6.8 Summary

Initial development and spread of the Conodonta conforms in pattern to that of a classical adaptive radiation, during which there was notable (and fairly rapid) elaboration of the skeletal apparatus and early establishment of major taxonomic groups. In the subsequent history of these groups, 20 cyclical fluctuations in evolutionary vigor and species-level diversity seem related to second-order eustatic events (although not always on a one-for-one basis) and to the ability of various conservative stem-stocks to answer the challenges of repeatedly offered opportunities through iterative, or repetitive, development of generally short-lived lineages that were strongly homeomorphic with their predecessors in features of their skeletal apparatuses. From the Late Pennsylvanian on, however, species-level diversity was low, and populations appear to have been at or below the maintenance level for long periods of time. With only a brief interruption in the Early Triassic, extinction intensity increased from the Early Permian to the end of the Triassic, when the phylum was reduced to a few species, probably represented by populations of small size and provincial distribution. Although studies of generic-level survivorship would not have led Permian observers to predict the extinction of conodonts before some time in the Jurassic, the phylum did not survive the Triassic. Reasons for extinction are not at all clear but may have included not only those built into the long post-Late Pennsylvanian decline, but also a dollop of bad luck in the Late Triassic when surviving basinal populations may have been unable to adapt to conditions produced by a worldwide drop in sea level.

References

Bergström, S. M. (1983). Biogeography, evolutionary relationships, and biostratigraphic significance of Ordovician platform conodonts. *Fossils and Strata* **15**, 35–58.

Broadhead, T. W. and McComb, R. (1983). Paedomorphosis in the conodont family Icriodontidae and the evolution of *Icriodus*. *Fossils and Strata* **15**, 149–154.

Chatterton, B. D. E., and Perry, D. G. (1977). Lochkovian trilobites and conodonts from northwestern Canada. *J. Paleont.* **51**, 772–796.

Clark, D. L. (1983). Extinction of conodonts. *J. Paleont.*: **57**, 652–661.

——— (1987). Conodonts: The final fifty million years. Pp. 165–174 *Palaeobiology of Conodonts* (ed. R. J. Aldridge). Ellis Horwood, Chichester, 180 pp.

Clark, D. L., Sweet, W. C., Bergström, S. M., Klapper, G., Austin, R. L., Rhodes, F. H. T., Müller, K. J., Ziegler, W., Lindström, M., Miller, J. F., and Harris, A. G. (1981). Conodonta. In *Treatise on Invertebrate Paleontology* (ed. R. A. Robison). Pt. W, Suppl. 2. Geol. Soc. America and Univ. Kansas, 202 pp.

Gould, S. J. (1977). *Ontogeny and Phylogeny*. The Belknap Press of Harvard Univ. Press, Cambridge and London, 501 pp.

Jeppsson, L. (1979). Conodont element function. *Lethaia* **12**, 153–171.

——— (1980). Function of the conodont elements. *Lethaia* **13**, 228.

Johnson, J. G., Klapper, G., and Sandberg, C. A. (1985). Devonian eustatic fluctuations in Euramerica. *Bull. Geol. Soc. Am.* **96**, 567–587.

Lipps, J. (1981). What, if anything, is micropaleontology? *Paleobiol.* **7**(2), 167–199.

Merrill, G. K., and Powell, R. J. (1980). Paleobiology of juvenile (nepionic?) conodonts from the Drum Limestone (Pennsylvanian, Missourian—Kansas City area) and its bearing on apparatus ontogeny. *J. Paleont.* **54**, 1059–1074.

Mosher, L. C. (1970). New conodont species as Triassic guide fossils. *J. Paleont.* **44**, 737–742.

Nicoll, R. S. (1987). Form and function of the Pa element in the conodont animal. Pp. 78–90 in *Palaeobiology of Conodonts* (ed. R. J. Aldridge). Ellis Horwood, Chichester, 180 pp.

Orchard, M. J. (1983). *Epigondolella* populations and their phylogeny and zonation in the Upper Triassic. *Fossils and Strata* **15**, 177–192.

Sandberg, C. A., and Dreesen, R. (1984). Late Devonian icriodontid biofacies models and alternate shallow-water conodont zonation. *Geol. Soc. Am. Spec. Paper* **196**, 143–178.

Sepkoski, J. J., Jr., and Raup, D. M. (1986). Periodicity in marine extinction events. Pp. 3–36 in *Dynamics of Extinction* (ed. D. K. Elliott). Wiley, New York, 294 pp.

Sloss, L. L. (1963). Sequences in the cratonic interior of North America. *Bull. Geol. Soc. Am.* **74**(2), 93–114.

Sweet, W. C. (1970). Uppermost Permian and Lower Triassic conodonts of the Salt Range and Trans-Indus ranges, West Pakistan. Pp. 207–275 in *Stratigraphic boundary problems, Permian and Triassic of West Pakistan* (ed. B. Kummel and C. Teichert). Univ. Kansas, Dept. Geol., Spec. Publ. 4.

——— (1985). Conodonts: Those fascinating little whatzits. *J. Paleont.* **59**(3),485–494.

Vail, P. R., Mitchum, R. M., Jr., and Thompson, S. III (1977). Seismic stratigraphy and global changes of sea level, pt. 4: Global cycles of relative changes of sea level. Pp. 83–95 in *Seismic Stratigraphy—Applications to Hydrocarbon Exploration* (ed. C. E. Payton). Am. Assoc. Petrol. Geol., Mem. 26.

Van Valen, L. (1973). Festschrift. *Science* **180**, 488.

Walliser, O. H. (1983). Geologic processes and global events. *Terra Cognita* **4**, 17–20.

——— (1984). Pleading for a natural D/C-Boundary. Pp. 241–246 in *The Devonian-Carbonifer-*

ous boundary (ed. E. Paproth and M. Streel). *Courier Forschungsinst. Senckenberg*, **67**.

Weddige, K. (1977). Die Conodonten der Eifel-Stufe im Typusgebiet und in benachbarten Faziesgebieten. *Senckenbergiana Lethaea*, **58**(4/5), 271–419.

Weddige, K., and Ziegler, W. (1979). Evolutionary patterns in middle Devonian conodont genera *Polygnathus* and *Icriodus*. *Geol. et Palaeont.* **13**, 157–164.

Ziegler, W., and Lane, H. R. (1987). Cycles in conodont evolution from Devonian to mid-Carboniferous. Pp. 147–163 in *Palaeobiology of Conodonts* (ed. R. J. Aldridge). Ellis Horwood, Chichester, 180 pp.

7. PALEOECOLOGY AND PALEOBIOGEOGRAPHY

7.1 Introduction

For many years after their discovery it was generally held that conodonts were cosmopolitan in their occurrence within the marine realm and that most species would turn out to be worldwide in their distribution. However, as more and more information was assembled, it gradually became clear that conodont species behaved like the species of most other animals. That is, there were lateral limits to their geographic distribution and decided, if broad, ecologic factors that determined where they lived within Paleozoic and Triassic seas. Furthermore, in 1958 Huckriede suggested that the development of Triassic conodonts followed different courses in various biogeographic provinces, and in 1959 Sweet, Turco, Warner, and Wilkie concluded that the distribution of conodonts in rocks of Late Ordovician age outlined two distinct faunal provinces.

Neither the ecology nor the biogeography of conodonts could be studied in a very meaningful way, of course, until a biologically sound taxonomy had been developed and fleshed out systematically. Nevertheless, the repeated discovery between 1926 and 1966 that conodonts are exceptionally useful biostratigraphically stimulated the assembly of comprehensive, geographically widespread collections through the known range of the group. Thus, with the development of the multielement taxonomy I have outlined in Chapter 5, much of the information needed for both paleoecologic and paleobiogeographic interpretation of conodont faunas has fallen into place.

Because conodonts are extinct and have no very close analogs in the living biosphere, features of their biologic associations and lithic occurrences must be used to reconstruct their habits, habitats and ecologic preferences. Information derived in such indirect ways is always subject to debate, revision, and periodic reinterpretation, of course, and this must be firmly understood as a preamble to the following discussion. Obviously, until we have satisfied our curiosity and reached some agreement as to these matters, conodonts will be less useful in reconstructing conditions in past depositional environments than are features of the rocks formed in those environments in interpreting the conodonts. Thus I approach a discussion of conodont ecology cautiously, with initial emphasis on features on which there is general agreement.

7.2 Mode of Life, or Habit, of Conodonts

It is common knowledge that elements representing the same or closely similar conodont species may be collected from sedimentary rocks that represent a wide variety of depositional environments. In addition, many conodont species have been shown to have had a very broad geographic distribution, and a few may have been cosmopolitan. Those facts, together with the elongate, wormlike body and distinctively finned tail of the few complete specimens known, support the generally held conclusion that the animals were probably nektic in habit and that the distribution of most species may have been largely independent of conditions on the bottom. Such an interpretation seems to be supported most strongly by the common occurrence of conodonts in certain types of black shales that yield fossils only of pelagic organisms because they accumulated under anoxic bottom conditions that inhibited development of a benthic fauna.

Conodonts occur in rocks with corals, brachiopods, echinoderms, and cephalopods, whose living representatives are all marine, but they have not been reported as autochthonous components of rocks known to have accumulated in nonmarine environments. Thus, as at least a first approximation, conodonts may be interpreted as pelagic, probably nektic, marine animals. Such a habit may also suggest that the animals were predators, a conclusion that is consistent with their possession of a complexly mineralized apparatus of tooth- and jawlike elements in the cephalic lobe. However, the worrisome question of how they might have manipulated the tooth- and jawlike elements that made up the cephalic apparatus may suggest caution in accepting this idea too rapidly.

But swimming predators can live in a great variety of places in the sea and under many different conditions. Some nibble on living animals or plants on the seafloor or on accumulations of carrion there and thus are nektobenthic in habitat. The fossil distribution of swimming animals with such habits should look very much like that of other groups of benthic organisms in that their remains should have a distribution closely similar to that of their benthic prey.

Other swimming predators, particularly small ones, are adapted to life at different levels in the water column, where they feed on phytoplankton, floating algae, or the larvae of other animals. The chaetognaths, which have many anatomic features in common with conodonts and are widely distributed at various levels in modern seas, are examples of swimming marine predators with this sort of adaptation.

In any consideration of life habits, of course, one should also keep in mind that many modern marine animals spend intervals of varying duration in their larval history as pelagic organisms and then settle down to take up either sedentary benthic or nektobenthic adult lives. Kozur (1976) and other authors have inferred such a mode of life for certain late Paleozoic and Triassic conodonts from the curiously biased or disjunct distribution of elements that are interpreted to represent parts of the skeletal apparatus of the same species. That is, if all individuals of every species of the Pennsylvanian idiognathodontid *Streptognathodus* had a skeletal apparatus like the ones preserved in the natural assemblages described from the Pennsylvanian of Illinois by Rhodes (1952), then it would be reasonable to expect that components of the apparatus of *Streptognathodus* should occur in collections of disjunct elements in about the same proportions as they occur in known natural assemblages. But this is not the case. In collection after collection from various levels in the Pennsylvanian and from different places in the world, the platformed Pa elements of *Streptognathodus* dominate the elements that represent other positions in the skeletal apparatus. It has been suggested that this disjunct, or biased, distribution, which leads to the common phenomenon of "platform over-representation" in some collections from late Paleozoic rocks, may indicate that the skeletal apparatuses of at least some conodont species may have differed in composition from one ontogenetic stage to another and that such differences may indicate a change from a pelagic mode of life in early ontogenetic stages to a nektobenthic or benthic mode in later ones.

If conodonts were adapted to some or all of these variations on a swimming mode of life, it will obviously take very careful study to determine which of the several species represented in a particular sample were nektobenthic nibblers, which might have been more at home in the several pelagic levels above the depositional site and which might have had different modes of life at different stages in their life history. If one adds to this the observation that, within a given depth zone in the water column there may also be lateral segregation, in a nearshore-offshore sense, then deciphering the life habits of individual conodont species becomes a challenging game, indeed. But it is a game that can be won through careful attention to the relationship of various types of conodonts to individual features of the depositional basin as deduced from the nature and arrangement of the strata that yield the conodonts.

7.3 Ecologic Models

Two general models, or hypotheses, have been devised to explain the distribution of conodonts in Paleozoic and Triassic marine rocks. The depth-stratification model, outlined in 1971 by Seddon and Sweet, began with the assumption that conodonts were pelagic and attempted to provide an explanation for puzzling discontinuities in their lateral distribution. The lateral-segregation model of Barnes and Fåhraeus (1975), on the other hand, attempts to explain the habit or mode of life of conodonts from discontinuities in their lateral distribu-

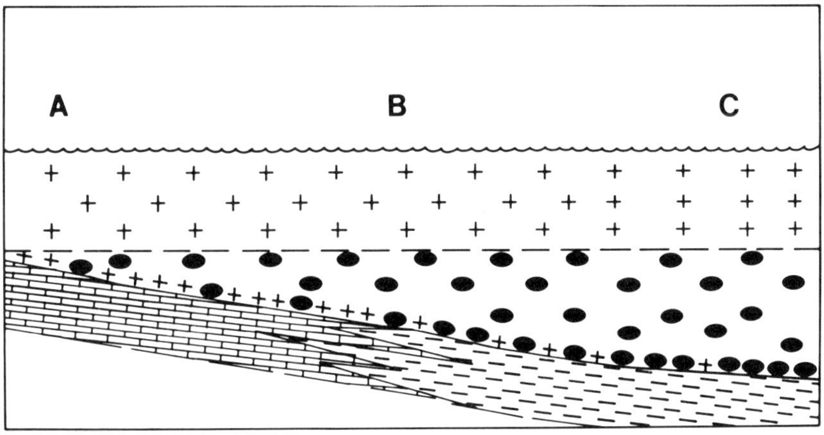

Fig. 7.1. Schematic explanation of the depth-stratification model of Seddon and Sweet (1971). Diagram shows two species; one indicated by crosses is uniformly distributed in a surficial water stratum; the other, indicated by filled ellipses, is uniformly distributed in a deeper-water stratum. Elements of both species would accumulate together on the seafloor, but between A and B those of the "cross" species would dominate, whereas between B and C the reverse would be true. Redrawn, with omissions, from Seddon and Sweet (1971).

tion in sedimentary rocks. To Barnes and Fåhraeus, the existence of laterally segregated biofacies was taken to indicate that most conodonts were benthic or nektobenthic organisms, whereas Seddon and Sweet found that laterally segregated biofacies (which are real) required a special explanation if conodonts were pelagic. Although the two models seem very different—and they certainly began with very different premises—neither has been generally accepted or rejected. They both bear summary discussion before we go on to more concrete matters.

7.3.1 The depth-stratification model

Seddon and Sweet (197l) pointed out that the Ordovician and Devonian conodonts they were studying in eastern North America and Australia, respectively, characterized distinct, laterally segregated biofacies instead of being more or less uniform in their lateral distribution, as one might expect the remains of truly pelagic organisms to be. To explain lateral segregation of conodonts in the benthic record, Seddon and Sweet proposed that, in their lifetimes, different conodonts inhabited different levels in the sea, in the manner of living chaetognaths. A schematic illustration of the depth-stratification model is given in Fig. 7.1.

Figure 7.1 considers just two pelagic conodont species, one represented by crosses, the other by filled ellipses. The cross species is assumed to have been widely and uniformly distributed in a stratum nearer to the surface than the ellipse species, which was also uniformly distributed, but was confined in its lifetime to a deeper-water layer. As individuals of each of these species died, their heavy, phosphatic skeletal remains would have settled to the bottom. We would expect to find fossil representatives of the cross species everywhere on the bottom. However, skeletal elements of the ellipse species would accumulate only seaward of the level at which the top of its depth zone "dragged bottom." Consequently, there might have been a tract of the bottom to the left of point A in Fig. 7.1 on which no specimens of the ellipse species came to rest and in which the fossil record was exclusively of the cross species. On the other hand, the ellipse species, a relatively rare component of collections made from the rocks that accumulated beneath point A, would gradually increase in abundance relative to the cross species in samples taken from rocks that were formed at greater and greater depth. At point B in Fig. 7.1, for example, the two species might be represented by approximately equal numbers of specimens; farther to the right, at point C, specimens of the ellipse species might be dominant, even though equal numbers of cross specimens were contributed throughout the area represented by the diagram.

If the distribution of conodonts at various levels in the water column was originally as uniform as is implied in Fig. 7.1, a gradual increase in diversity should characterize samples taken from progressively more seaward sites. That is, conodonts that inhabited the layer nearest the surface should be represented in every sample, whereas those that were restricted to the deepest levels would be expected in only a few. Experience shows that this is generally not the case, and the fact that shallow-water conodonts are generally not represented, or are very rare, in samples dominated by deeper-water forms has been identified as a fatal flaw by detractors of the depth-stratification model.

However, if conodonts led a depth-stratified existence but were not uniformly distributed throughout the water layer to which they were confined, contrasts might be increased between the biofacies that accumulated in even closely adjacent biotopes. For example, if the population density of the cross species in Fig. 7.1 was greater near the shore than farther offshore, or if that species was restricted to a nearshore, or marginal, environment, crosses and ellipses might not have accumulated together on the bottom, even though they occupied adjacent levels in the stratified water column. In addition, it is possible that different species might have occupied the same water layer, one in its nearshore portion, another in its more offshore parts. Druce (1973) recognized the likelihood that densities might have decreased offshore in each of several depth-stratified populations, and a similar feature was incorporated in a version of the depth-stratification model I followed in a report on Cincinnati Region Ordovician conodonts that was written in 1973 but not published until 1979.

7.3.2 The lateral-segregation model

Barnes and Fåhraeus (1975) have interpreted the laterally segregated distribution of conodonts to mean that during their lifetimes, most of these animals were benthic or nektobenthic in habit. Only a few species were interpreted as pelagic, all characterized by a very wide geographic distribution, essential lithofacies independence, and skeletal apparatuses composed of coniform elements. The lateral-segregation model, as this hypothesis has been termed, is merely an application of the simplicity principle to an interpretation of the distribution of conodonts in sedimentary deposits. As such, it has great appeal although, as Klapper and Barrick (1978) have shown, the laterally segregated record of conodonts can in no way be taken as proof of a benthic or nektobenthic mode of life.

Interpretations of conodont distribution are vastly simplified if the animals were nektobenthic or were components of the vagrant benthos. However, it is still difficult to explain the essentially cosmopolitan distribution of many species and the common occurrence of many others in phosphatic black shales, which presumably accumulated under anoxic bottom conditions that would have been hostile to benthic animal life. An increasingly compelling solution to these dilemmas is to abandon the probably simplistic notion that conodonts were all pelagic, benthic, or nektobenthic and to use in the solution of each problem the mode-of-life interpretation that best accords with the distributional facts.

7.4 Selected Studies of Conodont Ecology

There are as yet few ecologic generalizations that may safely be applied to conodonts as a group. However, the literature includes a number of elegant studies from which it may eventually be possible to draw such principles. Thus, before attempting to make generalizations about conodont ecology, I summarize several recent studies to determine if the conclusions reached in them suggest any common themes or relationships.

7.4.1 Ordovician paleoecology of Cincinnati Region

Conodonts are exceptionally abundant in the fossiliferous Middle and Upper Ordovician shales and limestones of the Cincinnati Region in Kentucky, Ohio, and Indiana (Fig. 7.2). The

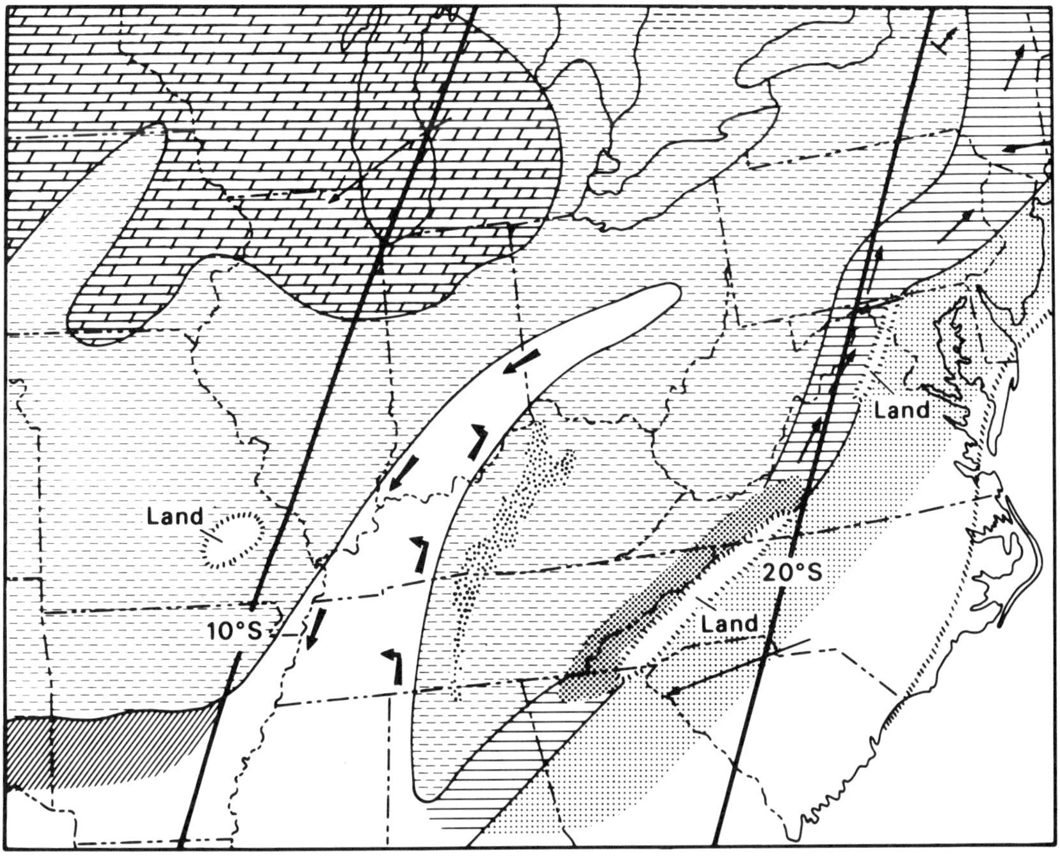

Fig. 7.2. Situation of the Cincinnati Region in the Middle and Late Ordovician. The Cincinnati Region is the tri-state tract in the center of the map. In general, it was part of a broad carbonate shelf to which terrigenous material was introduced from the southeast and that was bordered on the northwest by the distal extremity of a narrow, deep sag (white) that permitted connection with cold, phosphate-rich waters to the south. Approximate positions of 10° and 20° S paleolatitudes indicated by heavy lines; light arrows suggest prevailing-wind direction; heavier arrows in sag suggest wind-driven circulation pattern permitting upwelling of phosphate-rich water in Cincinnati Region. Adapted from Cressman (1973).

species represented are quite well known taxonomically, and their stratigraphic and geographic distribution has been worked out in considerable detail. Somewhat less detailed information is available on the petrology of Ordovician rocks in the Cincinnati Region, and there are some unanswered questions about the environments in which they accumulated. Nevertheless, enough is known to support generalizations about at least some of the factors that may have controlled distribution of the conodonts in the warm, low-latitude seas that covered this area in the Middle and Late Ordovician.

The Middle and Upper Ordovician rocks of the Cincinnati Region accumulated on a cratonic platform that was situated a few to as many as 20 degrees south of the Ordovician equator. Lithofacies are various mixtures of shale derived from orogenic sources far to the east and highly fossiliferous carbonate rocks generated on the spot, primarily from skeletal matter furnished by echinoderms, brachiopods, and bryozoans and reworked and fragmented in some environments by wave- or storm-generated currents. Fossils in most of the carbonate rocks have been transported, but probably not too far from their living sites.

The conceptually straightforward depositional framework of the Cincinnati Region was complicated by differential uplift and subsidence of segments of the platform along pre-

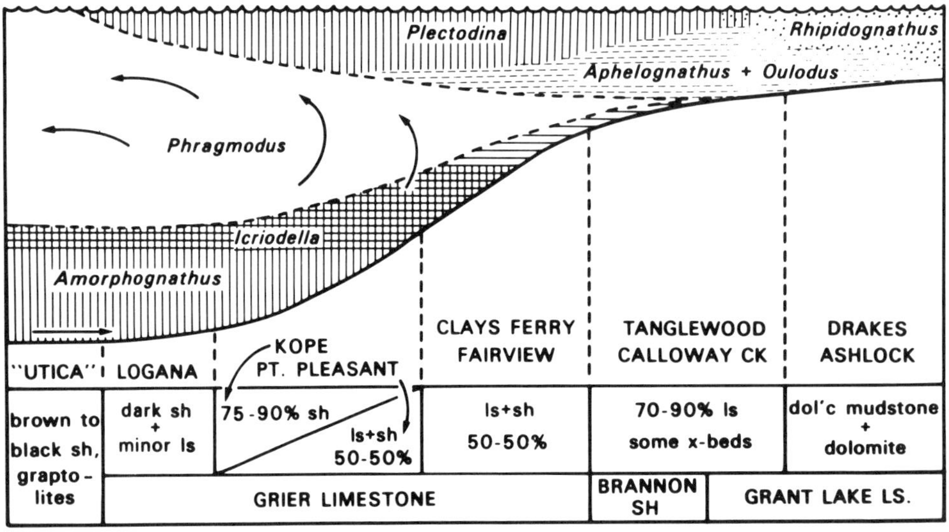

Fig. 7.3. Hypothetical transect of the Cincinnati Region, reconstructed from interpretations of depositional environments of Middle and Upper Ordovician rocks. Above are interpreted distributions of characteristic conodonts, inferred from the relative abundance and lateral distribution of their remains in the facies named below.

Ordovician structural lines (see Borella and Osborne, 1978). These movements evidently were responsible for the maintenance of substantial relief here and there and thus for rather rapid and seemingly erratic lateral variations in lithofacies at several stratigraphic levels. It is thus difficult to draw a profile of the Ordovician seafloor in the Cincinnati Region that would be valid for any very wide segment of the area or for any specific time. So, for purposes of general ecologic analysis, I have used Walther's law and prevailing estimates of the depth and energy regimes represented by Cincinnati Region lithofacies to arrange them along a hypothetical transect. That arrangement, which draws heavily on work by Cressman (1973) and data summarized by Pojeta (1979), is the basis for the summary profile in Fig. 7.3.

Near the deep end of the transect in Fig. 7.3 is the Logana Member of the mostly Middle Ordovician Lexington Limestone. At the shallow end is the Late Ordovician (Richmondian) Drakes Formation. The Logana is interbedded, brownish, petroliferous calcisiltite and brownish fissile shale, which Cressman (1973) postulates was deposited in a generally quiet, dysaerobic environment, probably below the compensation depth. The Logana contains lots of conodonts, and diversity is high. More than 95 percent of the specimens represent *Phragmodus undatus*, however; only a few are assignable to each of the other dozen species recognized.

The Drakes Formation, at the shallow end of the transect in Fig. 7.3, is sparsely glauconitic mudstone and fine-grained dolomite, with ripple marks, mud cracks and, in the upper part, several beds that are rich in colonial corals. The Drakes is a primarily quiet-water deposit that probably accumulated on broad, shallowly submerged flats. Bands of corals may have enclosed lagoons here and there on the Drakes flats, within which dolomites formed in water of somewhat elevated salinity (Hatfield, 1968). Dolomitic limestones in the Drakes are quite productive of conodonts, but at least half the specimens are customarily assignable to *Rhipidognathus symmetricus*, whereas the other half represent species of *Aphelognathus* and *Oulodus*. Diversity is very low in the Drakes, although single specimens of almost any species known from the Upper Ordovician of the Cincinnati Region may turn up here and there in isolated samples.

Between the bathymetric extremes marked in Fig. 7.3 by the Logana Member and Drakes Formation are two major groups of facies, which are similar in terms of the conodonts they contain but are mapped as differently named formations or members because they are composed of carbonates and shales in different proportions or because they develop particularly conspicuous internal structures.

In the late Middle Ordovician, before the arrival in the Cincinnati Region of much terrigenous detritus from the southeast, phosphatic limestones assigned to the Grier Member of the Lexington Limestone were the chief deposit in oxygenated, well-lit water somewhat, to considerably, shallower than that in which the Logana Member was deposited. Cressman (1973) estimates that the Grier accumulated in water less than 15 m deep, but the range may have been much greater than this for at the base the Grier grades both laterally and vertically into the deeper-, quieter-water Logana and, at the top into the Brannon or Tanglewood members, which formed in very shallow, more highly agitated water. Furthermore, the lower Grier shows a dominance by *Phragmodus undatus* nearly equal to that of the Logana, whereas *Plectodina* dominates in the upper Grier. The high phosphatic content of the Grier, a matter of special interest to Bourbon whiskey distillers, suggests that water above Grier depositional sites was supplied by upwelling with more phosphate than could be abstracted by plants. This adds to the suspicion that at least some of the carbonate rocks now included in the Grier were deposited below 15 m depth, perhaps not too far above the compensation depth.

Late in the Middle Ordovician considerable quantities of silt and clay spread into the Cincinnati Region from the southeast. Much of this bypassed or was washed through bathymetric highs to settle in thick beds in topographically lower portions of the seafloor (the elongate, southwestward-opening sag shown in Fig. 7.2). Thus the broad region characterized in the Middle Ordovician by carbonates of the Grier Member was partitioned in later Ordovician times into deeper-water areas in which the very shaly Kope and less shaly Point Pleasant formations accumulated, and somewhat shallower areas in which the much less shaly Clays Ferry and (later on) Fairview formations were deposited.

The Kope and Point Pleasant facies are both dominated by *Phragmodus undatus*, have only minor numbers of elements assignable to *Oulodus* or *Aphelognathus*, and never yield specimens of *Rhipidognathus*. In Clays Ferry and Fairview strata, on the other hand, *Phragmodus*, *Plectodina*, *Oulodus*, and *Aphelognathus* are almost equally represented, specimens are characteristically larger, and total collections are smaller. Furthermore, *Rhipidognathus* is represented by an occasional specimen in samples from the Clays Ferry (but not from the Fairview).

The Calloway Creek facies, entirely of Late Ordovician age, accumulated in probably somewhat deeper-water settings than the mudflat environments in which the Drakes Formation was deposited, and in quieter-water environments than those in which the conspicuously cross-bedded Tanglewood Member of the Lexington Limestone was deposited. In both the Calloway Creek and Tanglewood, *Oulodus* and *Aphelognathus* are the dominant conodonts; *Plectodina* is common, and *Phragmodus* is reduced to about 25 percent of the fauna. Specimens of *Rhipidognathus* may constitute as much as 5 percent of the collection from some units. The Grant Lake and Ashlock formations were deposited, in Late Ordovician time, at various sites slightly seaward of the Drakes flats; the Grant Lake represents wave-agitated shell-heap shoals, between and among which various facies of the Ashlock (or members of the Grant Lake) were deposited in sheltered lagoons. Grant Lake conodonts are mostly referable to *Plectodina*, *Oulodus*, and *Aphelognathus*, but specimens of *Rhipidognathus* are also common and, locally, may make up as much as 10 percent of a collection. *Rhipidognathus* may be as important in samples from the Ashlock as in samples from the very shallow-water Drakes, but the proportion of *Plectodina* is commonly greater. Otherwise, the two formations resemble one another lithologically and faunally.

It should also be noted that the Logana and Grier members of the Lexington Limestone, and the Late Ordovician Kope Formation, grade laterally into dark, fissile shales commonly logged by drillers as the "Utica," but is nowhere exposed at the surface in the Cincinnati Region. We have retrieved only a few conodonts from the Utica in the subsurface of the northwestern Cincinnati Region. These are all referable to *Amorphognathus*, a common component of a contingent of exotic species that is fairly abundantly represented in the Logana, the Kope, and deeper-water facies of the Grier Member of the Lexington but is largely unknown in shallower-water facies. This contingent, which includes *Icriodella*, *Rhodesognathus*, and *Periodon*, is characteristic of Ordovician rocks deposited at high southern latitudes. Schopf (1966) reported that a more diverse association of conodont species, including a substantial contribution by *Phragmodus undatus*, is represented in rocks of similar facies in the Upper Ordovician of New York.

To the information on Ordovician lithofacies and their conodonts summarized briefly in foregoing paragraphs I have added a certain amount of imagination to derive the conclusions on living distribution also indicated in Fig. 7.3. Because virtually all the species abundantly represented in Middle and Late Ordovician rocks of the Cincinnati Region were geographically widespread in tropical or subtropical environments, I have interpreted them in Fig. 7.3 as pelagic.

Although a pelagic interpretation is not required by the evidence for all the conodonts whose distribution is shown in Fig. 7.3, the common occurrence of specimens of *Amorphognathus* and its associates in black and nonblack shales strongly suggests a pelagic mode of life for the species they represent. The greater abundance of specimens of these species in rocks interpreted to have accumulated at the deeper-water sites in the Cincinnati Region (and elsewhere) further suggests that their distribution may have been defined primarily by that of a body of colder water, drawn into the Cincinnati Region as a bottom current from deeper-water sites to the west (now to the south). Upwelling of that cold bottom water might well have produced the phosphate now concentrated in various of the shallow-water facies marginal to the "sag" and might also have brought species of *Amorphognathus* and its associates in smaller numbers into shallower-water environments. However, if Cressman (1973) is right in concluding that the Logana Member (of the Lexington Limestone) formed on bottoms generally below the oxygen-minimum depth, it is likely that *Amorphognathus* and its associates lived in the cool water just above, not on the bottom itself.

The relative abundance of specimens representing *Phragmodus undatus* is maximum in rocks formed at what are interpreted as the deeper-water sites in the Cincinnati Region. The Logana Member of the Lexington, thought by Cressman (1973) to have accumulated below the oxygen-minimum depth, exhibits the highest relative abundance of *P. undatus* of all Cincinnati Region facies. Thus, in Fig. 7.3, I show *P. undatus* as the principal inhabitant of oxygenated, well-lit water above the colder bottom-water mass and "offshore" of the more shoal-water, high-energy, or tidal-flat areas dominated throughout the Middle and Late Ordovician by species of *Aphelognathus*, *Oulodus*, and *Rhipidognathus*.

Specimens of *Plectodina* (and the ozarkodinide "*Plectodina*") are components of collections from most Cincinnati Region lithofacies, but they diminish in relative importance in samples from rocks deposited at shallow nearshore sites in concert with an increase in relative abundance of specimens of closely related *Aphelognathus*. Thus, in Fig. 7.3 I show *Plectodina*, *Aphelognathus*, and *Oulodus* (an almost constant companion of the latter), as inhabitants of the same near-surface body of water, but with *Aphelognathus* and *Oulodus* progressively replacing *Plectodina* in an onshore direction. *Rhipidognathus*, as earlier noted, was certainly the dominant conodont in tidal-flat environments of the Cincinnati Region. I show it as pelagic in Fig. 7.3, but there is no compelling evidence for this at all.

Frequency and distributional data of the sort just summarized for Ordovician rocks of the Cincinnati Region suggest few profound conclusions as to factors other than water depth and energy that may have controlled distribution of living conodonts. To the west (north in the Ordovician) of the Cincinnati Region, species of *Panderodus* become relatively more important in the dominantly dolomitic Upper Ordovician strata, and *Aphelognathus* and *Oulodus* are the principal nonpanderodontide conodonts. *Rhipidognathus* is common in strata in

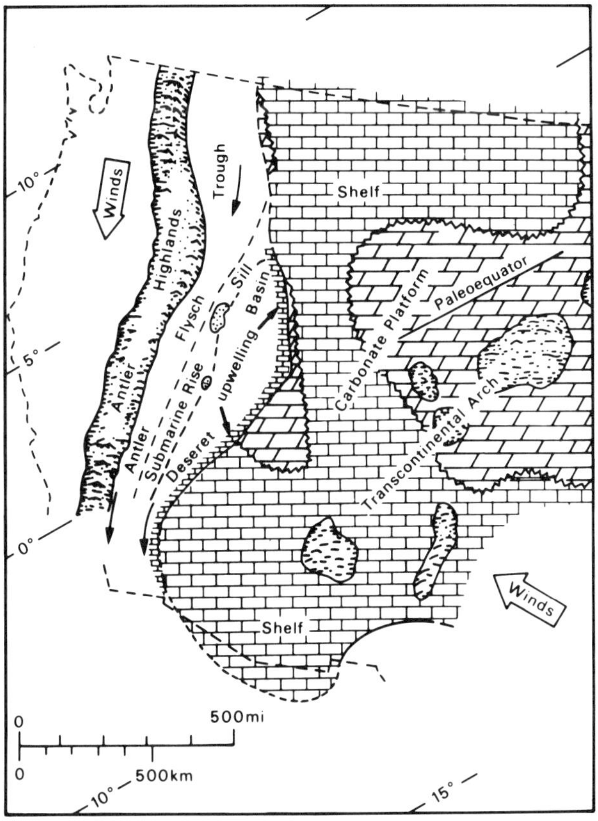

Fig. 7.4. Location of the Deseret Basin and adjacent features of the Mississippian of the western United States. Redrawn and simplified from Sandberg and Gutschick (1984).

northern Wyoming thought to have accumulated at sites marginal to those in which evaporites formed in the Williston Basin (Sweet, 1979). These features in the distribution of those conodonts square with the ones deduced from the record in the Cincinnati Region but offer little more as to the actual ecologic controls.

7.4.2 Mississippian paleoecology, western United States

Sandberg and Gutschick (1979, 1984) provide accounts of the distribution of conodonts in the Mississippian rocks of western Utah and adjacent Nevada that are rich in detail and of exceptional value in building up a general picture of conodont paleoecology. The rocks in question were deposited on and along the west edge of the broad carbonate platform of the continental interior, and in the Deseret Starved Basin, which was separated from the Antler Flysch Trough on the west by a narrow submarine rise, or sill. The general arrangement in the Deseret Basin and adjacent platform areas of Mississippian rocks of the Anchoralis-Latus Zone is shown in Fig. 7.4. From the distribution of conodonts in the several lithofacies shown in Fig. 7.5, Sandberg and Gutschick (1984) have also inferred the habitat of conodont species that represent the genera indicated in that figure.

Dark brown to black mudstones, phosphorites, and phosphatic shales are the deepest-water deposits recognized in the Mississippian transect summarized diagrammatically in Fig. 7.5. Sandberg and Gutschick (1984) conclude that these dark-colored rocks formed in cold, oxygen-deficient marine water at least 300 m deep along the axis of the Deseret Starved Basin.

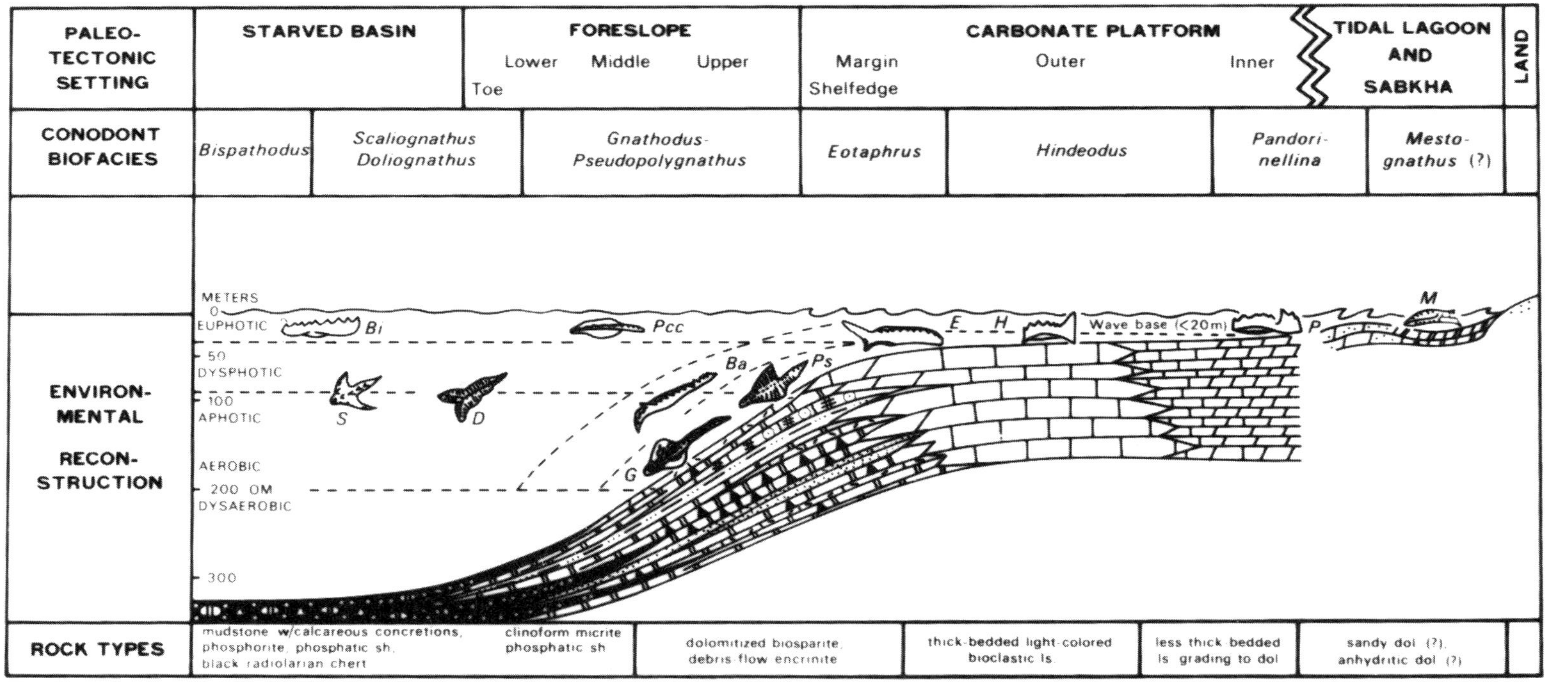

Fig. 7.5. Transect of Deseret Basin in mid-Mississippian time, showing living distribution of conodonts inferred from the distribution of conodont biofacies relative to distinctive lithofacies. Redrawn and simplified from Sandberg and Gutschick (1984).

The deep-water deposits intertongue eastward with thicker accumulations of dolomitized biosparite that include encrinites whose textures proclaim their formation as debris flows. These rocks, thought to have accumulated on the foreslope of the broad cratonic platform just east of the Deseret Basin, probably were deposited in the murky but well-aerated water of the aphotic zone; the encrinites, of course, are built of material that moved down the slope from the shelf edge or from sites higher on the slope itself.

East of the platform margin, rocks of the Anchoralis-Latus Zone are thick-bedded, light-colored, bioclastic limestones that are replaced farther eastward by more thinly bedded limestones. These, in turn, grade laterally into dolomites. Sandberg and Gutschick infer that the latter rocks should merge even farther to the east with sandy dolomites, perhaps with anhydrite or other evaporites. Those facies have not yet been sampled, however. In general, all these facies represent deposition in the relatively shallow-water environments of the carbonate platform. All those environments were certainly within the euphotic zone.

Representatives of *Bispathodus utahensis* Sandberg and Gutschick, 1984, and *Polygnathus communis* Branson and Mehl, 1934, have been recovered from all but the most nearshore facies. This near-ubiquity of occurrence suggests that both species were pelagic and inhabited the euryhaline environments of the euphotic zone.

Scaliognathus and *Doliognathus*, curiously specialized and biostratigraphically very useful prioniodinides of the family Bactrognathidae, are represented in the deepest-water deposits of the Deseret Basin as well as in the much shallower-water strata of the foreslope (Fig. 7.5). Such a distribution logically suggests that individuals of these genera were also pelagic, but lived at greater depths than did *Bispathodus utahensis* and *Polygnathus communis*. In short, species of these two genera were probably mesopelagic. Their depth zone may well have been defined by the well-oxygenated but still murky conditions of the dys- and aphotic zones.

Bactrognathus, a distinctive prioniodinide, is represented in samples taken from all parts of the reconstructed foreslope, including its presumably dysaerobic, deep-water toe. The absence of *Bactrognathus* specimens in axial deposits of the Deseret Basin, coupled with its occurrence in samples from both poorly and well-oxygenated slope environments, suggested to Sandberg and Gutschick (1984) that *Bactrognathus* was a swimmer, but that it confined its swimming to the aerated water just basinward of the slope floor. In other words, *Bactrognathus*, like its close relatives, *Scaliognathus* and *Doliognathus*, was mesopelagic, but its habitat may have hugged the bottom above the oxygen-minimum zone.

Gnathodus and *Pseudopolygnathus* are also represented by specimens from upper and middle slope deposits, but by very few from rocks that represent slope environments interpreted to have been below the oxygen-minimum level. Because of the latter fact, it seems unlikely that individuals of either genus ever strayed much above the bottom or out into water above the deepest parts of the basin. In short, species of *Gnathodus* and *Pseudopolygnathus* may well have been nektobenthic, with habitats that "bottomed out" basinward against the oxygen-minimum layer.

Eotaphrus is confined to rocks that formed high on the slope and in adjacent parts of the shelf edge; *Hindeodus* is commonly represented in strata deposited on the outer shelf; *Pandorinellina* appears to have characterized the inner shelf, in environments above wave base; and *Mestognathus*, a close relative of both *Pandorinellina* and the cavusgnathids, may well have been adapted to hypersaline lagoonal environments along the innermost margin of the platform. There is little in the reported distribution of conodonts assignable to any of these genera that provides a definitive answer to their mode of life. Their absence in slope or deep-basin deposits suggests they may not have been pelagic, in the manner of *Bispathodus* and *Polygnathus*. Possibly they were also nektobenthic.

It is always risky to generalize from observations made in limited areas, no matter how sophisticated the studies or convincing the conclusions. However, it may be noted that *Pandorinellina* is the acknowledged ancestor of the Cavusgnathidae and that *Cavusgnathus* (and

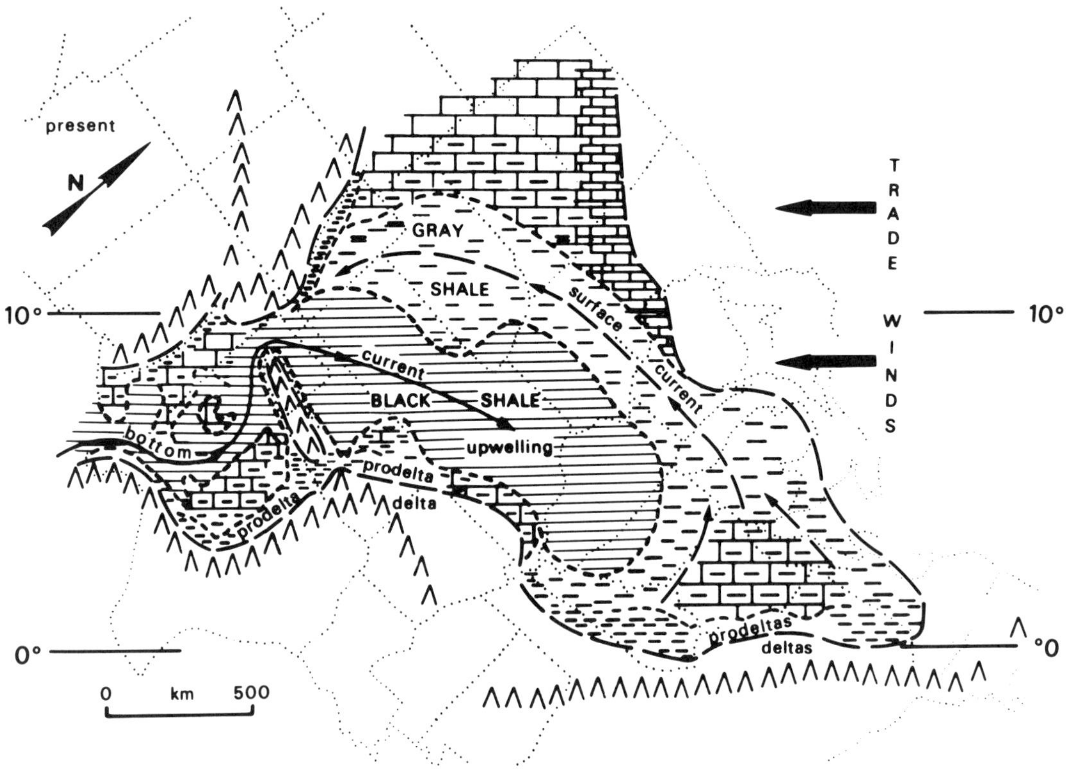

Fig. 7.6. Inferred environments of deposition of cyclic Pennsylvanian sedimentary deposits in the U.S. Midcontinent. Modified and redrawn from Heckel (1980).

its dominantly Pennsylvanian relative, *Adetognathus*) are elsewhere the dominant conodonts in rocks that represent very shallow-water, nearshore environments. In Pennsylvanian rocks, as will be noted in Section 7.4.3, *Hindeodus* seems to characterize environments that were farther offshore, and probably of more normal-marine character, than the inshore areas favored by *Cavusgnathus.* Thus there are parallels in other Mississippian and Pennsylvanian biofacies for several of the platform biofacies identified by Sandberg and Gutschick (1984) in the Mississippian of the western United States.

7.4.3 *Paleoecology of Pennsylvanian conodonts*

Conodonts are common fossils in Pennsylvanian rocks; their generic-level taxonomy is now reasonably well known, and their distribution defines a number of distinctive biofacies, which are identified, characterized, and have been named in important papers by Merrill and various coauthors (summarized in Merrill and Von Bitter, 1984). Factors that governed the formation and spatial differentiation of these biofacies in Pennsylvanian seas are outlined either directly or inferentially in comprehensive, well-documented reports by Heckel (1977, 1980), Heckel and Baesemann (1975), and Swade (1985). It would be misleading to suggest that there is general agreement on all matters respecting the ecology of Pennsylvanian conodonts, since there are clearly basic differences of opinion on certain matters. Not all these differences are as great as occasional stridency in the tone of the argument would suggest, however, and a combination of information from the two sets of reports provides valuable insight into the habits and habitats of Pennsylvanian conodonts and probably also of their ancestors and descendants.

Pennsylvanian rocks of midcontinental North America accumulated in a chain of foreland basins (Fig. 7.6) that were situated in an elongate area just north of the equator. This area was bordered on the north and northeast by the relatively stable cratonic interior and on what were then the south and northwest sides by active ranges of folded or fault-block mountains. The southern margin and the southeastern end of the foreland-basin chain received great quantities of terrigenous detritus from streams draining the Appalachian and Ouachita-Marathon foldbelts. This spread out to form a complex of laterally shifting alluvial plains and coalescing deltaic systems, which were only occasionally (and then briefly) transgressed by offshore marine environments. Farther west and north, the basin chain was more persistently occupied by the sea, sediments are of finer grain, and the record of conodonts and other groups of fossils is more nearly continuous.

In parts of the foreland-basin chain that were most persistently occupied by the sea, the Pennsylvanian rock record is made up of a succession of *cyclothems*. These are repeated sets of distinctive rock types, which record a number of transgressive-regressive cycles that are commonly regarded as the results of eustatic rise and fall of sea level. Heckel and Baesemann (1975) and Swade (1985) have related the distribution and frequency of conodonts to the lithic components of a number of the Pennsylvanian cyclothems in Iowa and Kansas, with the general results summarized in Fig. 7.7.

Note in Fig. 7.7 that sandy "outside shales" at the base of a typical Kansas-Iowa cyclothem initially record deposition in nonmarine environments but, in their upper parts, pick up a few fossils that indicate shallow, nearshore marine environments. The first conodonts to appear in these rocks represent *Adetognathus* and *Ellisonia*, but others referable to *Hindeodus*, *Aethotaxis*, and the *Idiognathodus-Streptognathodus* plexus follow shortly. In the "middle limestone," just above the "outside shale" in a typical cyclothem, the invertebrate fauna becomes more diverse and is much better represented. *Adetognathus* may continue to be a prominent member of the conodont fauna, but diversity commonly is greater than in shaly rocks below and dominance begins to be shared with conodonts of the *Idiognathodus-Streptognathodus* plexus.

In the typical Kansas-Iowa cyclothem, the middle limestone is followed by a meter or so of abundantly fossiliferous gray-brown shale, which is typically (but not invariably) introduced by (or includes) a distinctive component of fissile black, phosphatic shale. In this "core-shale" interval, conodont diversity may drop off, but frequency of occurrence commonly reaches a maximum and preservation is generally optimum. Representatives of *Gondolella* are almost completely restricted in their occurrence in Kansas and Iowa to the core-shale interval, and specimens of the prionidinide *Idioprioniodus* are more numerous in this facies than in any of the others in the Kansas-Iowa cyclothem.

In the basal part of the "upper limestone," which succeeds the core shale of a Kansas-Iowa cyclothem, maximum conodont diversity is reestablished and specimens of the *Idiognathodus-Streptognathodus* plexus continue to dominate in frequency. *Adetognathus*, represented by few, if any, specimens in the lower part of upper-limestone intervals, typically becomes more abundantly represented upward in those intervals and may regain a dominance in the uppermost part that carries over into the lower, marine part of the superjacent outside-shale unit.

As indicated in Fig. 7.7, the typical Kansas-Iowa Pennsylvanian cyclothem is viewed by most geologists as the record of a transgressive-regressive event. Thus, from the relationship between conodonts and the lithofacies that represent various stages in that transgressive-regressive history, it should be possible to reconstruct a general ecologic model. Such a model has been developed by Heckel and Baesemann (1975) and Swade (1985) and is shown diagrammatically in Fig. 7.8.

Heckel (1977) concludes that the phosphatic black shale component of the core-shale interval accumulated in anoxic bottom water beneath a thermocline and thus represents the deepest-water environment recorded by any of the lithofacies in a cyclothem of Kansas-Iowa type. From this conclusion it follows that conodonts commonly represented in core black shales were not only pelagic but were also segregated in depth zones in life. That is, *Gondolella*, *Idioprioniodus*, *Neognathodus*, and various members of the *Idiognathodus-Strep-*

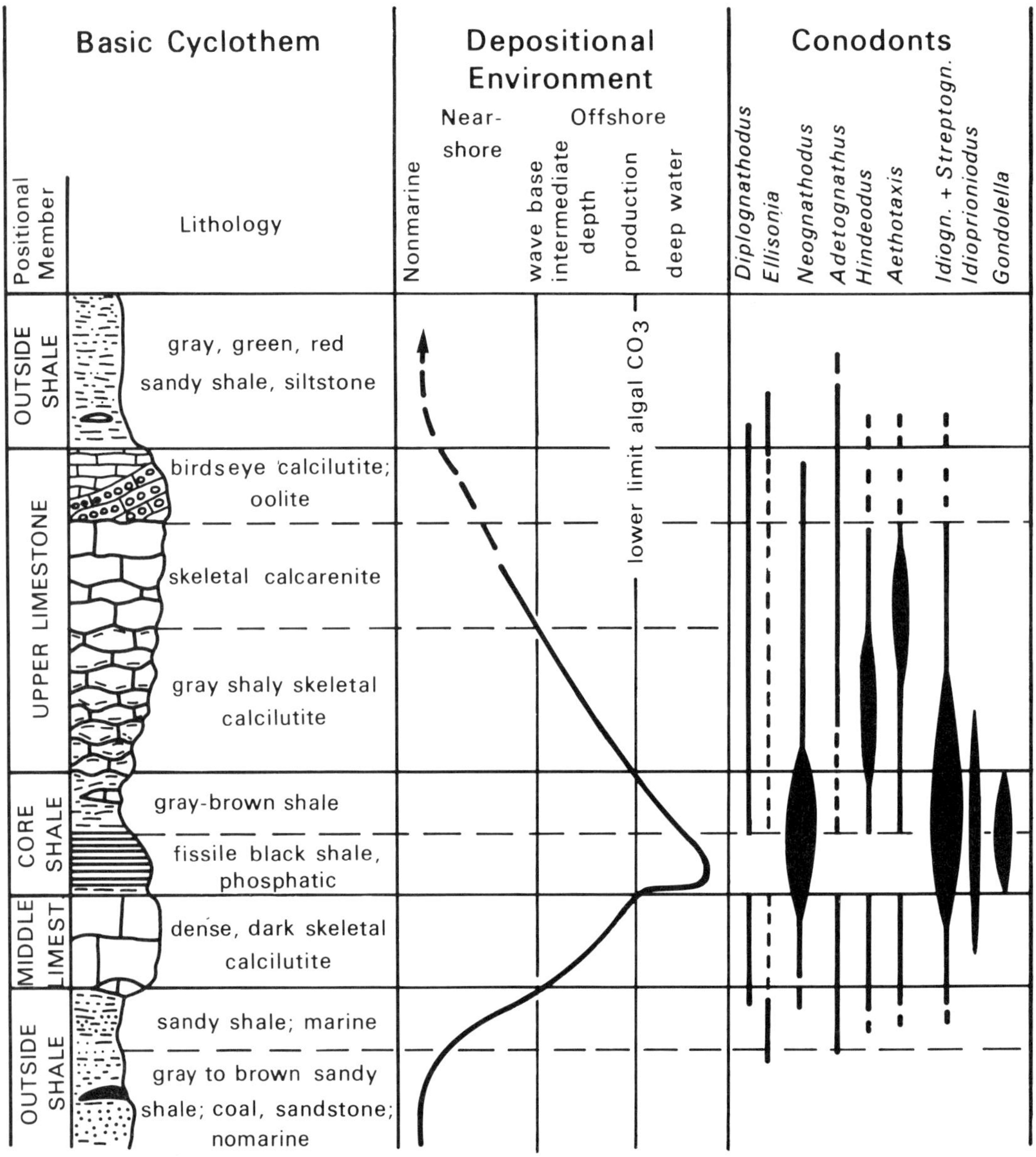

Fig. 7.7. An individual cyclothem of Kansas-Iowa type, with inferred environments of deposition of members and general distribution of principal types of conodonts. Modified and redrawn from Heckel (1977).

tognathodus plexus, which are common components of the core black-shale fauna, must have lived somewhere in the aerated water masses above the bottom; but *Gondolella*, which is restricted in many of its North American occurrences to the core-shale interval, must have been confined in life to a deeper part of the water column than the others. If, as is suggested in Fig. 7.8, anoxic conditions on the bottom were created by development and maintenance of a thermocline; and if the areal distribution of *Gondolella* is confined to the black- and gray-shale components of the core-shale interval, it follows that *Gondolella*'s depth zone may have been defined by the cold, dysaerobic water of the thermocline. Depth *per se* may not have been a factor that controlled *Gondolella*'s distribution, but it was probably the most readily decipherable of the several parameters that describe distribution of the cold, dense, oxygen-poor water of the thermocline, to which *Gondolella* was best adapted and in which *Idioprioniodus* also thrived. As Heckel (1977) noted, that water may also have been rich in settling organic matter, phosphate, and heavy metals, and these may also have played a part in defining *Gondolella*'s depth-zone habitat.

Adetognathus and *Ellisonia* were certainly at the other end of the ecologic spectrum in Pennsylvanian seas. The common occurrence of specimens of *Adetognathus*, for example, in rocks that represent obviously nearshore or marginal marine environments (like the outside shale of Fig. 7.7), in dolomitic strata that intertongue with evaporite sequences, and also (in greatly diminished abundance) as components of higher-diversity conodont faunas that include forms associated more commonly with normal marine conditions suggests that *Adetognathus* was euryhaline and adapted to life in well-oxygenated nearshore waters that were susceptible to wide variations in salinity. Merrill recognized the *Adetognathus* biofacies in 1962 and, in subsequent reports (Merrill, 1968; von Bitter, 1972), it has been suggested that ratios between specimens of *Adetognathus* and *Idiognathodus* or *Streptognathodus*, which dominate environments more offshore, might provide a quantitative sense of distance from shore and a measure of relative salinity.

Environments intermediate between the shallow, well-oxygenated nearshore waters with highly variable salinity and the much deeper, colder, more dysaerobic waters of the thermocline seem to have supported the most diverse conodont biotas in the Pennsylvanian. Throughout the Pennsylvanian, dominant conodonts in these intermediate environments were members of the *Idiognathodus-Streptognathodus* plexus, and rocks that accumulated in them have been assigned to a ubiquitous *Idiognathodus-Streptognathodus* biofacies (Merrill and von Bitter, 1984). But these intermediate environments were also the ones in which the closely related anchignathodontids

Fig. 7.8. Hypothetical cross sections of Pennsylvanian sea at maximum extent in the U.S. Midcontinent. Upper diagram relates sedimentary facies to inferred quasi-estuarine circulation pattern; lower diagram shows inferred living distribution of conodonts as reconstructed from their occurrences and frequencies in various sedimentary facies. Redrawn and slightly modified from Swade (1985).

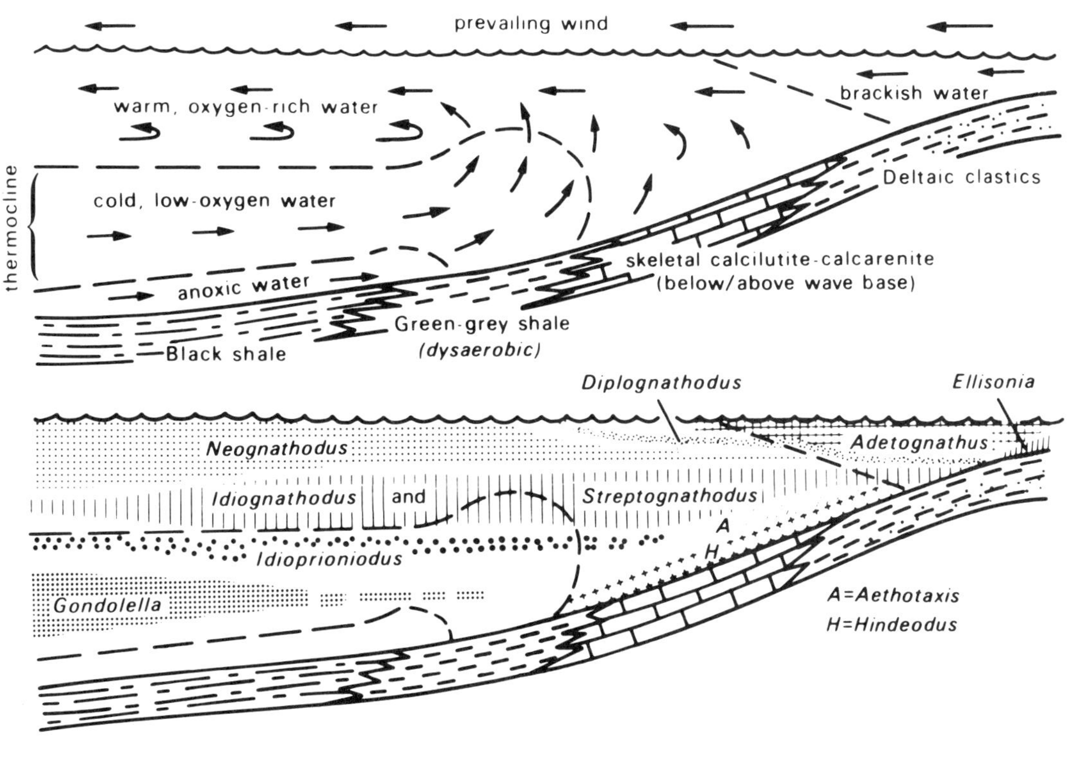

Hindeodus and *Aethotaxis* enjoyed their widest distribution and achieved their greatest frequency. In honor of the latter, Merrill and various coauthors (summarized by Merrill and von Bitter, 1984) have dubbed the biofacies that developed in this environment the *Aethotaxis* biofacies. Swade (1985), however, notes that *Hindeodus* is represented in more basinward lithofacies than *Aethotaxis* and also occurs in thin limestones interbedded with shales. Thus, the distribution of *Aethotaxis* and *Hindeodus* might be useful in dividing the intermediate, well-oxygenated, carbonate-substrate environment into clear- and turbid-water areas, the former providing the optimum environment for *Aethotaxis*, the latter for *Hindeodus*.

The principal point of contention between the two groups of writers who have considered the ecology of Pennsylvanian conodonts involves interpretation of the core shale of typical cyclothems. In the preceding paragraphs, I have followed the interpretation of Heckel (1977), which I find to be supported by the widest variety of facts. However, there are other interpretations. For example, a decade ago Merrill and Martin (1976) adopted a model in which a black shale component of the La Salle Member in northwest Illinois was interpreted as the transgressive record of a nearshore, flotant marsh, and thus of an environment very different in character than the deep-water, anoxic one visualized by Heckel.

If the flotant-marsh interpretation is extended to include all Pennsylvanian black shales, as Merrill and various coauthors apparently urge, then conclusions drawn from the Heckel-Baesemann and Swade models are either universally wrong or are applicable in the Pennsylvanian to only very limited parts of the system of foreland basins in the cratonic interior of North America. Heckel (1977), however, has pointed out that black shale may have accumulated under different conditions in Pennsylvanian basins and that all such rocks need not have the same environmental significance. He finds it difficult to interpret the phosphatic black shales of the Kansas-Iowa cyclothems as the record of shallow-water deposition in flotant marshes and suggests that black shales that record such conditions should lack sedimentary phosphate and commonly grade laterally into sands and conglomerates indicative of other nearshore or marginal environments. Even if phosphatic black shales overlie coals, they need not be interpreted as shallow-water, flotant-marsh deposits. Heckel notes, for example, that rapid transgression of coal-swamp sites cut off from terrigenous influx by distributary abandonment might have produced a low-pH or dysaerobic environment unfavorable to carbonate production, so that the coal-forming swamp could have foundered without further deposition. Subsequent deepening might have established the offshore, anoxic environment in which only fine-grained matter and phosphate-rich organic matter settled out, and this would be recorded by superposition of a phosphatic black shale directly on a coal. Acknowledging that such situations exist and that there is reasonable explanation for them seems better to me than insisting that all Pennsylvanian black shales be interpreted in the same terms.

7.5 Ecologic Generalizations

From the three studies summarized in Section 7.4 we learn that reasonable inferences as to the living distribution of conodonts can be made from careful studies of the relationship between the fossils and the sedimentary rocks in which they occur. This, of course, was implied in both the depth-stratification and lateral-segregation models, as proposed by Seddon and Sweet (1971) and Barnes and Fåhraeus (1975), respectively. However, we also learn that logical interpretations of living distribution require a combination of features of both models, together with the conclusion that the mode of life of conodonts may have varied from truly pelagic to nektobenthic. On the other hand, we have seen that it is rarely possible to identify the physical, chemical, or biological factors that actually controlled the distributions inferred. It is clear, for example, that *Mestognathus* and the cavusgnathids were, for millions of years, successful inhabitants of shallow, nearshore marine environments that were probably characterized by elevated temperatures and wide fluctuations in salinity, but it is not clear from the record what factors of those harsh environments controlled the distributions observed.

7.5.1 *Depth as a factor*

Many of us who have written about conodont paleoecology have discussed the distribution of various species in terms of water depth, not because depth *per se* is a significant ecologic factor, but because a sense of at least relative depth may be derived from the textures, structures, and other fossils in the sedimentary rocks from which our conodont collections have been made. Although depth itself may not have exerted a primary influence on conodont distribution, factors such as temperature, light penetration, light intensity, turbidity, energy, salinity, and water density fluctuate directly or inversely with depth, and one or a combination of these may have exerted the direct control on distribution. I suspect, for example, that temperature, not depth, was the important factor governing distribution of species of *Amorphognathus* and related balognathids. That is, those species are represented in high-latitude Ordovician rocks that formed in cool, relatively shallow water, whereas at low-latitude sites their remains are largely restricted to the deposits of deeper waters in which temperatures may have been comparable to those nearer the surface at high latitudes.

7.5.2 *Temperature as a factor*

Differences in temperature were probably responsible for development, in the Ordovician, of distinctive conodont faunas in the high- and low-latitude marine realms. This conclusion is suggested by the fact that rare invasions into higher latitudes of species characteristic of low latitudes were accompanied by accumulation of vaughnitic carbonate rocks that indicate concurrent invasions of warmer than usual water. Also, as noted in Section 7.5.1, the few typically high-latitude species that migrated from time to time in the Ordovician into the low-latitude realm became established there in deeper-water environments that were probably similar thermally to the high-latitude ones in which the species were most widely distributed.

As noted in Section 7.4.3, the distribution of *Idioprioniodus* and *Gondolella* in Pennsylvanian cyclothems in Kansas and Iowa suggests that species of those genera were confined in life to a relatively deep part of the water column and also that their depth zone may have been defined by the cold, dense dysaerobic water of the thermocline thought to be responsible for anoxic conditions on the bottom. Unfortunately, from study of the skeletal elements of cold-water Ordovician and Pennsylvanian species, I have been unable to isolate morphologic features by which other cold-water forms might be identified.

7.5.3 *Nearshore and offshore faunas*

All three of the models summarized in section 7.4 relate the distribution of conodonts to onshore-offshore transects reconstructed from the distribution and nature of the sedimentary-rock record. In all three models there is reasonably clear distinction between an offshore group of species that may be related to stable temperatures, "normal" marine salinities, and limited turbidity; and another group that apparently thrived in nearshore environments characterized presumably by harsher, more varied environmental conditions such as great fluctuations in temperature, salinity, and turbidity. It is of interest that conodonts typical of the offshore and nearshore ends of the environmental spectrum in all three models have a number of features in common.

In all three of the situations described in Section 7.4, species characteristic of nearshore environments built apparatuses of relatively large elements, with a reduced number of discrete denticles of circular cross section and little, if any, white matter. I also have the impression that elements of nearshore species are more variable in morphologic detail than those of offshore species, and their apparatuses are more variable in composition. Furthermore, note that Ordovician and Triassic rocks representing nearshore environments have produced all of the "fibrous" elements on which Branson and Mehl (1944) based their concept of the suborder Neurodontiformes. Finally, it is clear that at any time species diversity was considerably lower in nearshore than in offshore environments and, because it is my experience that there are far fewer elements in a kilogram of nearshore rock than in a comparable weight of offshore rock, I judge that conodont populations may also have been relatively small in nearshore environments.

In the Ordovician, Mississippian and Pennsylvanian situations described in section 7.4, conodonts that were evidently adapted to offshore environments formed relatively small elements with considerable white matter. Ramiform elements tend to have delicate processes set with many compressed, laterally confluent denticles, and the variety of platformed pectiniform elements seems to have been much greater than in nearshore environments. It is also my experience that species diversity was higher in offshore than in nearshore environments and that populations were larger.

Although the qualities of conodonts characteristic of nearshore and offshore environments seem to have persisted, at least in a general way, through the Paleozoic and Triassic, it would be foolhardy to treat any of these characters as invariable indices to such environments. As an example, I merely recall that elements built by species of the prioniodinide *Idioprioniodus* are stout, largely hyaline, and characterized by short processes that bear just a few discrete denticles of circular cross section. This combination of features would surely suggest a nearshore, shallow-water environment, but we have seen in Section 7.4.3 that *Idioprioniodus* was an associate of *Gondolella* in what have been interpreted as relatively deep-water environments that were probably well offshore.

7.5.4 Phyletic changes in ecological distribution

In the three examples summarized in Section 7.4 only genera were considered. Lest this convey the impression that the distribution of all species of every conodont genus has the same ecologic significance, I should cite several examples in which this is not the case.

In Section 7.4.1, I related the distribution of *Phragmodus* in Middle and Upper Ordovician rocks of the Cincinnati Region to deeper parts of the basin of deposition. The oldest known species of *Phragmodus*, however, were clearly members of nearshore faunas. They formed stout skeletal elements with rounded denticles and limited amounts of white matter. Their descendants, in the later Middle Ordovician, apparently "went to sea" and built apparatuses characterized by offshore features. Development of *Plectodina* involves a similar history. Early Middle Ordovician species are represented in nearshore associations, and the elements they formed have many nearshore features. Later species, including the primitive ozarkodinides referred to "*Plectodina*," clearly were denizens of more offshore environments. Thus, it would surely be dangerous to use either *Phragmodus* or *Plectodina* elsewhere in the Ordovician as invariable guides to the environments indicated by their Late Ordovician distribution.

In short, it is a reasonable conclusion from information such as that just mentioned that, for safety, the game of conodont paleoecology must be played on the species level. To be sure, stocks such as the Icriodontidae—or even the Prioniodinida—seem to have had a general affinity through time for nearshore, shallow-water environments. However, even at this level of generalization, one must keep in mind that by Mississippian time, the Prioniodinida had developed a group of offshore explorers (the Bactrognathidae), and from mid-Pennsylvanian time on, offshore and nearshore branches of this venerable order were firmly established in the Gondolellidae and Ellisoniidae, respectively.

7.6. Paleobiogeography

In the introduction to this chapter I noted that as early as 1958 and 1959 studies of Triassic and Ordovician conodonts, respectively, had led different investigators to conclude that during at least certain intervals in their history the geographic distribution of conodonts may have been distinctly provincial. These suggestions, which were met intially with considerable skepticism, have subsequently become well-established parts of conodont lore, although it has also been shown that at other times in their history, conodonts were remarkably cosmopolitan in their geographic distribution.

The subject of paleobiogeography is currently of great interest in geological circles because of the light it may shed on the position of major crustal plates in the past. For this reason, a number of detailed studies of conodont paleobiogeography have been published in the last few years. Most of these have been brought up to date fairly recently (e.g., Clark, ed., 1984), so I provide only a brief summary here.

7.6.1 Late Cambrian and Ordovician paleobiogeography

Miller (1984) notes that both classes of conodonts appear to have originated in the Late Cambrian in warm, low- to midlatitude seas. By earliest Ordovician time, however, representatives of both groups had spread into higher-latitude seas, and *Cordylodus* (a cavidont) had become essentially cosmopolitan. A pair of major eustatic events in this Late Cambrian-earliest Ordovician interval are thought to have been influential in directing the development of the Conodonta so that, by late in the Tremadocian, very different faunas had evolved in low- and higher-latitude seas.

The faunal differentiation outlined by Miller continued through the Ordovician, when the marine realm was clearly divided into two major biogeographic units (Sweet and Bergström, 1974, 1984). I now prefer to regard these major units as warm- and cold-water realms, although they were originally described as the North Atlantic and North American Midcontinent provinces. Descendants of the pioneer lineage of the Conodonti, the *Teridontus* stock, figured prominently in faunas of both the warm- and cold-water realms and, as Bergström and I pointed out in 1974, there was a not-surprising parallelism in development. Only a few cosmopolitan species existed, mostly representatives of *Drepanoistodus* and *Panderodus*, but also including the cavidonts *Dapsilodus*, *Walliserodus*, and *Ansella*.

By the Late Ordovician, and possibly before, distribution of conodonts had come to define a number of biogeographic provinces in the warm- and cold-water realms. For example, a combination of R- and Q-mode cluster analysis of data from 26 sites in North America (Sweet and Bergström, 1984) permitted discrimination of an equator-straddling *Red River Province* and a somewhat higher-latitude *Ohio Valley Province* in the Late Ordovician of the warm-water realm. Lateral differences in contemporaneous conodont faunas also permitted recognition of *British*, *Baltoscandic*, and *Mediterranean* provinces in the cold-water marine realm, in which conodonts may have lived in subpolar seas well above 60 degrees south of the equator.

Late in the Ordovician, as noted in Section 6.3.2, conodont species diversity declined dramatically, and this seems to have led, by the end of that period, to virtual elimination of the fauna characteristic of the cold-water realm. Stocks that survived into the Silurian (i.e., the Icriodellidae, Prioniodinidae, Spathognathodontidae, and Panderodontidae) are either ones that were well established in low- to midlatitude seas of the Ordovician or are surviving relatives (e.g., Distomodontidae) of ones that were.

7.6.2 Later Paleozoic and Triassic paleobiogeography

Reported Silurian conodont faunas are surprisingly cosmopolitan. However, 46 new species are described from the Lower and Middle Silurian of New South Wales in a recent report by Bischoff (1986), which may suggest that Silurian cosmopolitanism may be partly a result of faulty knowledge of Silurian conodont faunas. It is also of interest to note that Silurian conodonts have been collected largely from rocks thought to have been deposited equatorward of the 40th parallels (Charpentier, 1984).

Klapper and Johnson (1980), whose comprehensive review is authoritative, confirm that conodonts appear to have been restricted in the Early and Middle Devonian to more or less tropical areas. Early in the Devonian, endemism was relatively high, but later in the period, with increased marine transgression of cratonic areas, endemism decreased and the number of cosmopolitan species increased. Although Klapper and Johnson noted that most conodont genera were uniformly distributed, they logged a number of differences, at the species level, between the conodont faunas of various epeiric seas of the Early and Middle Devonian. It was their choice not to formalize the endemic units as provinces because such a procedure would have amounted to "... no more than naming each separate Devonian epeiric sea."

Conodonts continued to be confined through the Late Devonian to a low-latitude belt bounded north and south by the 40th parallels. The almost universal applicability of the biozonal scheme devised by Ziegler (1962) in Upper Devonian rocks deposited in offshore, normal marine environments attests to the wide geographic distribution of Late Devonian species and the absence of well-marked provincialism. Charpentier's (1984) analyses are marred by reliance on a data base composed of form species, but they seem to confirm a lack of provincialism in the Late Devonian.

The distribution of Carboniferous conodonts shows essentially no provincialism at the generic level, and so little attention has been given to the multielement taxonomy of most Carboniferous species that distinctions suggested in the literature are difficult to evaluate biogeographically. Charpentier's (1984) analysis of the distribution of Mississippian and Pennsylvanian form species reached the logical conclusion that, with Late Carboniferous closure of the Rheic Ocean, different faunas may have developed on opposite sides of Pangaea. However, he noted that data were so few as to make any conclusions suspect. Interestingly, von Bitter, Sandberg, and Orchard (1986) note a surprisingly cosmopolitan distribution for the Mississippian genus *Mestognathus*, despite the fact that it was apparently adapted to the harshest and most restricted of inshore environments.

There is too little information on the occurrence of Permian conodonts to judge if their distribution was provincial. Nicoll (1976) reported that, despite their abundance in the Carboniferous and Triassic, conodonts were absent from the Permian of Australia. He attributed this to climatic change caused by extensive glaciation in the Late Carboniferous and Permian. Charpentier's (1984) analysis, although based on few data, suggests provincial differences between faunas on the east and west coasts of Pangaea. No formal provinces have been established or characterized, however.

Huckriede's 1958 report on Triassic conodonts was the first to recognize the possibility of provinciality in their distribution and one of the first to suggest provincialism as a feature in the distribution of conodonts of any age. Subsequent authors have recognized a Tethyan Province on the east side of Pangaea, a Germanic (or Muschelkalk) Province in central Europe, and a North American Province on the west side of Pangaea. Charpentier's (1984) analysis, however, suggests that endemism diminished in the Early Triassic and that Middle and Late Triassic seas were occupied by increasingly cosmopolitan conodont faunas. Because provincial faunas of the Triassic have been diagnosed primarily at the species level, broad identifications (or misidentifications) may have played a considerable role in supporting either a concept of provincialism or one of increasing cosmopolitanism. Until there is greater agreement than exists today on the taxonomy of Triassic conodonts, I doubt that further arguments about biogeographic distributions will be worthwhile.

References

Barnes, C. R., and Fåhraeus, L. E. (1975). Provinces, communities, and the proposed nektobenthic habit of Ordovician conodontophorids. *Lethaia* **8**, 133–149.

Bischoff, G. C. O. (1986). Early and Middle Silurian conodonts from midwestern New South Wales. *Courier Forschungsinst. Senckenberg* **89**, 1–137.

Borella, P. E., and Osborne, R. H. (1978). Late Middle and early Late Ordovician history of the Cincinnati Arch Province, central Kentucky to central Tennessee. *Geol. Soc. Am. Bull.* **89**(10), 1559–1573.

Branson, E. B., and Mehl, M. G. (1934). Conodont studies no. 4. *Univ. Missouri Studies* **8**, 265–349.

——— (1944). Conodonts. Pp. 235–246 in *Index Fossils of North America* (ed. H. W. Shimer and R. R. Shrock). Wiley, New York.

Charpentier, R. R. (1984). Conodonts through time and space: Studies in conodont provincialism. *Geol. Soc. Am. Memoir* **196**, 11–32.

Clark, D. L., ed. (1984). Conodont biofacies and provincialism. *Geol. Soc. Am. Memoir* **196**, 1–340.

Cressman, E. R. (1973). Lithostratigraphy and depositional environments of the Lexington Limestone (Ordovician) of central Kentucky. *U. S. Geol. Survey Prof. Paper 768*, 1–59.

Hatfield, C. B. (1968). Stratigraphy and paleoecology of the Saluda Formation (Cincinnatian) in Indiana, Ohio, and Kentucky. *Geol. Soc. America Spec. Paper 95*, 1–34.

Heckel, P. H. (1977). Origin of phosphatic black shale facies in Pennsylvanian cyclothems of mid-continent North America. *Am. Assoc. Petrol. Geol. Bull.* **61**(7), 1045–1068.

——— (1980). Paleogeography of eustatic model for deposition of mid-continent Upper Pennsylvanian cyclothems. Pp.197–215 in *Paleozoic paleogeography of the west-central United States* (ed. T. D. Fouch and E. R. Magathan). Rocky Mountain Sec., Soc. Econ. Paleont. Mineral.

Heckel, P. H., and Baesemann, J. F. (1975). Environmental interpretation of conodont distribution in Upper Pennsylvanian (Missourian) megacyclothems in eastern Kansas. *Am. Assoc. Petrol. Geol. Bull.* **59**, 486–509.

Huckriede, R. (1958). Die Conodonten der Mediterranen Trias und ihr stratigraphischer Wert. *Paläont. Z.* **32**, 141–175.

Klapper, G., and Johnson, J. G. (1980). Endemism and dispersal of Devonian conodonts. *J. Paleont.* **54**, 400–455.

Kozur, H. (1976). Paleoecology of Triassic conodonts and its bearing on multielement taxonomy. *Geol. Assoc. Canada Spec. Paper 15*, 313–324.

Merrill, G. K. (1968). Allegheny (Pennsylvanian) conodonts. Unpubl. Ph.D. thesis, Louisiana State Univ., Baton Rouge, 184 pp.

Merrill, G. K., and Martin, M. D. (1976). Environmental control of conodont distribution in the Bond and Matoon formations (Pennsylvanian, Missourian), northern Illinois. *Geol. Assoc. Canada Spec. Paper 15*, 243–271.

Merrill, G. K., and von Bitter, P. H. (1984). Facies and frequencies among Pennsylvanian conodonts: Apparatuses and abundances. *Geol. Soc. Am. Spec. Paper 196*, 251–261.

Miller, J. F. (1984). Cambrian and earliest Ordovician conodont evolution, biofacies, and provincialism. *Geol. Soc. Am. Spec. Paper 196*, 43–68.

Nicoll, R. S. (1976). The effect of Late Carboniferous-Early Permian glaciation on the distribution of conodonts in Australia. *Geol. Assoc. Canada Spec. Paper 15*, 273–278.

Pojeta, J., Jr. (1979). The Ordovician paleontology of Kentucky and nearby states—Introduction. *U. S. Geol. Surv. Prof. Paper 1066*, A1-A48.

Rhodes, F. H. T. (1952). A classification of Pennsylvanian conodont assemblages. *J. Paleont.* **26**, 886–901.

Sandberg, C. A., and Gutschick, R. C. (1979). Guide to conodont biostratigraphy of Upper Devonian and Mississippian rocks along the Wasatch Front and Cordilleran hingeline. *Brigham Young Univ. Geol. Studies* **26**, 107–133.

——— (1984). Distribution, microfauna, and source-rock potential of Mississippian Delle Phosphatic Member of Woodman Formation and equivalents, Utah and adjacent states. Pp. 135–178 in *Hydrocarbon source rocks of the Greater Rocky Mountain Region* (ed. J. Woodward, F. F. Meissner, and J. L. Clayton). Rocky Mountain Assoc. Geologists, Denver.

Schopf, T. J. M. (1966). Conodonts of the Trenton Group (Ordovician) in New York, southern Ontario and Quebec. *N. Y. State Mus. Sci. Serv. Bull.* **405**, 1–105.

Seddon, G., and Sweet, W. C. (1971). An ecologic model for conodonts. *J. Paleont.* **45**, 869–880.

Swade, J. W. (1985). Conodont distribution, paleoecology, and preliminary biostratigraphy of the upper Cherokee and Marmaton groups (upper Desmoinesian, middle Pennsylvanian) from two cores in south-central Iowa. *Iowa Geol. Survey, Tech. Information Ser.* **14**, 1–71.

Sweet, W. C. (1979). Conodonts and conodont biostratigraphy of post-Tyrone Ordovician rocks of the Cincinnati Region. *U. S. Geol. Surv. Prof. Paper 1066-G*, G1–G26.

——— (1979). Late Ordovician conodonts and biostratigraphy of the western Midcontinent Province. *Brigham Young Univ. Geol. Studies* **26**, 45–85.

Sweet, W. C., and Bergström, S. M. (1974). Provincialism exhibited by Ordovician conodont faunas. *Soc. Econ. Paleont. Mineral. Spec. Publ.* **21**, 189–202.

——— (1984). Conodont provinces and biofacies of the Late Ordovician. *Geol. Soc. Am. Spec. Paper 196*, 69–87.

Sweet, W. C., Turco, C. A., Warner, E., Jr., and Wilkie, L. C. (1959). The American Upper Ordovician Standard. I. Eden conodonts from the Cincinnati Region of Ohio and Kentucky. *J. Paleont.* **33**, 1029–1068.

von Bitter, P. H. (1972). Environmental control of conodont distributon in the Shawnee Group (Upper Pennsylvanian) of eastern Kansas. *Univ. Kansas Paleont. Contrib.* **59**, 1–105.

von Bitter, P. H., Sandberg, C. A., and Orchard, M. J. (1986). Phylogeny, speciation, and palaeoecology of the Early Carboniferous (Mississippian) conodont genus *Mestognathus*. *Roy. Ontario Mus., Life Sci. Contrib.* **143**, 1–115.

Ziegler, W. (1962). Taxionomie und Phylogenie oberdevonischer Conodonten und ihre stratigraphische Bedeutung. *Abh. Hess. Landesamtes Bodenforsch.* **38**, 1–166.

8. THE PHYLUM CONODONTA

8.1 A Personal Bias

By now you are familiar with my bias concerning the biologic affinities of the conodonts. That bias, signaled by the title of this chapter, is to treat the group as a taxonomic entity of the highest rank, a phylum. However, it would be unfair to leave the impression that everyone shares my bias or that the conodonts have never been allied with, considered part of, or assigned to some other biologic group. The mission of this final chapter is thus to review the evidence for and evaluate previous opinions as to the biologic affinities of conodonts, and thereby provide some insight not only into my reasons for regarding them as an independent and long-extinct phylum, but also into those emphasized by others, whose conclusions on biologic affinities differ from mine.

8.2 Summary of Conodont Characters

Mark Twain might have been writing about opinions on the biologic affinities of conodonts when he wondered: Where else but in science might such a wholesale return of speculation result from such a trifling investment of fact? Speculation is fun, and it is an important component of scientific methodology, but it is always important to keep track of its factual basis. And, as has been emphasized repeatedly in previous parts of this book, the sorts of facts about conodonts that are useful in evaluating speculations (or hypotheses) about their biologic affinities are few in number and mostly of fairly recent discovery.

The four Scottish Namurian specimens described by Briggs et al. (1983) and Aldridge et al. (1986) and a single Silurian *Panderodus* specimen from Wisconsin (Mikulic et al., 1985a, 1985b; Smith et al., 1987) provide the only direct evidence we have as to the shape, size, and soft-body anatomy of conodonts. I have provided a summary of that evidence in Chapter 3.

Direct evidence as to the composition of the skeletal apparatus of some conodonts is provided by more than 500 natural assemblages and fused clusters; and evidence on the shape and internal structure of mineralized elements thought to represent as many as 4000 species is available in literally millions of discrete specimens in extant collections. All the rest is really speculation or, if you will, interpretation.

As noted in Chapter 3, the four Scottish specimens, although preserved in slightly different ways and thought to represent at least two different genera, are all compressions in which originally organic material is preserved as a brownish crystalline substance with overlying patches of pale blue mineral. These compressed and altered specimens are the remains of elongate, wormlike, dominantly soft-bodied organisms, 40 to 42 mm long and 1.2 to 1.8 mm wide. One end of one specimen, interpreted as a cephalic lobe, is slightly expanded, bilobed, and includes an essentially complete natural assemblage of conodont elements, which are oriented with their long axes transverse to that of the specimen as a whole. The opposite end, interpreted as the tail and shown more or less clearly in two specimens, tapers to a bluntly rounded point and bears marginal sets of short, posteriorly directed, fringelike impressions interpreted to be the rays of fins surrounding the periphery of the tail.

A pair of dark spots in the posterior part of the cephalic lobe of one Scottish specimen may represent eyes, and a series of posteriorly opening, V-shaped impressions characterize what are interpreted to be the sides of at least the posterior parts of all four specimens. Although Briggs et al. (1983) were appropriately cautious in comparing these latter features with the chevron-shaped myotomes or muscle blocks of the lancet, *Branchiostoma*, and fish, Aldridge et al. (1986) describe these features as "somites" without further justification and this evidently figured prominently in their ultimate determination that the affinities of the conodonts are with the chordates.

Also indicated with varying degrees of fidelity on all four Scottish specimens is a pair of subparallel longitudinal lines, which extend from a point (in one specimen) not far behind the cephalic lobe to a point near the posterior end of the tail. Aldridge et al. (1986) point out that these lines might represent ". . . a septum dividing the body somites longitudinally, a mesentery separating compartments of the body cavity, a major longitudinal blood vessel, a stiffening rod or notochord, a nerve chord, the gut." They note that the gut is the one of these structures most likely to be preserved; however, they also note that if the pairs of longitudinal lines do represent the outlines of the gut, the fact that they extend almost to the posterior end of the tail in two specimens suggests that the animal lacked a postanal segment, or true tail.

As noted in Chapter 2, the internal structure of conodont elements indicates they are composed of two parts, crown and base, which must have been completely enveloped by (or embedded in) secretory tissue at least during times of growth. Apatite crystallites in lamellae of crowns are oriented with their prism surfaces parallel to the direction of growth. Bases are less densely mineralized, and the framework of organic material may be thicker. In a few specimens, the basal portion may include a plug of bonelike material within which Barskov, Moskalenko, and Starostina (1982) have identified structures interpreted as osteoblasts.

Among the other characters that have been considered in speculations on the biologic affinities of conodonts, I should also mention their geographic and stratigraphic distribution and the cosmopolitan occurrence of many species in rocks representing a wide variety of sedimentary facies. Conodonts are common indigenous fossils in rocks that contain the remains of animals such as brachiopods and echinoderms, which are exclusively marine at present and apparently always have been. They have been found only as reworked or redeposited specimens in rocks that might be interpreted as nonmarine. The first true conodonts are from Upper Cambrian rocks, the last from strata of latest Triassic age. Conodonts have been recovered from virtually all types of sedimentary rocks, and records of conspecific specimens from rocks representing many different depositional envionments are common. Especially significant (to me, at least) is the fact that most conodont genera are represented somewhere in the world in black shale deposits, which commonly lack remains of benthic organisms but in many places include well-preserved specimens of pelagic ones.

8.3 Conodonts as Invertebrates

In the 130 years since their first formal recognition, the conodonts have been referred at one time or another and with various degrees of seriousness to most of the major phyla of invertebrate animals. The hypotheses represented by many of these referrals are easily dismissed, but some others are not so readily discounted and should at least be mentioned in this opinion poll.

8.3.1 Arthropod and Annelid connections

Even before Pander's original diagnosis of conodonts was published in 1856, Murchison, Barrande, and Carpenter had registered the opinion that the tiny fossils were the tip-ends of some parts of the trilobite carapace, and thus were arthropod, not chordate, fragments. And, not long after Pander's monograph appeared, Owen (1860) suggested that conodonts might be analogous with the ". . . spines, hooklets, or denticles, of naked molluscs or Annelids."

Although no one seems to have followed up on the interpretation of conodont elements as parts of a trilobite carapace, two authors (Harley, 1861; Simpson, in Newberry, 1875) suggested that conodont elements might be crustacean remains. These suggestions have not proved particularly appealing, largely, I suspect, because the variety of element types of most natural assemblages is not duplicated in the array of spines that rim crustacean carapaces.

Owen's suggestion that conodonts might be a group of the Annelida received serious attention from Zittel and Rohon (1886) and influenced the thinking of many other paleontologists (e.g., Scott, 1934; DuBois, 1943; Rhodes, 1954) until the middle 1950s, when hypotheses of annelid affinities seem to have been abandoned. Indeed, through the years, the "annelid hypothesis of conodont affinities" has been the primary rival of those favoring the chordates.

Rhodes (1954) lists and comments on the views of Zittel and Rohon, Scott, and DuBois that conodonts are annelids. His summary includes the following points.

- The form and structure of conodont elements are similar to those of annelid jaw components, and conodont elements lack the pulp cavity typical of vertebrate teeth.
- The general arrangement, number, and variation in form of the component elements in natural assemblages resemble those in fossil and recent assemblages known to represent jaw apparatuses of polychaete annelids.
- Paired elements of these assemblages seem to represent lateral pairs and not the upper and lower forms that characterize most vertebrates.
- The lack of wear on most conodont elements suggests that they did not function as true teeth. The "grasping jaw components" of worms, however, rarely show signs of wear.
- Some polychaete annelids secrete calcium phosphate in the form of tubes. If the buccal cavity and pharynx of these worms form by invagination of the embryonic ectoderm and are thus stomodael in origin, it seems possible that annelid jaws are physiologic homologs of the external cuticle, or tube. It thus seems less difficult than formerly to postulate an annelid with jaws made largely of calcium phosphate instead of chitin.

Although Rhodes (1954) concluded his discussion of the arguments just listed with the view that collectively they offer strong evidence in favor of an annelid affinity for conodonts, he also pointed out (as others have done) that a major objection is that living annelids are apparently unable to secrete calcium phosphate internally. A more telling objection was presented by Müller (in Clark et al., 1981), who showed that scolecodonts, the fossil jaws of polychaete annelids, differ fundamentally from conodonts in internal structure.

8.3.2 Molluscan connections

In the same place that he suggested the possibility that conodont elements might represent an annelid group, Owen (1860) also raised the possiblity that they might be the radular teeth of some group of mollusks. This suggestion has not appealed to many paleontologists, but it was supported by Morse (in Newberry, 1875), James (1884), and Loomis (1936), all of whom based their conclusions on gross morphologic comparison of the radular elements of living gastropods and certain coniform conodont elements.

Rhodes (1954), in countering the argument that conodont elements might be the radular teeth of gastropods, notes that most types of conodont elements lack morphologic counterparts among the radular teeth of living gastropods and that conodonts are not found in many deposits that contain abundant gastropod shells. He also notes that conodont elements are phosphatic, whereas the radular teeth of snails and other mollusks are entirely organic ("chitinous"), and that radular teeth are almost invariably arranged in symmetrical rows and that they commonly show considerable evidence of wear, whereas conodont elements do not. Müller (in Clark et al., 1981) added what was then the knockout punch, suggested by Yochelson (1975), that most living mollusks are not equipped physiologically to secrete very much phosphate and hence are unlikely to represent the major group to which conodonts are most closely related.

Very recently, Tillier and Cuif (1986) have revived the hypothesis that conodonts might be mollusks by suggesting that many features of the Scottish Carboniferous specimens invite comparison with those of various caudofoveate aplacophorans. In suggesting such a comparison, Tillier and Cuif cite the fact that living aplacophorans are about the same size as the Scottish Carboniferous conodonts; that in either the larvae or adults of some Aplacophora, the oral shield is divided into two symmetrical parts; and that the pedal groove of living Solenogastres and the midventral suture of certain caudofoveates have positions reminiscent of the axial lines of the Scottish conodont specimens. They also suggest that the chevron-shaped impressions, which are such striking features of the Scottish specimens, might correspond either to the dorso-ventral muscle series of the Solenogastres or to the chevronlike arrangement of dermal spicules that characterizes the dorsal side of other aplacophoran mollusks. According to Tillier and Cuif, features interpreted as fin rays in the tail region of two Scottish specimens might also be duplicated in certain prochaetodermatids by very long spicules in the posterior region.

Tillier and Cuif further suggest that the cephalic apparatus of the Scottish conodonts may be analogous in arrangement and even in mineralogy to the radular teeth and, particularly, the buccal mandibles of the Aplacophora. They note, as I have in a previous paragraph, that the principal reason for rejecting a molluscan affinity for conodonts is the argument that mollusks are incapable of secreting phosphatic hard parts. However, Tillier and Cuif have subjected the radular teeth and buccal mandibles of *Chevroderma turnerae* (a prochaetodermatid caudofoveate) to both X-ray microdiffraction and microprobe analysis and have discovered that both teeth and mandibles include substantial amounts of calcium phosphate, as is probably also the case with the radular teeth of at least certain scaphopods. (Gastropod radulae studied thus far seem to lack phosphate, and the external spicules of the Aplacophora are aragonitic.)

The features just summarized led Tillier and Cuif to comment that if conodonts were aplacophoran mollusks, it would be normal for the phosphatic buccal mandibles, at least, to be fossilized even though the aragonitic dermal spicules might recrystallize and be unrecognizable as conodont elements. They therefore regard the Conodonta as a potential class of the Mollusca, most closely related to one or another of the groups included in the paraphyletic Aplacophora.

8.3.3 Connections with other invertebrates

Müller (in Clark et al., 1981) lists a large number of additional invertebrate groups that have been considered appropriate, for one reason or another, as taxonomic receptacles for the conodonts. In addition to the arthropods, annelids, and mollusks, which have already been discussed, these include the lophophorata (or tentaculata) and such disparate groups of wormlike animals as various aschelminthes, gnathostomulids, and chaetognaths. In fact, about the only major invertebrate phyla that have not been suggested as the parental group for the conodonts are the Archaeocyathida (which were largely or entirely extinct before conodonts appeared), and the Porifera, Bryozoa, Echinodermata, and Hemichordata.

Lindström (1973) and Conway-Morris (1976) have either directly or indirectly suggested that conodonts might be an extinct group of the invertebrate superphylum termed Lophophorata or Tentaculata by various authors and composed at present of the phyla Brachiopoda, Bryozoa, and Phoronida. Conway-Morris's suggestion is based on his interpretation of a curious little fossil from the famous Middle Cambrian Burgess Shale of British Columbia as a vagrant lophophorate, and on his interpretation of impressions in the supposed lophophore of that organism as conodont elements. Unfortunately, the conical elements supposed to have supported lobes of the lophophore of the creature Conway-Morris named *Odontogriphus omalius* have all been dissolved away, and the fossil is from rocks far older than those yielding the oldest specimens I would regard as conodonts. Thus, in addition to questioning the real significance of a vagrant, or swimming, lophophorate, I disqualify *Odontogriphus* as a serious contributor to our understanding of conodont affinities.

Lindström (1973), on the other hand, pointed out features that suggest conodont elements were internal (not external) structures; noted a general tendency in the evolution of conodonts toward an increase in surface area of elements; and suggested that pits in the upper surfaces of certain conodont elements might have been the sites of muscle attachment. These observations led him to suggest that conodont elements might have served as internal supports for a muscle-operated tentacular apparatus, or lophophore, and this led to the further suggestion that

> ancestors of the conodont stock and those of the brachiopod stock were closely related at some relatively late

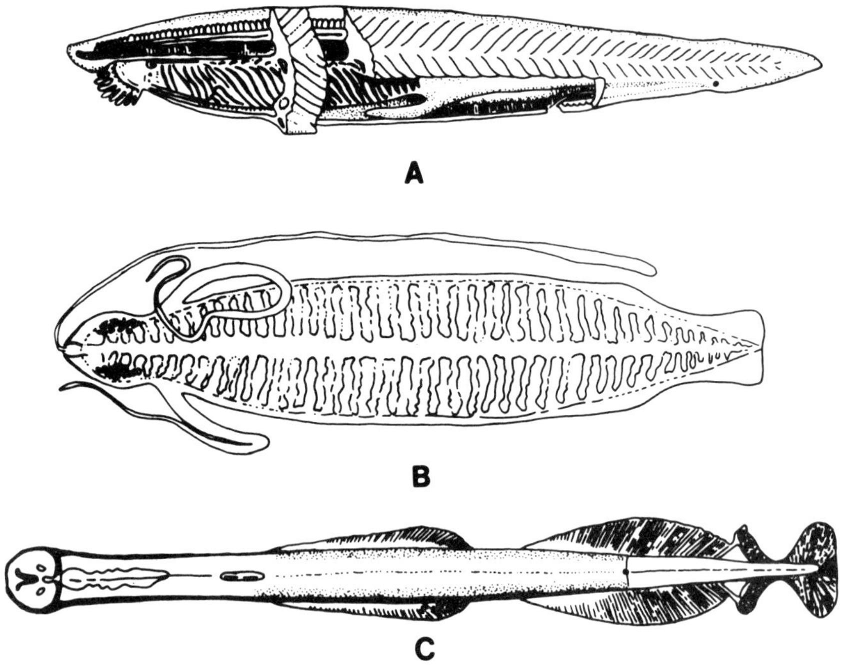

Fig. 8.1. Representatives of several groups that have been considered as repositories for the Conodonta. (A) *Branchiostoma* (or "Amphioxus"), a living acraniate (Chordata); (B) *Nectonemertes*, a pelagic nemertine (Rhynchocoela); and (C) *Sagitta*, a pelagic arrow worm (Chaetognatha). (A) Redrawn from various sources; (B) and (C) redrawn from Hyman (1951, 1959).

stage of the Precambrian, and that the fundamental differences evolved because the brachiopods, after a planktonic larval stage, became sessile, whereas most conodonts remained planktonic through life.

Lindström carefully took into account all features of conodonts known in 1973, but would, I suspect, be the first to comment that the hypothetical model he deduced from those features needs to be squared with features of the Scottish Namurian impressions, which were not described until 1983. Those specimens lack any feature that I would interpret as a lophophore and include others (e.g., repeated, chevron-shaped structures; ray-supported fins), which, if correctly interpreted (e.g., as myotomes and caudal fins, respectively) would indicate a group of animals very different in organization than any known lophophorate. However, it will be noted later in this chapter that interpretation of some features of the Scottish specimens is also rather "iffy," so Lindström's reconstruction cannot be too summarily rejected.

Several authors have also suggested that conodonts might be related to living arrow worms of the phylum Chaetognatha, and this zoologically isolated group briefly became the category of choice when Szaniawski (1982) demonstrated a striking similarity in structure between certain late Cambrian and early Ordovician coniform elements (now assigned to *Phakelodus*) and the grasping spines of *Sagitta* (Fig. 8.1C), a living chaetognath. I was particularly attracted to the possibility that the conodonts might turn out to be Paleozoic and early Mesozoic relatives of the Chaetognatha because in 1971 George Seddon and I had used the habits and distribution of that group in building a model to explain certain anomalies we perceived in the distribution of Ordovician and Devonian conodonts. And, with the appearance of Szaniawski's elegant study, many concluded that we had finally discovered the conodonts' roots.

There are numerous problems, however, with assigning the conodonts to the Chaetognatha, despite attractive similarities in distribution and near identity in internal structure of

Phakelodus and *Sagitta* elements. I am not certai: at any of these problems is fatal, but they .ate caution. For example, the grasping .ies of all living chaetognaths are entirely organic in composition, whereas all specimens that I regard as skeletal elements of conodonts are thoroughly phosphatized. Furthermore, although Bengtson (1976) has outlined the steps by which the transition from organisms with elements like those of *Phakelodus* to those with elements with the internal structure of all known conodonts might have been made, that transition has not yet been objectively documented. Thus it is my perhaps ultraconservative conclusion that *Phakelodus* and other organisms with skeletal elements of similar internal structure (the Paraconodontida of the *Treatise* classification) are not conodonts. If that is the case, then identity in structure of *Phakelodus* and *Sagitta* elements has no bearing on the question of conodont affinities, although it may suggest that roots of the Chaetognatha are to be found in the Paraconodontida, a group formerly assigned to the Conodonta on the basis of gross morphology of skeletal elements.

8.4 Conodonts as Chordates

Only a few students of conodonts have taken a firm stand regarding the biologic affinities of condonts, and most of them favor an assignment in or near the phylum Chordata. It is thus of interest to look critically at the evidence on which that assignment has been based.

8.4.1 The opinions of Pander

Pander (1856), who coined the name and provided the first diagnoses and descriptions of conodonts, included those diagnoses and descriptions in a monograph that dealt primarily with fossil fishes. Throughout that pioneer monograph, Pander treated conodonts as the hard parts, possibly the teeth and jaws, of a group of extinct fishes, primarily, it seems to me because that is what they looked like to him. Pander did not seriously consider assigning the conodonts to any group other than the Chordata, but he noted that Murchison, Barrande, and Carpenter had examined some specimens and concluded that they were "... fragments of the hard, crustaceous ends of the segments of some trilobite." In apparent deference to that opinion, which he did not attempt to refute, Pander remarked somewhat wistfully that (my translation) "... Complete conviction that these remains are really teeth of extinct fishes could be reached only if similar structures could be demonstrated in living animals of the same class."

8.4.2 Newberry, Hinde, Huxley and Myxine

Newberry (1875), on his own account, and Hinde (1879), on the advice of Huxley, concluded apparently from shape and luster alone that the conodont elements they examined were so closely similar to the teeth of the hagfish, *Myxine* (Fig. 8.2), as to be indistinguishable. In fact, Huxley is reported to have in-

Fig. 8.2. *Myxine glutinosa*, a hagfish (or slime eel), represents the chordate division to which many believe conodonts to be most closely related. Redrawn from Müller (1836).

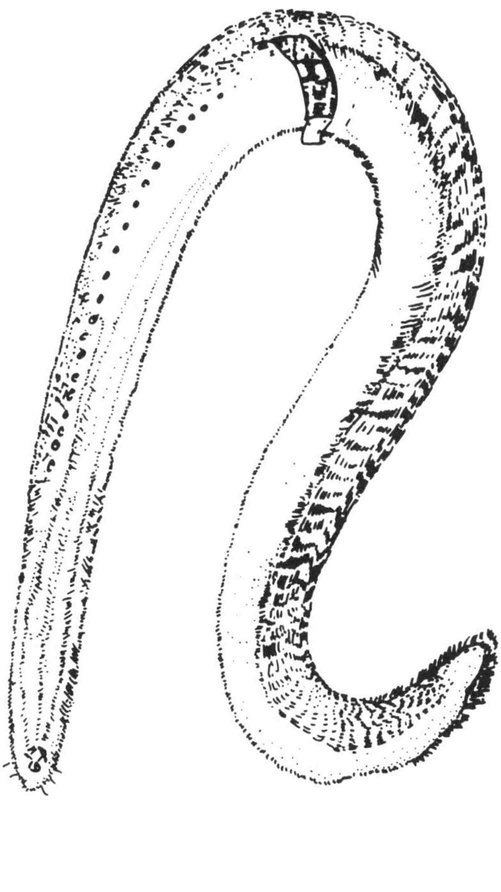

formed Hinde that "... it would be difficult to prove that they did not belong to fishes of this order..." It is probably not necessary to point out, however, that such a statement can hardly substitute for objective demonstration that conodont elements and the horny "teeth" of hagfishes are homologous. Internally, at least, they are quite different in structure (Fig. 8.4A and 8.4B); and, of course, the elements of conodonts are phosphatic, whereas those of hagfish are entirely organic. Also, Pander described internal structures in conodont elements that were clearly unlike those of hagfishes (*Myxine*) or any other fish with which he was familiar. Would that Huxley had been so careful!

8.4.3 *Macfarlane and the nemertinean connection*

Macfarlane (1923), in a curious book intended to show that the vertebrates evolved from nemertinean worms (Rhynchocoela) (Fig. 8.1B) in freshwater environments, grouped the conodonts together in an order Conodontes, which was assigned to a more comprehensive category (the Malacodermata) to which he also relegated the hagfishes (order Myxinoides) and lampreys (order Petromyzontes). Macfarlane supported his conclusion that conodonts and cyclostomes (i.e., hagfishes and lampreys) were closely related by comparing crude line drawings of conodont elements traced from papers by Hinde and Newberry with similarly generalized drawings of hagfish and lamprey teeth. Nothing was mentioned about internal structure or mode of growth, and the only difference noted between conodonts and cyclostome teeth was that the former are calcareous, whereas the latter are chitinous. Confidence in Macfarlane's assessment of conodont affinities is not increased by noting that he also asserted that the Devonian black shales from which conodonts had been collected in New York and Ohio were nonmarine. In short, Macfarlane's most memorable contribution may be the formal establishment of an order-group category for conodonts, the first suprageneric group proposed for these animals. Note, however, that even though the factual basis for Macfarlane's conclusions on conodont affinity is either questionable or erroneous, those conclusions are precisely the ones reached by a rather different route in one of the most recent papers on conodont affinity (e.g., Aldridge et al., 1986).

8.4.4 *The belief of Ulrich and Bassler*

Ulrich and Bassler (1926) also expressed the "belief" that conodonts are the teeth and dermal plates of primitive fishes, not necessarily all of the same group. It is clear, however, that they reached this conclusion primarily (if not entirely) because of a gross similarity in external shape among conodont elements, the teeth and dermal plates of sharks, and the teeth of the hagfish, *Myxine glutinosa*. I can find no evidence that they took any account of the substantial differences in internal structure described and pondered over in detail by Pander.

8.4.5 *Discovery and interpretation of basal bony material*

Kirk (1929) reported that conodont elements occur with the well-known ostracoderms in the Middle Ordovician Harding Sandstone of central Colorado. He also noted that some of the conodont elements were attached at the base to a substance that appeared similar in color and luster to the associated ostracoderm bone. These observations led Kirk to the opinion that some conodont elements might have been mouth parts of ostracoderms, but he was unwilling to extend this conclusion to include all conodonts. However, many years later, Barskov, Moskalenko, and Starostina (1982) described spongy masses in the basal plates of several elements assigned to *Coleodus*, a genus that is also represented abundantly in the Harding Sandstone. These masses, together with structures that appear to correspond in size and morphology to osteoblasts, osteocytes, and osteoclasts, suggested to Barskov and his coauthors that the basal plates were bone and that the specimens were vertebrate remains.

Branson and Mehl (1933) concurred in the interpretation of conodont elements as the teeth of primitive fishes and claimed to have conclusive evidence, which was never presented. They did note, however, as Kirk had done a few years earlier (and as I did many years later), that many of the elements they collected from the Middle Ordovician Harding Sandstone of central Colorado were attached at the base to a substance that ". . . appears bony but does not have the structure of ordinary bone." Branson and Mehl's (1933) comment that ". . . the materials to which the conodonts are attached could not be from annelids or from anything but vertebrates . . ." takes on considerable importance because it was the conclusion of a pair of trained vertebrate paleontologists; but it can hardly be accepted as evidence of affinity in the absence of further description, illustration, or histologic analysis. Possibly they were referring to material such as that described from Siberia by Barskov, Moskalenko, and Starostina (1982), but this will always be uncertain.

8.4.6 Composition and growth of elements

Ellison (1944), and Hass and Lindberg (1948) cleared up long-standing confusion about the chemical composition of conodont elements and pointed out that the phosphatic minerals they identified were the same as those in fossil and recent bones and teeth. However, Gross (1954), a student of fossil fishes, made the important determination that conodont elements are not composed of dentine and lack features that could be interpreted as either a pulpa or dentine channels. Furthermore, Gross prepared thin sections of the Pa elements of *Ozarkodina murchisoni* (Pander) and determined from the arrangement of their lamellae that the elements must have grown by accretion of new material to their outer surfaces, not by addition of new lamellae on the inner surface of the element, as Pander (1856) had concluded was the case. Gross was thus able to assert with some confidence that conodont elements were not the teeth or dermal scales of vertebrates, nor were they components of the vertebrate endoskeleton, which should show evidence of ossification about a core of spongy, cartilagenous tissue. Gross concluded from his studies that conodonts were primarily soft-bodied chordates assignable to a distinct branch of the Agnatha. It is ironic, I suppose, that his detailed studies led Gross to conclusions very similar to those advanced by Huxley, Macfarlane, and Ulrich and Bassler on little evidence other than comparative gross morphology.

8.4.7 Interpretations of the Scottish Carboniferous specimens

The most recent discussions of conodont affinities (Briggs et al., 1983; Aldridge et al., 1986) are concerned primarily with appropriate interpretation of the four Scottish Carboniferous specimens, to which I have already made repeated reference. Briggs et al. compared the first-discovered and best-preserved Scottish specimen with the cephalochordate *Branchiostoma* (Fig. 8.1A), the ammocoete larva of lampreys, and arrow worms of the invertebrate phylum Chaetognatha (Fig. 8.1C). Their conclusion was that neither a chordate nor a chaetognath model ". . . provides a satisfactory answer to the question of conodont affinity."

Following description of three additional specimens in 1986, however, Aldridge and his colleagues concluded that conodonts represent a group of jawless, craniate chordates separate from previously described groups but perhaps most closely related to the hagfishes (Myxinoidea). A similar conclusion is expressed, but just in passing, by Jefferies (1986), who notes that the hypothetical animal "s" he proposes as the first "crown vertebrate" resembles the Scottish Carboniferous conodont animal described by Briggs et al. (1983). However, in the nearly three pages Jefferies devotes to listing the characters of this hypothetical creature, I find only six that might be more or less objectively represented in (or by) the Scottish specimens (marine habitat, eel-shaped body, head-trunk-tail, rasping teeth, ?lensless paired eyes, ?somites). I suggest that, as Jefferies intimates, the resemblance between hypothetical animal "s" and the Scottish conodonts may be so striking because the latter were used as a general model for the former.

In reaching their conclusions on conodont affinities, Aldridge et al. (1986) and Jefferies (1986) clearly identify with Pander, Huxley, Newberry, Macfarlane, and Ulrich and Bassler, although their reasoning is somewhat different. Because the Scottish specimens lack any evidence of jaws or a bony skeleton, it is the opinion of Aldridge and his coauthors that

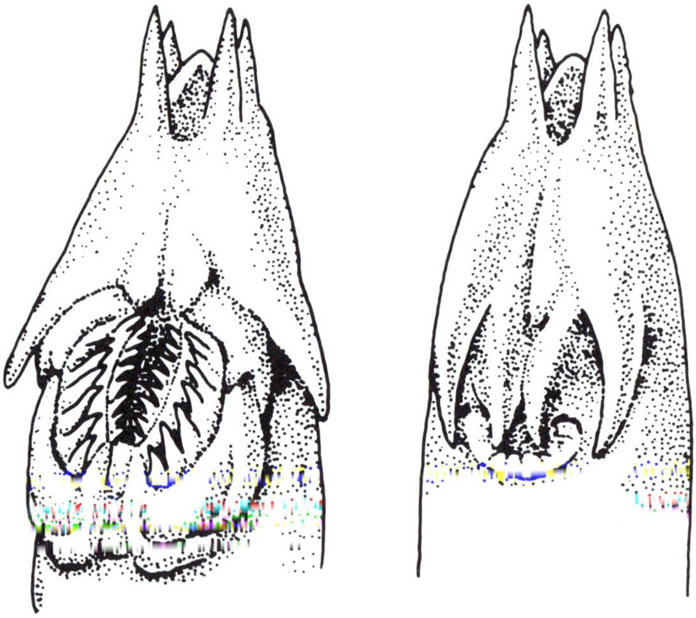

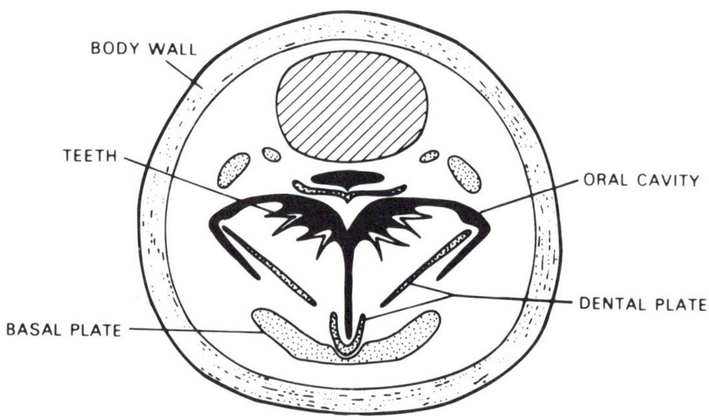
BODY WALL
TEETH
ORAL CAVITY
DENTAL PLATE
BASAL PLATE

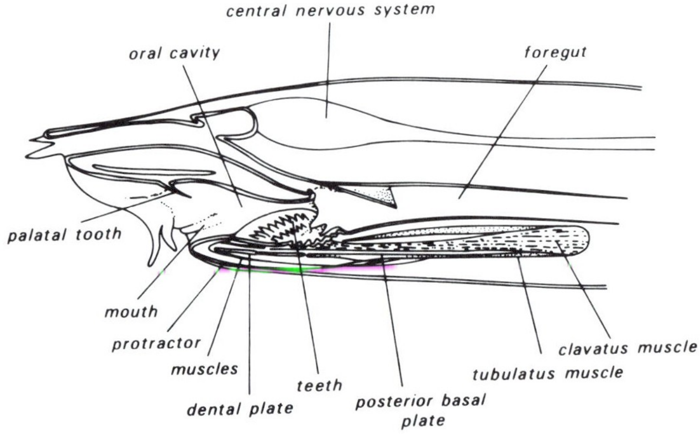
central nervous system
oral cavity
foregut
palatal tooth
mouth
protractor
muscles
dental plate
teeth
posterior basal
plate
tubulatus muscle
clavatus muscle

> the chordate feature to which the conodont apparatus shows the greatest similarity is the lingual structure of hagfishes . . . in which the rows of teeth are not opposed when at rest, but are everted by protractor muscles in such a way that the action is bilateral rather than vertical.

Since this is also the chordate feature to which Huxley, Newberry, Macfarlane, and Ulrich and Bassler assigned conodont elements, and a feature of Jefferies' (1986) hypothetical animal "s," the first crown vertebrate, a brief review of its characters is in order.

8.4.8 The lingual apparatus of Myxine

Figure 8.3 includes ventral views of the head of *Myxine glutinosa*, a typical hagfish, drawn to show the toothed lingual structure everted (left, above) and retracted (right, above). The sagittal and transverse sections below these two views provide schematic information on the arrangement of supporting plates and principal muscles. According to Dawson (1963), the lingual structure of *Myxine* has four primary components. Above the mouth there is a single, backwardly directed *palatal tooth*, which is median in position and firmly attached to an overlying plate of palatal cartilage. The laterally compressed, slitlike mouth is flanked by flaps of mucous membrane, which are continuations of the cephalic epidermis and fold into and line the oral cavity. Embedded in the mucous membrane on either side of the oral cavity are two longitudinal rows of seven to nine yellowish-brown, horny teeth, which are conical in shape and joined at their bases to form a *tooth comb.* When the lingual apparatus is retracted (Fig. 8.3, upper right), the tooth combs lie flat, and apices of their component denticles point inward and posteriorly.

The lingual teeth of *Myxine* are firmly attached to a cartilaginous *dental plate* (Fig. 8.3), which is a troughlike structure of V-shaped cross section that lies just below the mucous membrane of the oral cavity. Its longitudinally keeled underside moves forward and backward in the grooved upper surface of a cartilaginous *basal plate*, which supports the floor of the mouth and is the rigid, immovable site of attachment of the jaw muscles.

Contraction of *protractor muscles* attached to its anterior end causes the dental plate to slide forward in the grooved upper surface of the basal plate. As it does so, its sides flatten and the rows of lingual teeth above it are raised, spread laterally, and everted from the mouth (Fig. 8.3, upper left). Dawson (1963) likens this sequence of events to the opening of a book. Contraction of muscles attached to the posterior end of the dental plate effects retraction of the plate, closing of the book, and approximation of the teeth in a grasping movement. Repeated protraction and retraction of the lingual apparatus alternately spreads and approximates the rows of teeth and is effective in tearing food and conducting it into the mouth. During these ripping motions, the lone palatal tooth may serve as a gaff by impaling chunks of foodstuff and thereby keeping them from slipping out of the mouth during protractive movements of the dental plate.

Individual lingual teeth of *Myxine glutinosa* (Fig. 8.4A) are complex structures made up of an outer horn cap separated by an epithelial layer from an inner cone that includes distinctive, goblet-shaped *pokal cells.* In the center of each tooth is a mesodermal *pulp cavity.* The base of the outer horny cap is embedded in a groove in the mucous membrane of the oral cavity, except laterally, where it joins adjacent teeth. Dawson (1963) notes that several authors believe that growth of the horn cap takes place by cell division and keratinization in this basal groove. Cells in the epithelial layer between the apex of the horn cap and that of the inner *pokal-cell cone* appear pulled apart, as if they had been stretched between a horn cap that grew more rapidly than the underlying pokal-cell cone. The latter, composed of white, opaque, resilient ectodermal cells, is peculiar to the Myxinoidea and surrounds a small pulp, which contains blood vessels and nerve fibers. Various authors have speculated that the pokal-cell cone might be a replacement tooth,

←

Fig. 8.3. *Myxine glutinosa.* Top two views show ventral side of head, with lingual apparatus extended (left) and retracted (right). Central and bottom figures are transverse and sagittal sections of the head to show elements and musculature of the lingual apparatus. Redrawn from Dawson (1963).

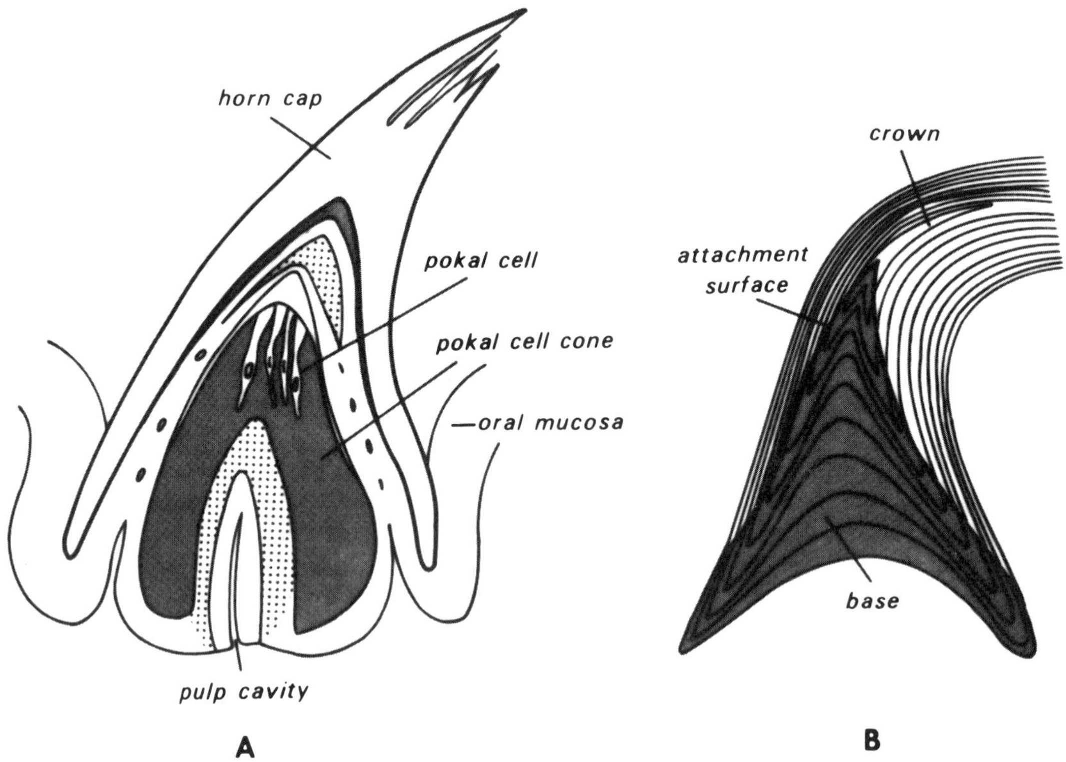

Fig. 8.4. (A) Lingual tooth of *Myxine* in longitudinal section. From Dawson (1963). (B) Typical coniform element of a conodont, also in longitudinal section. From Müller and Nogami (1971).

but this is not at all certain. Dawson (1963) summarizes her discussion of *Myxine* teeth by noting that "... it is most likely that there is no phylogenetic connection between these teeth and calcified teeth and that they are an individual adaptation to a parasitic mode of life."

In Fig. 8.4 I compare a lingual tooth of *Myxine glutinosa* with a coniform conodont element. There are profound and obvious differences in internal structure and, presumably, in mode of development. Whatever the significance of the pokal-cell cone in the lingual teeth of *Myxine*, there is no counterpart in complete conodont elements, which were stoutly scleritized and also enlarged during ontogeny by apposition of successive mineralized lamellae to their outer surfaces. The side-to-side grasping action of the lingual apparatus of *Myxine* is, indeed, an intriguing model for functional, or mechanical, interpretation of the cephalic apparatus of conodonts, as Aldridge, Smith, Norby, and Briggs (1987) and Smith et al. (1987) have shown. However, the Scottish and American Silurian specimens, which are the most completely preserved conodonts collected thus far, lack any trace of the cartilaginous palatal, dental, and basal plates that are required to make the apparatus function in *Myxine*. I am not so sure that this is attributable just to the vicissitudes of preservation, either. That is, two of the Scottish specimens preserve impressions at the posterior end of what have been interpreted as fin rays, which should have been cartilagenous, but neither the American Silurian specimen nor the best preserved of the Scottish Carboniferous specimens preserves anything in the head region that might be interpreted as cartilage.

Further of course, there are only three likely types of action in a buccal apparatus designed to tear, shred, rasp, and ultimately ingest food—up-and-down chewing; lateral grasping (à la *Myxine*); and in-and-out rasping (*à la* lampreys and the molluscan radula). I join Dawson (1963) in suspecting that apparatuses of any of these types might have been developed independently many times. If so, similarity in mechanical action, even if real, may not imply very much about biologic affinities.

8.4.9 Summary of features indicating chordate affinities

In summary, evidence taken by previous authors to suggest chordate (or vertebrate) affinities for conodonts consists of (1) similarity in chemical composition of conodont elements and the mineralized tissues of vertebrates; (2) the shapes of conodont elements, which suggest that they functioned as teeth or jaws; (3) the presence within the basal plates of several Siberian Ordovician specimens of structures such as those that characterize vertebrate bone; (4) the chevron-shaped impressions in the posterior half of several Scottish Carboniferous specimens, which have been interpreted to be myotomes comparable to those of *Branchiostoma* (=*Amphioxus*); and (5) implied similarity of function between the cephalic apparatus of conodonts and the lingual apparatus of *Myxine* and other hagfishes.

With respect to compositional similarity, it has been noted repeatedly that phosphate mineralization is widespread within the animal kingdom and hence is not, of itself, evidence of anything more than an enzymatic system that has not been completely modified to permit accumulation of carbonate minerals. Note, in this regard, the recent determination by Tillier and Cuif (1986) that the buccal elements of certain aplacophoran and scaphopodan mollusks are phosphatic.

With respect to shape, I need only point out that teeth, grasping spines, copulatory hooklets, and the like, have similar shapes in whatever animal body they are formed because those are the shapes that do the job, not because all organisms with teeth, grasping spines, and copulatory hooklets are closely related. In addition to those formed in various ways by different groups of chordates, structures that are similar to chordate teeth in form and function are found in distantly related groups such as mollusks, arthropods, trematodes, nemertineans, gnathostomulids, and chaetognaths. It is thus no surprise that someone, sometime, has considered one or another of those groups as either the repository for the Conodonta or its sister group.

Without in any way minimizing the possible significance of the bonelike features reported by Barskov, Moskalenko, and Starostina (1982) from the basal plate of several Middle Ordovician elements from Siberia, it should be noted that no such features were noted in comparable parts of any of the conodont elements studied by Ørvig (1951), Gross (1954), Müller and Nogami (1971), or in the closely similar elements of *Archaeognathus* recently analyzed in detail by Klapper and Bergström (1984). The affinities of *Coleodus*, to which the Siberian specimens have been assigned, are unknown among conodonts; and the Siberian fossils are different enough from the type of *Coleodus* to raise at least modest doubt about their identification. In short, the features noted by Barskov and his coauthors may indicate that the basal plates of the specimens in which they found them are bone. But it is not entirely clear that the specimens are conodont elements! Nor have those features been seen in the basal plates of elements representing other species.

Ørvig (1951) noted that some of the conodont elements in his collection from the Silurian of Estonia and the middle Devonian of Ohio have "a base consisting of a substance which differs . . . from . . . material of the tooth-like cusps . . . but in this substance one is not concerned with any kind of bone tissue." He also pointed out that the Middle Ordovician Harding Sandstone of Colorado, which yields elements of *Coleodus* and a great variety of other types of conodonts, also yields specimens assigned to the ostracoderms *Astraspis* and *Eriptychius* that exhibit practically all the types of hard tissue known in vertebrates. To my mind, this statement is significant because it suggests that conodont elements with superficially bonelike materials at the base and specimens composed of real bone occur in collections from the same beds and that they may be readily distinguished by one with a trained eye.

Finally, anyone familiar with the elegant illustration of *Myxine glutinosa* in Müller's (1836) famous monograph on the hagfishes (Fig. 8.2) is bound to be struck by the many features that seem to be shared with the best preserved of the Scottish Carboniferous specimens illustrated by Briggs et al. (1983). If scale is ignored (a typical *Myxine* is about 30 cm long and 2 cm wide), one is surely struck by the similarity between the V-shaped myotomes of *Myxine* and the V-shaped impressions in the posterior half of the Scottish specimen. However, one must also keep in mind that even if the V-shaped impressions in the latter specimen are correctly identified as myotomes (and Aldridge et al.,1986, surely waffle on this matter), an interpretation is not acceptable as objective evidence. I submit, for example (but without serious conviction), that the serially repeated V-shaped impressions might just as well be those of the numerous diverticula that branch from the gut of nemertineans (Fig. 8.1B), a group of acoelomate invertebrates that has been proposed by numerous authors (e.g., Hubrecht, 1883, 1887; Macfarlane, 1923; Willmer, 1974) as the sister group of the Chordata. Or, as Tillier and Cuif (1986) have suggested, they might represent the serial dorso-ventral musculature or V-shaped dorsal junction of sets of dermal spicules of aplacophoran mollusks.

Actually, one should keep in mind that three basic features unite the Chordata. At one stage or another in their ontogeny, chordates have a notochord, a series of pharyngeal clefts ("gill slits"), and a distinctively tubular nervous system that develops by infolding of the dorsal surface of the embryonic body. Other features, such as myotomes, paired pineal eyes, lateral-line systems, and the like, are developed to various degrees by many chordates but have apparently been modified or lost by others. Surely positive evidence, or strong indirect indication of at least one of the features (notochord, dorsal nerve cord, gill slits) that typifies all chordates should be identified in specimens such as the Scottish or American Silurian ones before conodonts are assigned so positively and enthusiastically to the Chordata. To be sure, all such features have a very low preservation potential. However, elements of the gill pouches are the most prominent features shown in the fossil lamprey recently described by Lund and Janvier (1986), even though that fossil lacks any trace of notochord, myotomes, or teeth.

8.5. L'Envoi

For more than a century we have been trying to stuff the Conodonta into the Chordata or into one or another of the invertebrate phyla, guided, evidently, by the often-stated conviction that most of the major organizational grades we recognize today as phyla arose late in the Precambrian or early in the Paleozoic and that none has been added (or subtracted) since about the middle of the Ordovician. However, an alternate view might be that in the late Precambrian-early Paleozoic heyday of metazoan radiation there were many more such organizational grades tried out than have survived, and that the "taxonomic distance" between those grades may have been much less than that which now separates the survivors.

We have learned to live with the idea (albeit somewhat uncomfortably) that "arthropod" is a grade of organization reached independently a number of times; that there are grades of organization even among phyla; and that at least one phylum-level group with a confusing combination of poriferan and cnidarian characters, the Archaeocyatha, evidently failed to survive the Cambrian. It has been suggested recently (Seilacher, 1983) that the bizarre members of Ediacara-type faunas, which have been assigned without obvious conviction to several animal phyla with living representatives, may instead represent additional organizational grades (phyla, or even an entire kingdom!) that failed to survive, and the Tommotiids and Conulata may also be such groups. And it has several times been suggested that even the venerable Chordata is possibly no more than a shaky confederation of invertebrates and vertebrates whose living representatives may record independent achievement of the "chordate" grade, but blend on several sides into other groups of phylum rank (e.g., the Echinodermata, one of the arthropodan phyla, or even the Rhynchocoela).

In brief, more than a century of trying to make conodonts something else has done little more than emphasize their uniqueness. Their skeletal elements lack essential features of the chordate groups with which they have commonly been compared, and what little hard information we have on soft-part anatomy speaks to us in several tongues. Thus, I am comfortable in considering the Conodonta an extinct phylum of small, solitary marine animals, with compressed or depressed vermiform bodies that were soft except for a variously differentiated assemblage of phosphatized epithelial elements in the cephalic lobe. The elements by which conodonts are represented in the fossil record may have functioned as grasping spines and buccal or pharyngeal teeth. And the cosmopolitan and broadly facies-independent distribution of many species suggests they were pelagic animals, probably nektonic, although others, of more limited distribution and facies dependence, may have been nektobenthic or even benthic. Conodonts were probably a minor phylum, never particularly conspicuous components of Paleozoic or Triassic marine faunas. Their ultimate extinction at the end of the Triassic was probably the cumulative result of a long decline through the Carboniferous and Permian and a gradual reduction in genomic diversity. In spite of their probable rarity in Paleozoic and Triassic seas, conodonts are abundant and common as fossils right up to the top of the Triassic, probably because they were blessed with heavy, chemically resistant skeletal elements that provided them a place in posterity denied to other less suitably endowed groups.

References

Aldridge, R. J., Briggs, D. E. G., Clarkson, E. N. K. and Smith, M. P. (1986). The affinities of conodonts—new evidence from the Carboniferous of Edinburgh, Scotland. *Lethaia* **19**, 1–14.

Aldridge, R. J., Smith, M. P., Norby, R. D., and Briggs, D. E. G. (1987). The architecture and function of Carboniferous polygnathacean conodont apparatuses. Pp. 63–75 in *Palaeobiology of Conodonts* (ed. R. J. Aldridge). Ellis Horwood, Chichester, 180 pp.

Barskov, I. S., Moskalenko, T. A., and Starostina, L. P. (1982). Novye dokazatel'stva prinadlezhnosti konodontoforid k pozvonochnym. *Pal. Zhur.* **1982**, 80–86. [English translation in *Paleontological Journal* **1982**, 82–90.]

Bengtson, S. (1976). The structure of some Middle Cambrian conodonts, and the early evolution of conodont structure and function. *Lethaia* 7, 185–206.

Branson, E. B., and Mehl, M. G. (1933). Conodont Studies No. 1. *Univ. Missouri Studies* **8**, 5–72.

Briggs, D. E. G., Clarkson, E. N. K., and Aldridge, R. J. (1983). The conodont animal. *Lethaia* **16**,1–14.

Clark, D. L., Sweet, W. C., Bergström, S. M., Klapper, G., Austin, R. L., Rhodes, F. H. T., Müller, K. J., Ziegler, W., Lindström, M., Miller, J. F., and Harris, A. G. (1981). Conodonta. Pt. W, Suppl. 2 in *Treatise on Invertebrate Paleontology* (ed. R. A. Robison), Geol. Soc. America and Univ. Kansas, 202 pp.

Conway-Morris, S. (1976). A new Cambrian lophophorate from the Burgess Shale of British Columbia. *Palaeontology* **19**, 199–222.

Dawson, J. A. (1963). The oral cavity, the 'jaws' and the horny teeth of Myxine glutinosa. Pp. 231–255 in *The Biology of Myxine* (ed. A. Brodal and R. Fange). Universitetsforlaget, Oslo.

Dubois, E. P. (1943). Evidence on the nature of conodonts. *J. Pal.* **17**, 155–159.

Ellison, S. P., Jr. (1944). The composition of conodonts. *J. Pal.* **18**, 133–140.

Gross, W. (1954). Zur Conodonten-Frage. *Senckenbergiana Lethaea* **35**, 73–85.

Harley, J. (1861). On the Ludlow bone-bed and its crustacean remains. *Q. J. Geol. Soc. London* **17**, 542–552.

Hass, W. H., and Lindberg, M. L. (1946). Orientation of crystal units of conodonts. *J. Pal.* **20**, 501–504.

Hinde, G. J. (1879). On conodonts from the Chazy and Cincinnati group of the Cambro-Silurian, and from the Hamilton and Genesee-shale divisions of the Devonian in Canada and the United States. *Q. J. Geol. Soc. London* **35**, 351–369.

Hubrecht, A. A. W. (1883). On the ancestral form of the chordate. *Q. J. Microsc. Soc.* **23**, 349–368.

——— (1887). The relation of the Nemertea to the Vertebrata. *Q. J. Microsc. Sci.* **27**, 605–644.

Hyman, L. H. (1951). *The Invertebrates: Platyhelminthes and Rhynchocoela—The acoelomate Bilateria.* V. II. Mc Graw-Hill, New York, 550 pp.

——— (1959). *The Invertebrates: Smaller Coelomate Groups.* V. 5. Mc Graw-Hill, New York, 783 pp.

James, U. P. (1884). On conodonts and fossil annelid jaws. *Cincinnati Soc. Nat. Hist. J.* **7**, 143–149.

Jefferies, R. P. S. (1986). *The Ancestry of the Vertebrates.* Cambridge Univ. Press, Cambridge, 376 pp.

Kirk, S. R. (1929). Conodonts associated with the Ordovician fish fauna of Colorado—A preliminary note. *Am. J. Sci.* Ser. 5, **18**(108), 493-496.

Klapper, G., and Bergström, S. M. (1984). The

enigmatic Middle Ordovician fossil *Archaeognathus* and its relations to conodonts and vertebrates. *J. Pal.* **58**(4), 949–976.

Lindström, M. (1973). On the affinities of conodonts. Pp. 85–102 in *Conodont Paleozoology* (ed. F. H. T. Rhodes). Geol. Soc. America Spec. Paper 141.

Loomis, F. G. (1936). Are conodonts gastropods? *J. Pal.* **10**, 663-664.

Lund, R., and Janvier, P. (1986). A second lamprey from the Lower Carboniferous (Namurian) of Bear Gulch, Montana (U.S.A.). *Geobios* **19**(5), 647-652.

Macfarlane, J. M. (1923). *Fishes the source of petroleum.* Macmillan, New York, 451 pp.

Mikulic, D. G., Briggs, D. E. G., and Kluessendorf, J. (1985a). A Silurian soft-bodied biota. *Science* **228**, 715-717.

——— (1985b). A new exceptionally preserved biota from the Lower Silurian of Wisconsin, USA. *Phil. Trans. Roy. Soc. London*, Ser. B, **311**, 78–85.

Müller, J. (1836). *Vergleichende Anatomie der Myxinoiden.* **I**. Berlin, 120 pp.

Müller, K. J., and Nogami, Y. (1971). Über den Feinbau der Conodonten. *Mem. Fac. Sci. Kyoto Univ.* ser. Geol. Mineral., **38**, 1–88.

Newberry, J. S. (1875). Descriptions of fossil fishes. *Ohio Geol. Survey*, Rept. 2, Pt. 2, 41–46.

Ørvig, Tor (1951). Histologic studies of placoderms and fossil elasmobranchs. I: The endoskeleton, with remarks on the hard tissues of lower vertebrates in general. *Arkiv for Zool.* **2**(2), 321–454.

Owen, Richard (1860). *Palaeontology or a systematic summary of extinct animals and their geological relations.* Adams and Black, Edinburgh, 463 pp.

Pander, C. H. (1856). Monographie der fossilen Fische des silurischen Systems der russisch-baltischen Gouvernements. *Konigl. Akad. Wiss., St. Petersburg*, 91 pp.

Rhodes, F. H. T. (1954). The zoological affinities of the conodonts. *Biol. Rev.* **29**, 419–452.

Scott, H. W. (1934). The zoological relationships of the conodonts. *J. Pal.* **8**, 448–455.

Seddon, G., and Sweet, W. C. (1971). An ecologic model for conodonts. *J. Pal.* **45**, 869–880.

Seilacher, A. (1983). Precambrian metazoan extinctions. *Geol. Soc. Am., Abstracts with Programs* **15**, 683.

Smith, M. P., Briggs, D. E. G., and Aldridge, R. J. (1987). A conodont animal from the lower Silurian of Wisconsin, USA, and the apparatus architecture of panderodontid conodonts. Pp. 91–104 in *Palaeobiology of Conodonts* (ed. R. J. Aldridge). Ellis Horwood, Chichester, 180 pp.

Sweet, W. C. (1955). Conodonts from the Harding Formation (Middle Ordovician) of Colorado. *J. Pal.* **29**, 226–262.

Szaniawski, H. (1982). Chaetognath grasping spines recognized among Cambrian protoconodonts. *J. Pal.* **56**, 806–810.

Tillier, S. and Cuif, J.-P. (1986). L'animal-conodonte est-il un Mollusque Aplacophore? *C. R. Acad. Sci. Paris* **303** ser. II, 627–632.

Ulrich, E. O., and Bassler, R. S. (1926). A classification of the toothlike fossils, conodonts, with descriptions of American Devonian and Mississippian species. *U. S. Nat. Mus. Proc.* **68**, Art. 12, 63 pp.

Willmer, E. N. (1974). Nemertines and possible ancestors of the vertebrates. *Bio. Rev.* **49**, 321–363.

Yochelson, E. L. (1975). Discussion of early Cambrian "molluscs'. *J. Geol. Soc. London* **131**, 661–662.

Zittel, K. A., and Rohon, J. V. (1886). Über Conodonten. *Bayer. Akad. Wiss.*, Sitzungsber. Mat.-Phys. Kl., **16**, 108–136.

APPENDIX A. A SUMMARY CLASSIFICATION OF THE CONODONTA

Phylum Conodonta Pander, 1856
- Class Cavidonti, new
  - Order Proconodontida, new
    - Family Proconodontidae Lindström, 1970
      - *Eoconodontus* Miller, 1980
      - *Proconodontus* Miller, 1969
    - Family Cordylodontidae Lindström, 1970
      - *Cambrooistodus* Miller, 1980
      - *Cordylodus* Pander, 1856
      - *Iapetognathus* Landing, 1982
    - Family Fryxellodontidae Miller, 1981
      - *Fryxellodontus* Miller, 1969
    - Family Pygodontidae Bergström, 1981
      - *Nericodus* Lindström, 1955
      - *Polonodus* Dzik, 1976
      - *Pygodus* Lamont and Lindström, 1957
  - Order Belodellida, new
    - Family Belodellidae Khodalevich and Tschernich, 1973
      - *Belodella* Ethington, 1959
      - *Coelocerodontus* Ethington, 1959
      - *Dvorakia* Klapper and Barrick, 1983
      - *Stolodus* Lindström, 1955
      - *Walliserodus* Serpagli, 1967
    - Family Ansellidae Fåhraeus and Hunter, 1985
      - *Ansella* Fåhraeus and Hunter, 1985
      - ?*Hamarodus* Viira, 1975
    - Family Dapsilodontidae, new
      - *Besselodus* Aldridge, 1982
      - *Dapsilodus* Cooper, 1976
- Class Conodonti Branson, 1938
  - Order Protopanderodontida, new
    - Family Protopanderodontidae Lindström, 1970
      - *Glyptoconus* Kennedy, 1980
      - *Monocostodus* Miller, 1980
      - *Oneotodus* Lindström, 1955
      - *Parutahconus* Landing, 1982
      - *Protopanderodus* Lindström, 1971
      - ?*Pseudooneotodus* Drygant, 1974
      - *Scabbardella* Orchard, 1980
      - *Semiacontiodus* Miller, 1969
      - *Staufferella* Sweet, Thompson, and Satterfield, 1975
      - ?*Strachanognathus* Rhodes, 1955
      - *Teridontus* Miller, 1980
      - *Tropodus* Kennedy, 1980
      - *Utahconus* Miller, 1980
      - *Variabiloconus* Landing, Barnes, and Stevens, 1986
    - Family Clavohamulidae Lindström, 1970
      - *Clavohamulus* Furnish, 1938
      - *Hirsutodontus* Miller, 1969
      - *Serratognathus* Lee, 1970
      - ?*Paraserratognathus* Yang, 1986
    - Family Acanthodontidae Lindström, 1970
      - *Acanthodus* Furnish, 1938
      - *Cornuodus* Fahråeus, 1966
      - *Drepanodus* Pander, 1856
      - *Parapaltodus* Stouge, 1984
      - *Scalpellodus* Dzik, 1976
      - *Ulrichodina* Furnish, 1938
    - Family Drepanoistodontidae Fåhraeus and Nowlan, 1978
      - *Drepanoistodus* Lindström, 1971
      - *Paltodus* Pander, 1856
      - *Paroistodus* Lindström, 1971
  - Order Panderodontida, new
    - Family Panderodontidae Lindström, 1970
      - *Belodina* Ethington, 1959
      - *Culumbodina* Moskalenko, 1973
      - *Neopanderodus* Ziegler and Lindström, 1971
      - *Panderodus* Ethington, 1959
      - *Parabelodina* Sweet, 1979
      - *Parapanderodus* Stouge, 1984
      - *Plegagnathus* Ethington and Furnish, 1959
      - *Pseudobelodina* Sweet, 1979

Order Prioniodontida Dzik, 1976
- Family Oistodontidae Lindström, 1970
  - Subfamily Oistodontinae Lindström, 1970
    - *Oelandodus* van Wamel, 1974
    - *Oistodus* Pander, 1856
    - *Protoprioniodus* McTavish, 1973
  - Subfamily Tripodontinae, new
    - *Rossodus* Repetski and Ethington, 1983
    - *Tripodus* Bradshaw, 1969
  - Subfamily Juanognathinae Bergström, 1981
    - *Histiodella* Harris, 1962
    - *Juanognathus* Serpagli, 1974 (incl. *Aurilobodus* An, 1983)
- Family Prioniodontidae Bassler, 1925
  - *Oepikodus* Lindström 1955
  - *Prioniodus* Pander, 1856
- Family Balognathidae Hass, 1959
  - *Amorphognathus* Branson and Mehl, 1933
  - *Baltoniodus* Lindström, 1971
  - *Complexodus* Dzik, 1976
  - *Gamachignathus* McCracken, Nowlan, and Barnes, 1980
  - *Noixodontus* McCracken and Barnes, 1982
  - *Rhodesognathus* Bergström and Sweet, 1966
  - *Sagittodontina* Knüpfer, 1967
- Family Icriodellidae, new
  - *Icriodella* Rhodes, 1953
  - *Pedavis* Klapper and Philip, 1971
  - *Sannemannia* Al-Rawi, 1977
  - *Steptotaxis* Uyeno and Klapper, 1980
- Family Distomodontidae Klapper, 1981
  - *Coryssognathus* Link and Druce, 1972
  - *Distomodus* Branson and Branson, 1947
  - *Rotundacodina* Carls and Gandl, 1969
- Family Icriodontidae Müller and Müller, 1957
  - *Antognathus* Lipniagov, 1978
  - *Icriodus* Branson and Mehl, 1938
  - *Latericriodus* Müller, 1962
  - *Pelekysgnathus* Thomas, 1949
- Family Polyplacognathidae Bergström, 1981
  - *Cahabagnathus* Bergström, 1983
  - *Eoplacognathus* Hamar, 1966
  - *Polyplacognathus* Stauffer, 1935
- Family Multioistodontidae Harris, 1964
  - *Leptochirognathus* Branson and Mehl, 1943
  - *Multioistodus* Cullison, 1938
  - *Neomultioistodus* Harris and Harris, 1965
  - *Paraprioniodus* Ethington and Clark, 1982
  - *Pteracontiodus* Harris and Harris, 1965j
- Family Plectodinidae, new
  - "*Aphelognathus*" (new genus for species with pastinate P elements)
  - *Plectodina* Stauffer, 1935
  - *Scyphiodus* Stauffer, 1935
  - ?*Tangshanodus* An, 1983
- Family Cyrtoniodontidae Hass, 1959
  - ?*Bryantodina* Stauffer, 1935
  - *Phragmodus* Branson and Mehl, 1933
- Family Rhipidognathidae Lindström, 1981
  - *Appalachignathus* Bergström and others, 1974
  - *Bergstroemognathus* Serpagli, 1974
  - *Rhipidognathus* Branson, Mehl, and Branson, 1951
- Family Periodontidae Lindström, 1970
  - *Microzarkodina* Lindström, 1971
  - *Periodon* Hadding, 1913

Order Prioniodinida, new
- Family Chirognathidae Branson and Mehl, 1944
  - *Chirognathus* Branson and Mehl, 1933
  - ?*Curtognathus* Branson and Mehl, 1933
  - *Erismodus* Branson and Mehl, 1933
  - *Erraticodon* Dzik, 1978
- Family Prioniodinidae Bassler, 1925
  - *Apatognathus* Branson and Mehl, 1934
  - *Erika* Murphy and Matti, 1982

Hibbardella Bassler, 1925
Idioprioniodus Gunnell, 1933
Kladognathus Rexroad, 1958
Oulodus Branson and Mehl, 1933
Prioniodina Bassler, 1925
Pristognathus Stone and Furnish, 1959
?*Ligonodina* Bassler, 1925
?*Lonchodina* Bassler, 1925
Family Bactrognathidae Lindström, 1970
Bactrognathus Branson and Mehl, 1941
Doliognathus Branson and Mehl, 1941
Dollymae Hass, 1959
Eotaphrus Pierce and Langenheim, 1974
Embaysgnathus Metcalfe, 1981
Scaliognathus Branson and Mehl, 1941
Staurognathus Branson and Mehl, 1941
Family Ellisoniidae Clark, 1972
Ellisonia Müller, 1956
Furnishius Clark, 1959
Gladigondolella Müller, 1962
Hadrodontina Staesche, 1964
Merrillina Kozur and Mock, 1974
Pachycladina Staesche, 1964
Sweetina Wardlaw and Collinson, 1986
Family Gondolellidae Lindström, 1970
Carinella Budurov, 1973
Cypridodella Mosher, 1968
Epigondolella Mosher 1968
Gondolella Stauffer and Plummer, 1932
Misikella Kozur and Mock, 1974
Mosherella Kozur, 1972
Neogondolella Bender and Stoppel, 1965
Neospathodus Mosher, 1968
Platyvillosus Clark, Sincavage, and Stone 1964
Pseudofurnishius van den Boogaard, 1966
Xaniognathus Sweet, 1970
Order Ozarkodinida Dzik, 1976
Family Spathognathodontidae Hass, 1959
Amydrotaxis Klapper and Murphy, 1980
Ancyrodella Ulrich and Bassler, 1926
Ancyrodelloides Bishoff and Sannemann, 1958
Aphelognathus Branson, Mehl, and Branson, 1951
Bispathodus Müller, 1962
Eognathodus Philip, 1965
Lochriea Scott, 1942
Mehlina Youngquist, 1945
Ozarkodina Branson and Mehl, 1933
Pandorinellina Müller and Müller, 1957
?"*Plectodina*" (new genus for species without pastinate P elements)
Polygnathoides Branson and Mehl, 1933
Pseudopolygnathus Branson and Mehl, 1934
Rhachistognathus Dunn, 1966
Scaphignathus Helms, 1959
Tortodus Weddige, 1977
Vogelgnathus Norby and Rexroad, 1985
Yaoxianognathus An, 1985
Family Kockelellidae Klapper, 1981
Ancoradella Walliser, 1964
Kockelella Walliser, 1957
Family Pterospathodontidae Cooper, 1977
Apsidognathus Walliser, 1964
Astropentagnathus Mostler, 1967
Aulacognathus Mostler, 1967
Carniodus Walliser, 1964
Johnognathus Mashkova, 1977
Pterospathodus Walliser, 1964
Family Polygnathidae Bassler, 1925
Ancyrognathus Branson and Mehl, 1934
?*Ancyrolepis* Ziegler, 1959
Polygnathus Hinde, 1879
Polylophodonta Branson and Mehl, 1934
Family Palmatolepidae, new
Klapperina Lane, Müller, and Ziegler, 1979
Mesotaxis Klapper and Philip, 1972

Palmatolepis Ulrich and Bassler, 1926
Schmidtognathus Ziegler, 1966
Family Elictognathidae Austin and Rhodes, 1981
Alternognathus Ziegler and Sandberg, 1984
Siphonodella Branson and Mehl, 1944
Family Gnathodontidae, new
Gnathodus Pander, 1856
Protognathodus Ziegler, 1969
Family Idiognathodontidae Harris and Hollingsworth, 1933
Declinognathodus Dunn, 1966
Idiognathodus Gunnell, 1931
Idiognathoides Harris and Hollingsworth, 1933
Neognathodus Dunn, 1970
Streptognathodus Stauffer and Plummer, 1932
Family Anchignathodontidae Clark, 1972
Aethotaxis Baesemann, 1973
Hindeodus Rexroad and Furnish, 1964
?*Subbryantodus* Branson and Mehl, 1934
Family Sweetognathidae Ritter, 1986
Diplognathodus Kozur and Merrill, 1975
Homoiranognathus Ritter, 1986
Iranognathus Kozur, Mostler, and Rahimi-Yazd, 1976
Isarcicella Kozur, 1975
Neostreptognathodus Clark, 1972
Rabeignathus Kozur, 1978
Sweetognathus Clark, 1972
Family Cavusgnathidae Austin and Rhodes, 1981
Adetognathus Lane, 1967
Capricornognathus Austin and Mitchell, 1975
Cavusgnathus Harris and Hollingsworth, 1933
Cloghergnathus Austin and Mitchell, 1975
Clydagnathus Rhodes, Austin and Druce, 1969
Patrognathus Rhodes, Austin and Druce, 1969
Taphrognathus Branson and Mehl, 1941
Family Mestognathidae Austin and Rhodes, 1981
Mestognathus Bischoff, 1957
Laterignathus Aristov and Alekseev, 1976
Order Unknown
Family Coleodontidae Branson and Mehl, 1944
Archaeognathus Cullison, 1938
Coleodus Branson and Mehl, 1933
?*Mixoconus* Sweet, 1955
Neocoleodus Branson and Mehl, 1933
?*Stereoconus* Branson and Mehl, 1933

APPENDIX B. STRATIGRAPHIC RANGE CHARTS

In charts on the next 14 pages I indicate the stratigraphic ranges of 562 conodont species in a biostratigraphic framework that includes 156 conodont-defined biozones and, in the Lower and Upper Carboniferous, several intervals for which there are currently no widely accepted conodont-biozonal schemes. In choosing which of the nearly 5000 named conodont species to include in the following charts, I followed no particular guidelines. By and large I have included species whose first and/or last occurrences define boundaries in the biozonal framework, and I have omitted species with exceptionally long ranges and/or poorly known or morphologically nondescript apparatuses. Representative specimens of many of the species included are illustrated in Chapter 5.

Mark Kleffner, a doctoral student at The Ohio State University, generously abstracted information on the ranges of important Silurian species from the much more inclusive composite standard section he has assembled by graphic means. In compiling other charts in this appendix I have supplemented my own records with information from sources listed below. Thus authors of those reports must be credited for most of the hard work that went into building the charts in which their data are used. Of course, they must also share some of the blame for any errors!

References

Higgins, A. C., and Austin, R. L., editors (1985). *A Stratigraphical Index of Conodonts.* Ellis Horwood, Chichester, 263 pp.

Klapper, G., and Johnson, J. G. (1980). Endemism and dispersal of Devonian conodonts. *J. Pal.* **54**(2), 400–455.

Klapper, G., and Ziegler, W. (1979). Devonian conodont biostratigraphy. *Spec. Papers in Palaeont. 23*, 199–224.

Kovacs, S., and Kozur, H. (1980). Stratigraphische Reichweite der wichtigsten Conodonten (ohne Zahnreihenconodonten) der Mittel- und Obertrias. *Geol. Paläont. Mitt. Innsbruck* **10**(2), 47–78.

Lane, H. R., Sandberg, C. A., and Ziegler, W. (1980). Taxonomy and phylogeny of some Lower Carboniferous conodonts and preliminary standard post-*Siphonodella* zonation. *Geol. et Palaeont.* **14**, 117–168.

Lane, H. R., and Straka, J. J., (1974). Late Mississippian and early Pennsylvanian conodonts, Arkansas and Oklahoma. *Geol. Soc.America Spec. Paper 152*, 144 pp.

Ritter, S. M. (1986). Taxonomic revision and phylogeny of post-Early Permian crisis bisseli-whitei Zone conodonts with comments on Late Paleozoic diversity. *Geol. et Palaeont.* **20**, 139–165.

Wardlaw, B. R., and Collinson, J. W. (1986). Paleontology and deposition of the Phosphoria Formation. *Contr. Geol. Univ. Wyoming* **24**(2), 107–142.

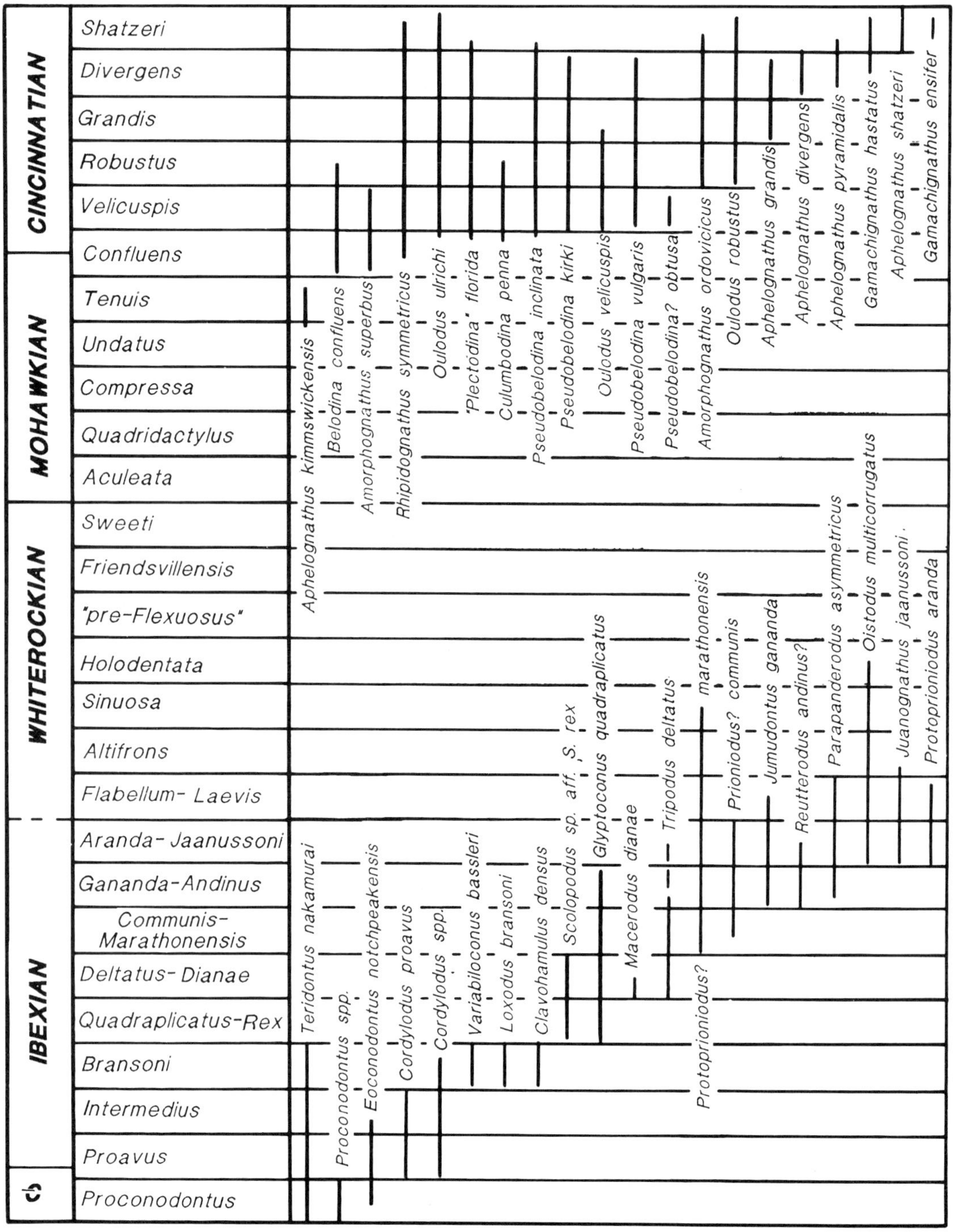

Chart 1. Ranges of selected Late Cambrian and Ordovician conodont species in low-latitude provinces *(left and right pages).*

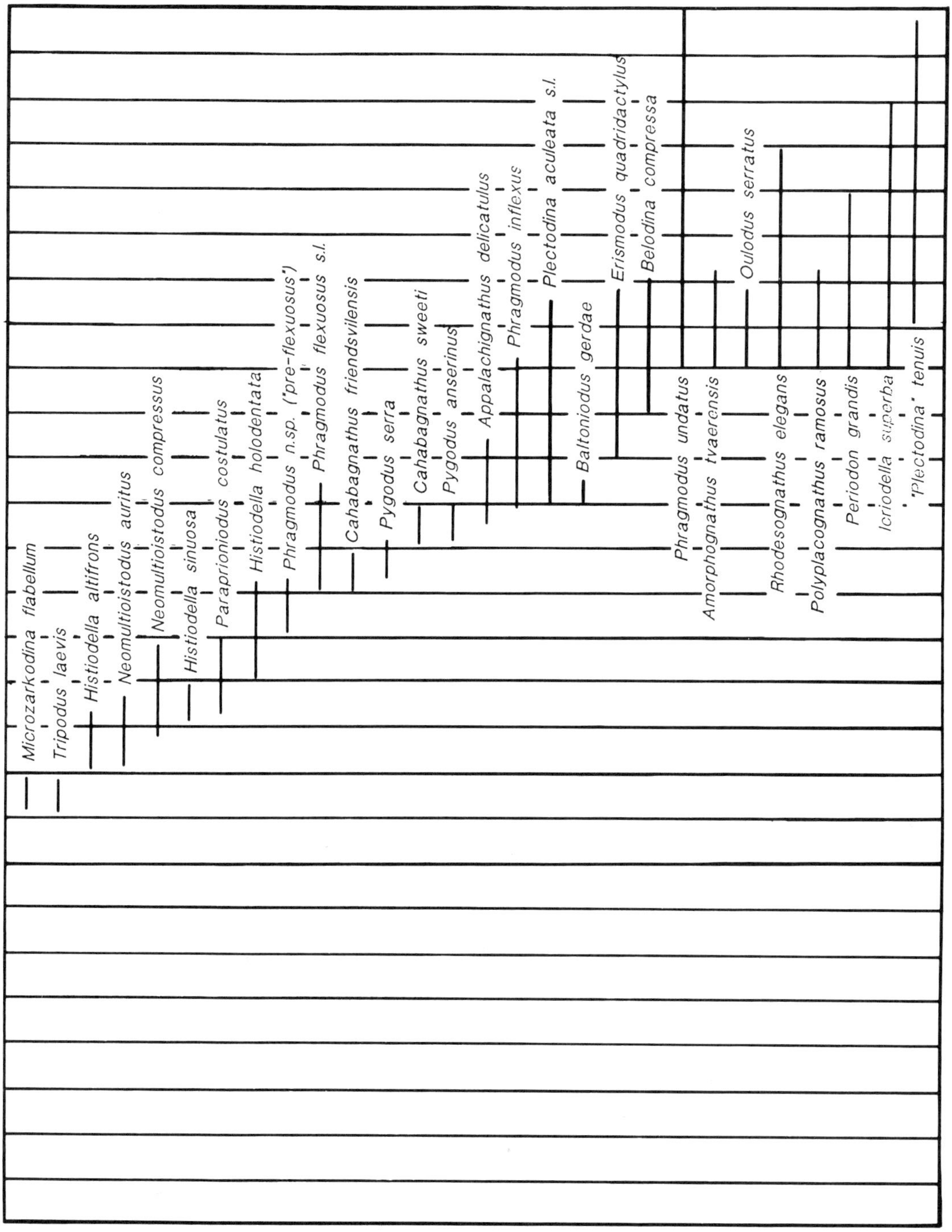
Microzarkodina flabellum
Tripodus laevis
Histiodella altifrons
Neomultioistodus auritus
Neomultioistodus compressus
Histiodella sinuosa
Paraprioniodus costulatus
Histiodella holodentata
Phragmodus n.sp. ("pre-flexuosus")
Phragmodus flexuosus s.l.
Cahabagnathus friendsvilensis
Pygodus serra
Cahabagnathus sweeti
Pygodus anserinus
Appalachignathus delicatulus
Phragmodus inflexus
Plectodina aculeata s.l.
Baltoniodus gerdae
Erismodus quadridactylus
Belodina compressa
Phragmodus undatus
Amorphognathus tvaerensis
Oulodus serratus
Rhodesognathus elegans
Polyplacognathus ramosus
Periodon grandis
Icriodella superba
"Plectodina" tenuis

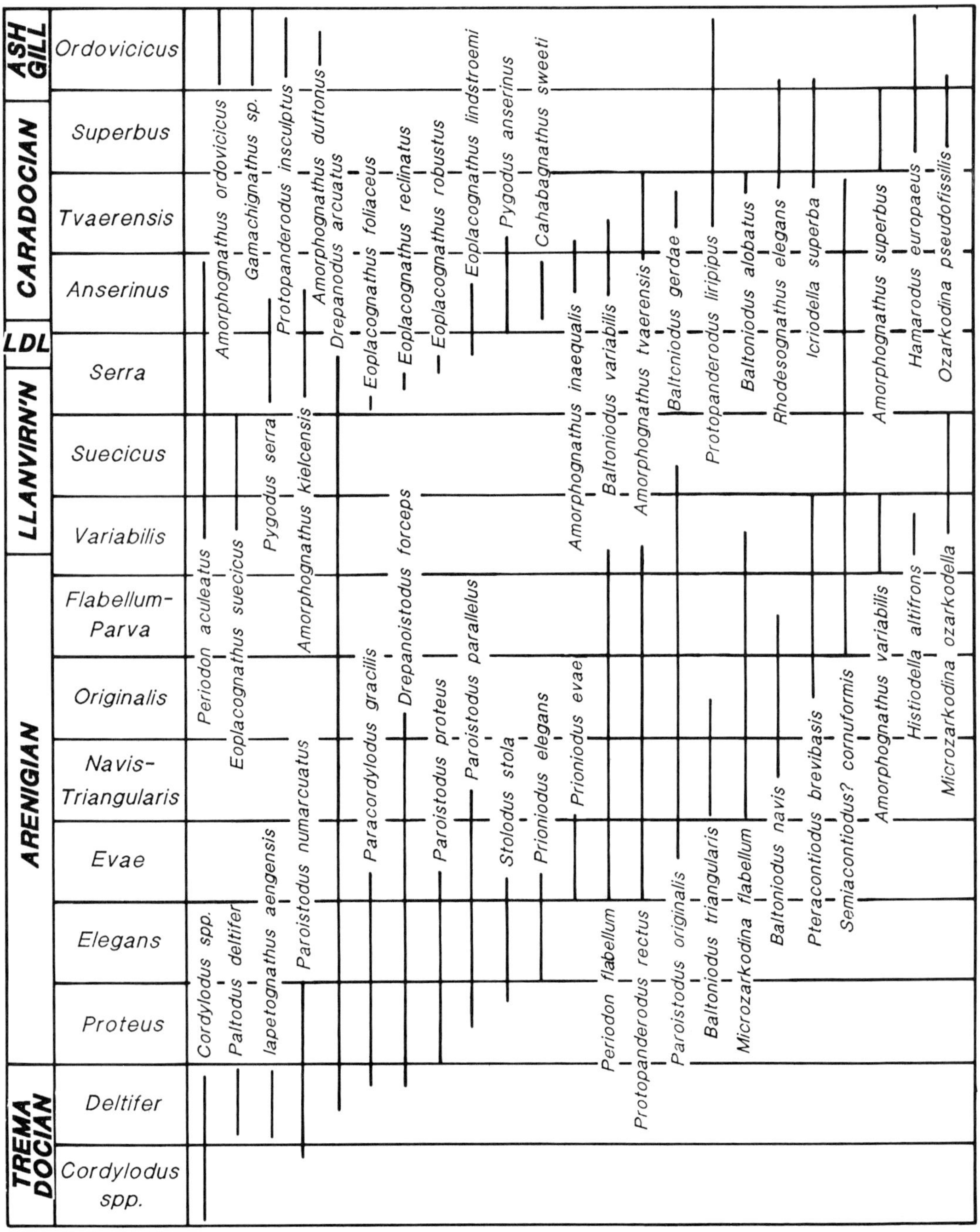

Chart 2. Ranges of selected conodont species in Ordovician high-latitude provinces.

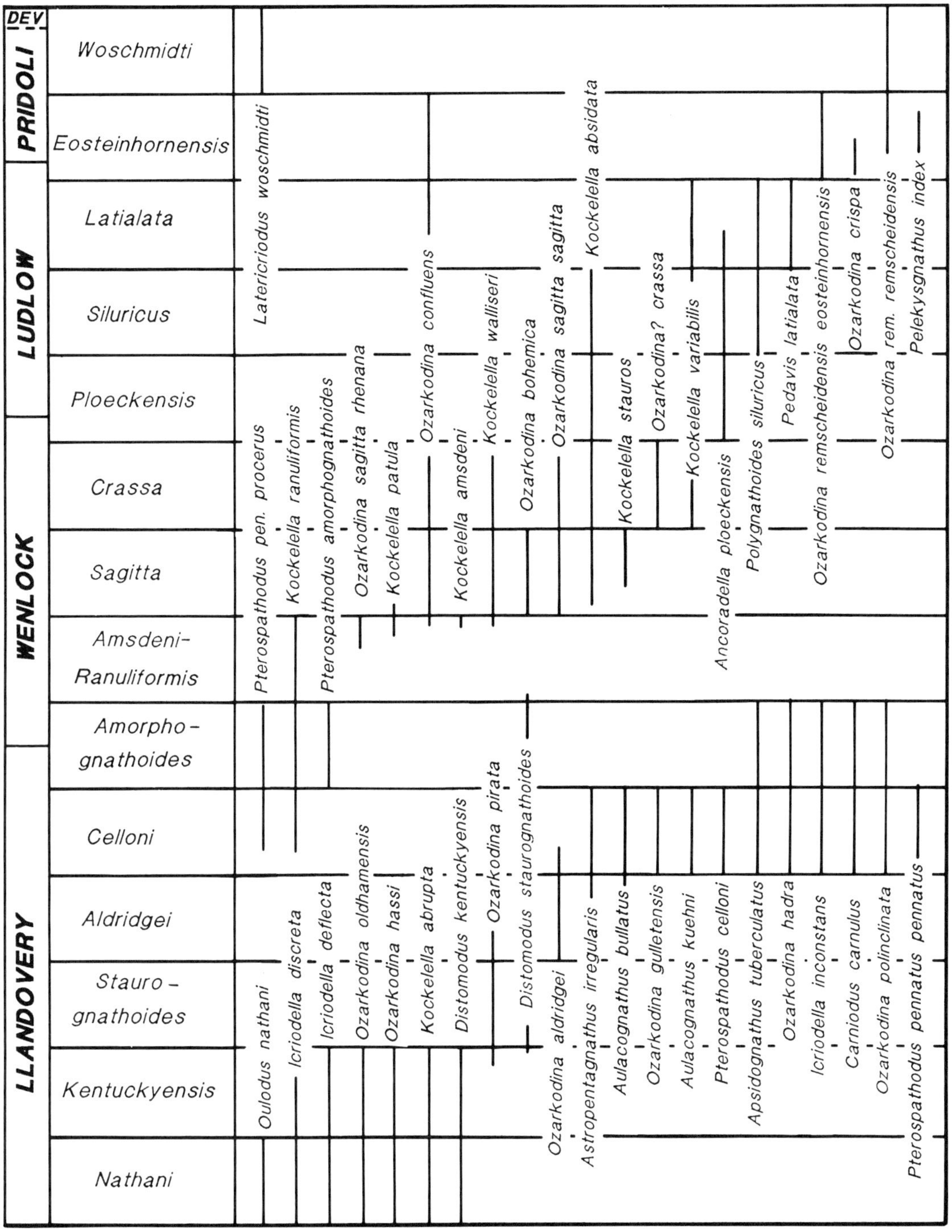

Chart 3. Ranges of selected Silurian conodont species.

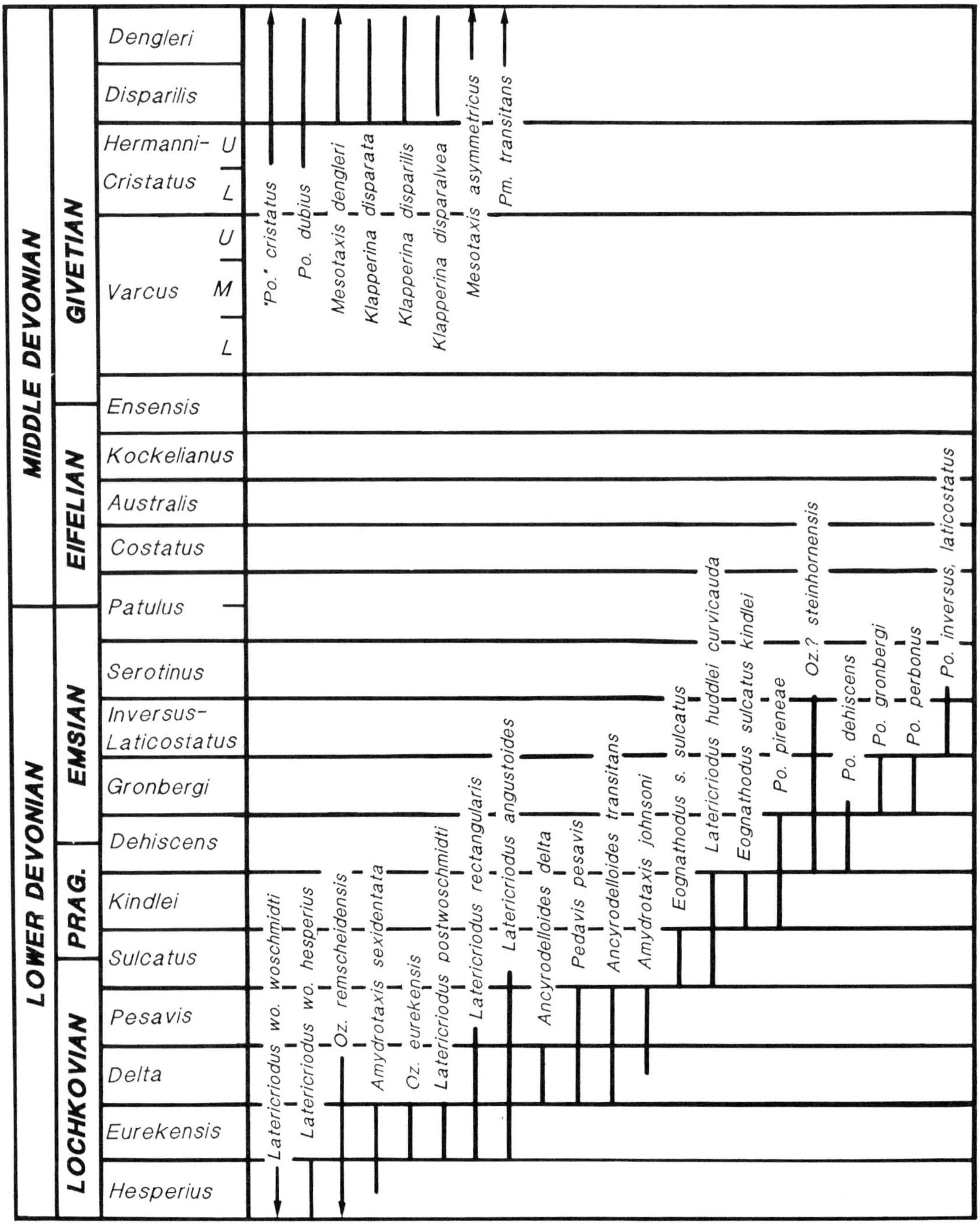

Chart 4. Ranges of selected Lower and Middle Devonian conodont species *(left and right pages).*

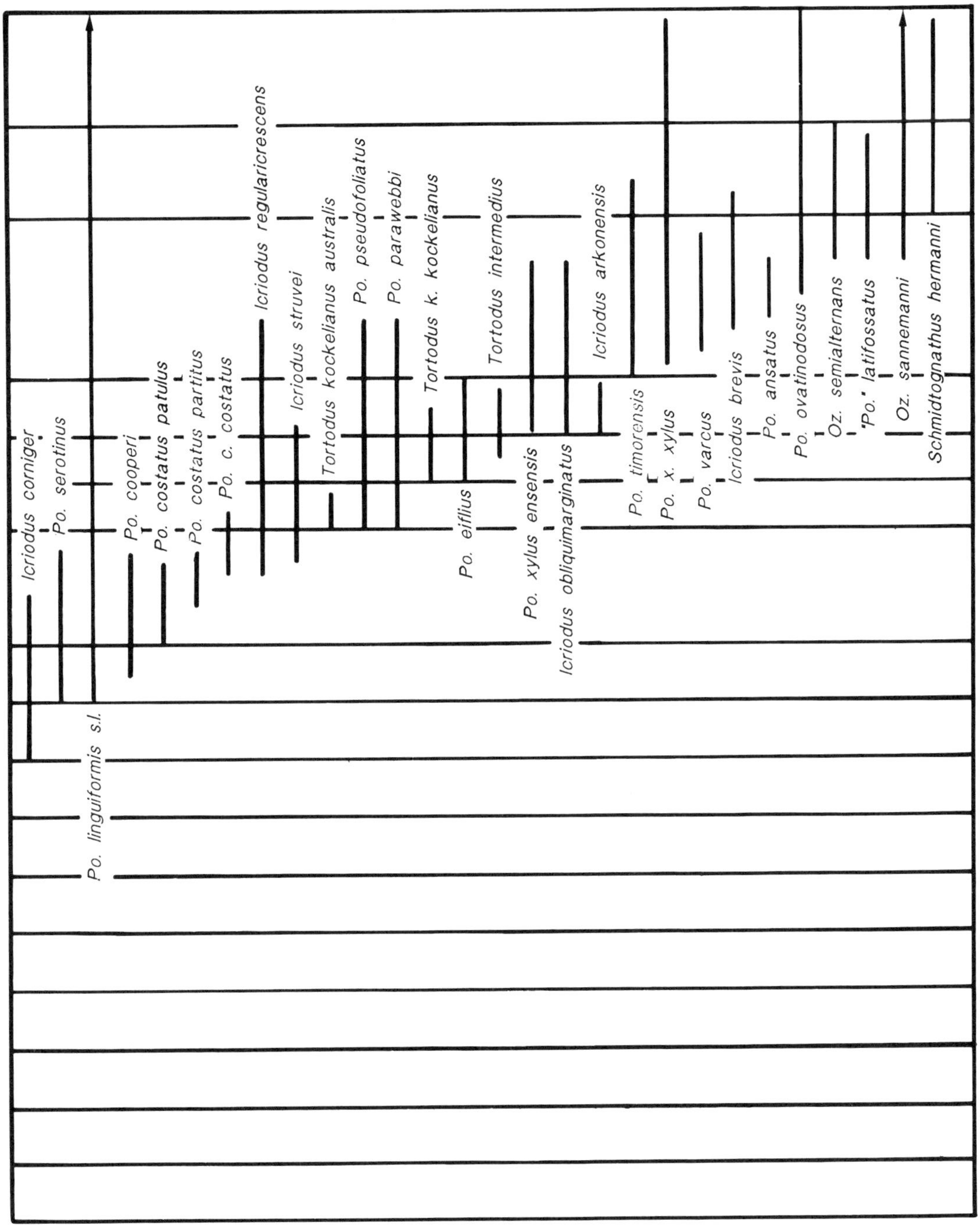

Icriodus corniger
Po. serotinus
Po. linguiformis s.l.
Po. cooperi
Po. costatus patulus
Po. costatus partitus
Po. c. costatus
Icriodus regularicrescens
Icriodus struvei
Tortodus kockelianus australis
Po. pseudofoliatus
Po. parawebbi
Tortodus k. kockelianus
Po. eiflius
Tortodus intermedius
Po. xylus ensensis
Icriodus obliquimarginatus
Icriodus arkonensis
Po. timorensis
Po. x. xylus
Po. varcus
Icriodus brevis
Po. ansatus
Po. ovatinodosus
Oz. semialternans
'Po.' latifossatus
Oz. sannemanni
Schmidtognathus hermanni

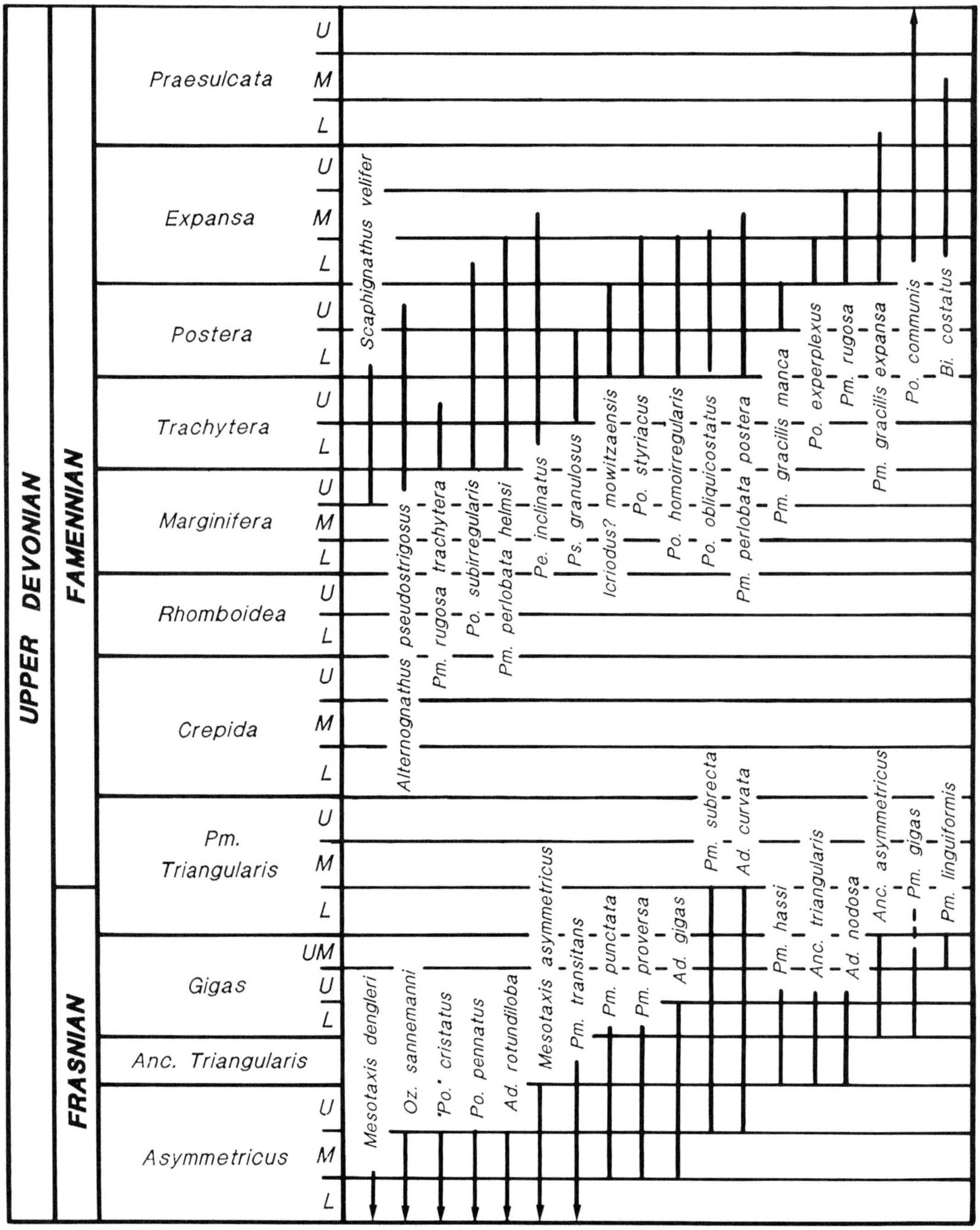

Chart 5. Ranges of selected Late Devonian conodont species *(left and right pages).*

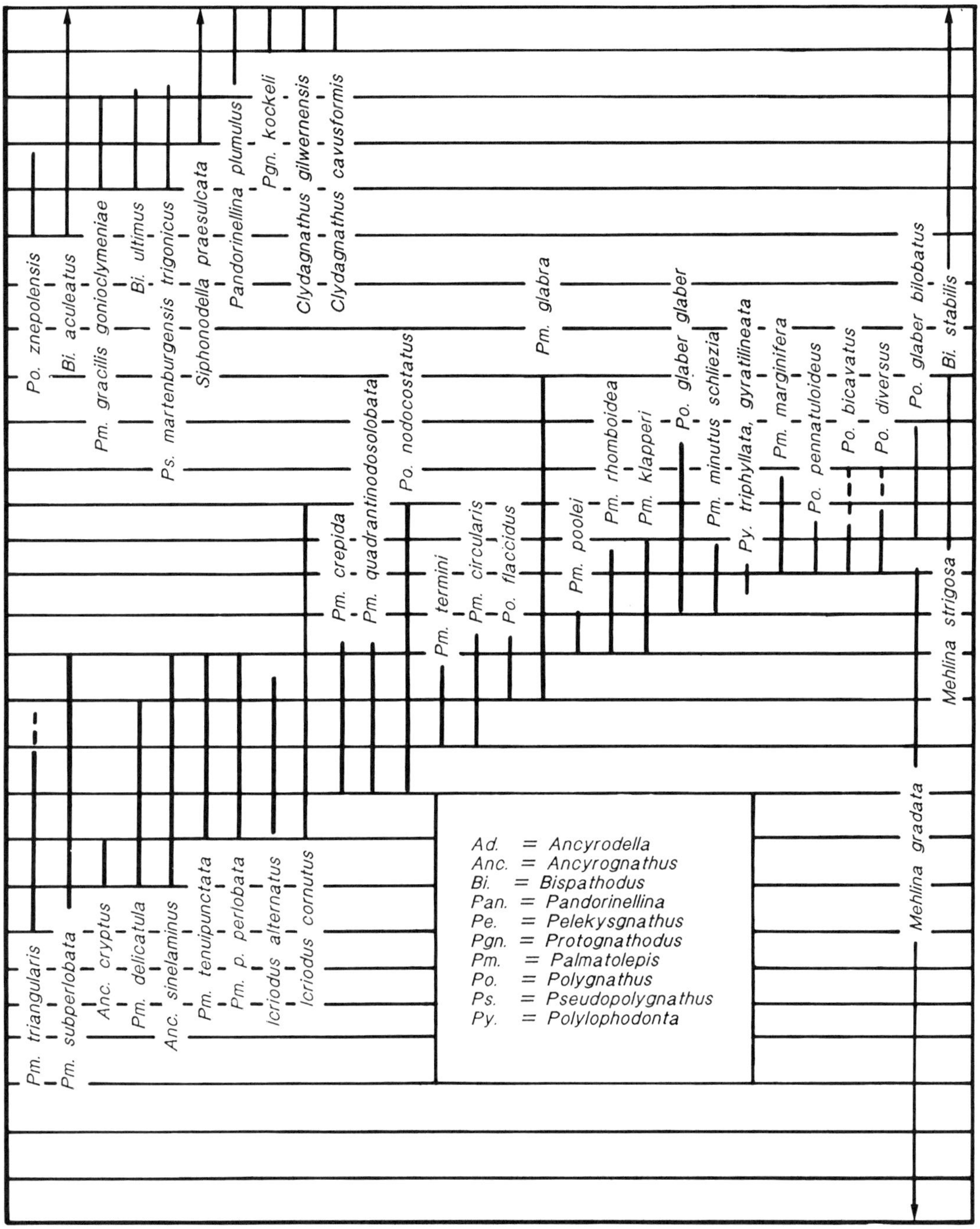

Po. znepolensis
Bi. aculeatus
Pm. gracilis gonioclymeniae
Bi. ultimus
Ps. martenburgensis trigonicus
Siphonodella praesulcata
Pandorinellina plumulus
Pgn. kockeli
Clydagnathus gilwernensis
Clydagnathus cavusformis
Pm. triangularis
Pm. subperlobata
Anc. cryptus
Pm. delicatula
Anc. sinelaminus
Pm. tenuipunctata
Pm. p. perlobata
Icriodus alternatus
Icriodus cornutus
Pm. crepida
Pm. quadrantinodosolobata
Po. nodocostatus
Pm. termini
Pm. circularis
Po. flaccidus
Pm. glabra
Pm. poolei
Pm. rhomboidea
Pm. klapperi
Po. glaber glaber
Pm. minutus schleizia
Py. triphyllata, gyratilineata
Pm. marginifera
Po. pennatuloideus
Po. bicavatus
Po. diversus
Po. glaber bilobatus
Bi. stabilis
Mehlina strigosa
Mehlina gradata
Ad. = Ancyrodella
Anc. = Ancyrognathus
Bi. = Bispathodus
Pan. = Pandorinellina
Pe. = Pelekysgnathus
Pgn. = Protognathodus
Pm. = Palmatolepis
Po. = Polygnathus
Ps. = Pseudopolygnathus
Py. = Polylophodonta

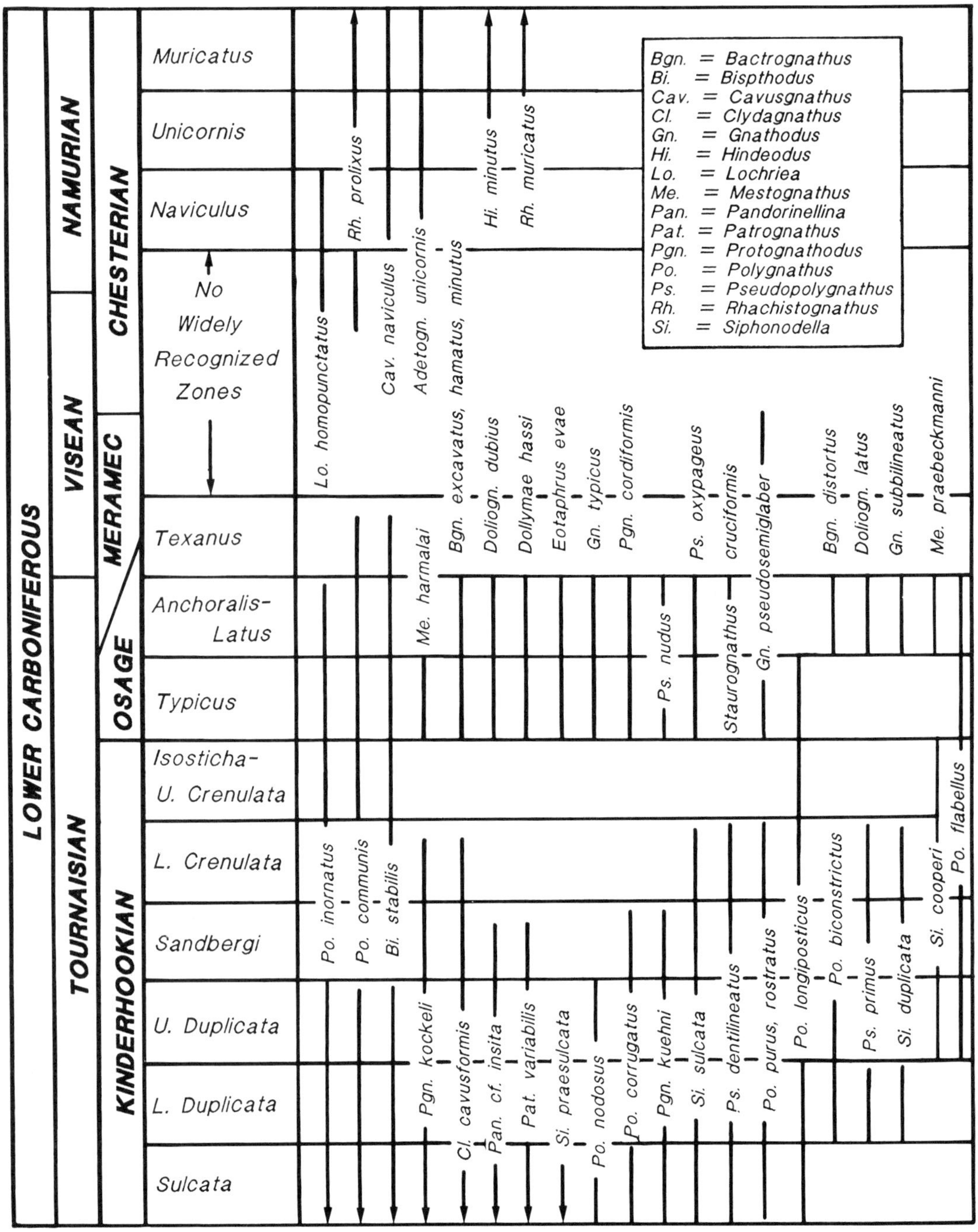

Chart 6. Ranges of selected Early Carboniferous (Mississippian) conodont species *(left and right pages).*

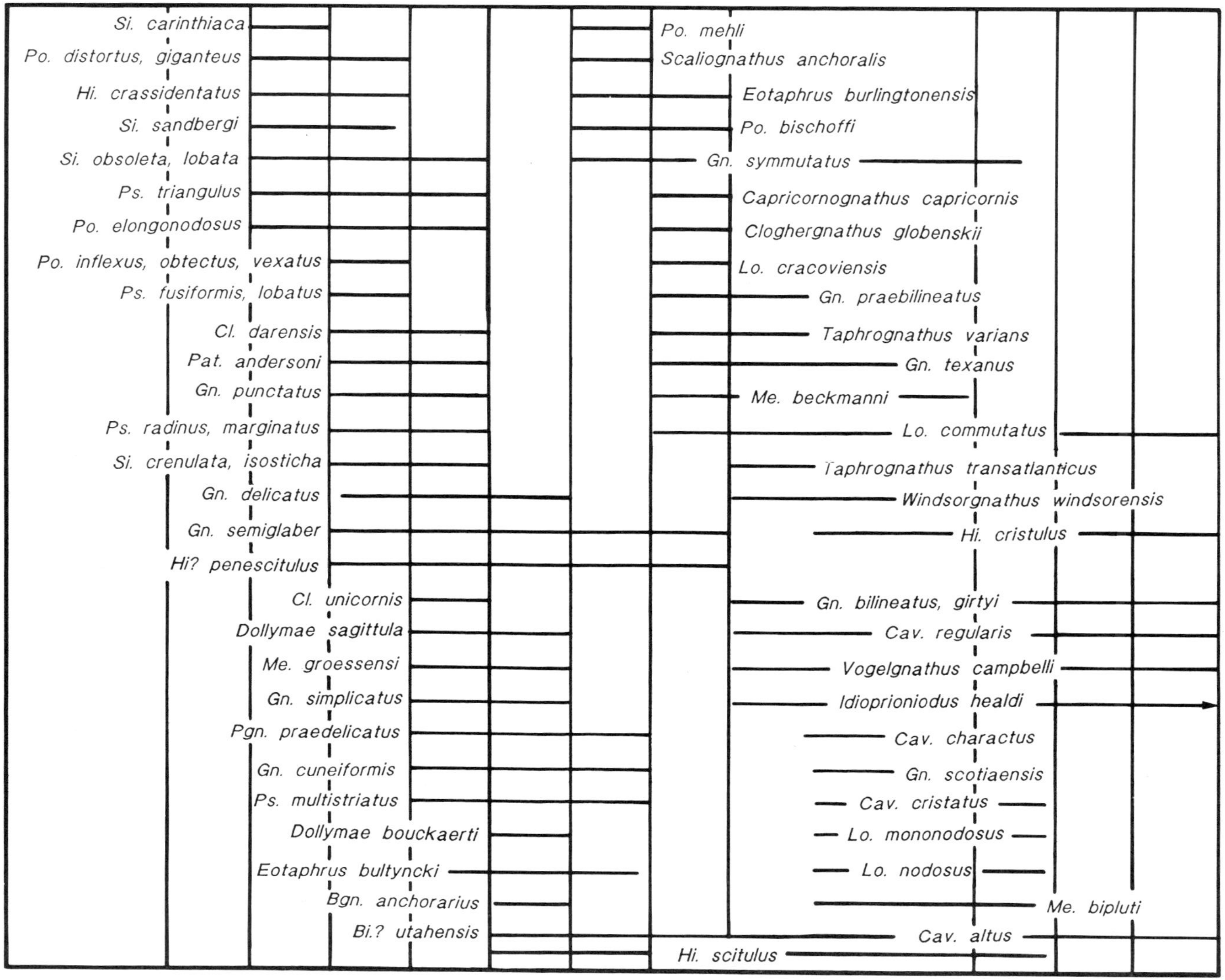
Si. carinthiaca
Po. distortus, giganteus
Hi. crassidentatus
Si. sandbergi
Si. obsoleta, lobata
Ps. triangulus
Po. elongonodosus
Po. inflexus, obtectus, vexatus
Ps. fusiformis, lobatus
Cl. darensis
Pat. andersoni
Gn. punctatus
Ps. radinus, marginatus
Si. crenulata, isosticha
Gn. delicatus
Gn. semiglaber
Hi? penescitulus
Cl. unicornis
Dollymae sagittula
Me. groessensi
Gn. simplicatus
Pgn. praedelicatus
Gn. cuneiformis
Ps. multistriatus
Dollymae bouckaerti
Eotaphrus bultyncki
Bgn. anchorarius
Bi.? utahensis
Po. mehli
Scaliognathus anchoralis
Eotaphrus burlingtonensis
Po. bischoffi
Gn. symmutatus
Capricornognathus capricornis
Cloghergnathus globenskii
Lo. cracoviensis
Gn. praebilineatus
Taphrognathus varians
Gn. texanus
Me. beckmanni
Lo. commutatus
Taphrognathus transatlanticus
Windsorgnathus windsorensis
Hi. cristulus
Gn. bilineatus, girtyi
Cav. regularis
Vogelgnathus campbelli
Idioprioniodus healdi
Cav. charactus
Gn. scotiaensis
Cav. cristatus
Lo. mononodosus
Lo. nodosus
Me. bipluti
Cav. altus
Hi. scitulus

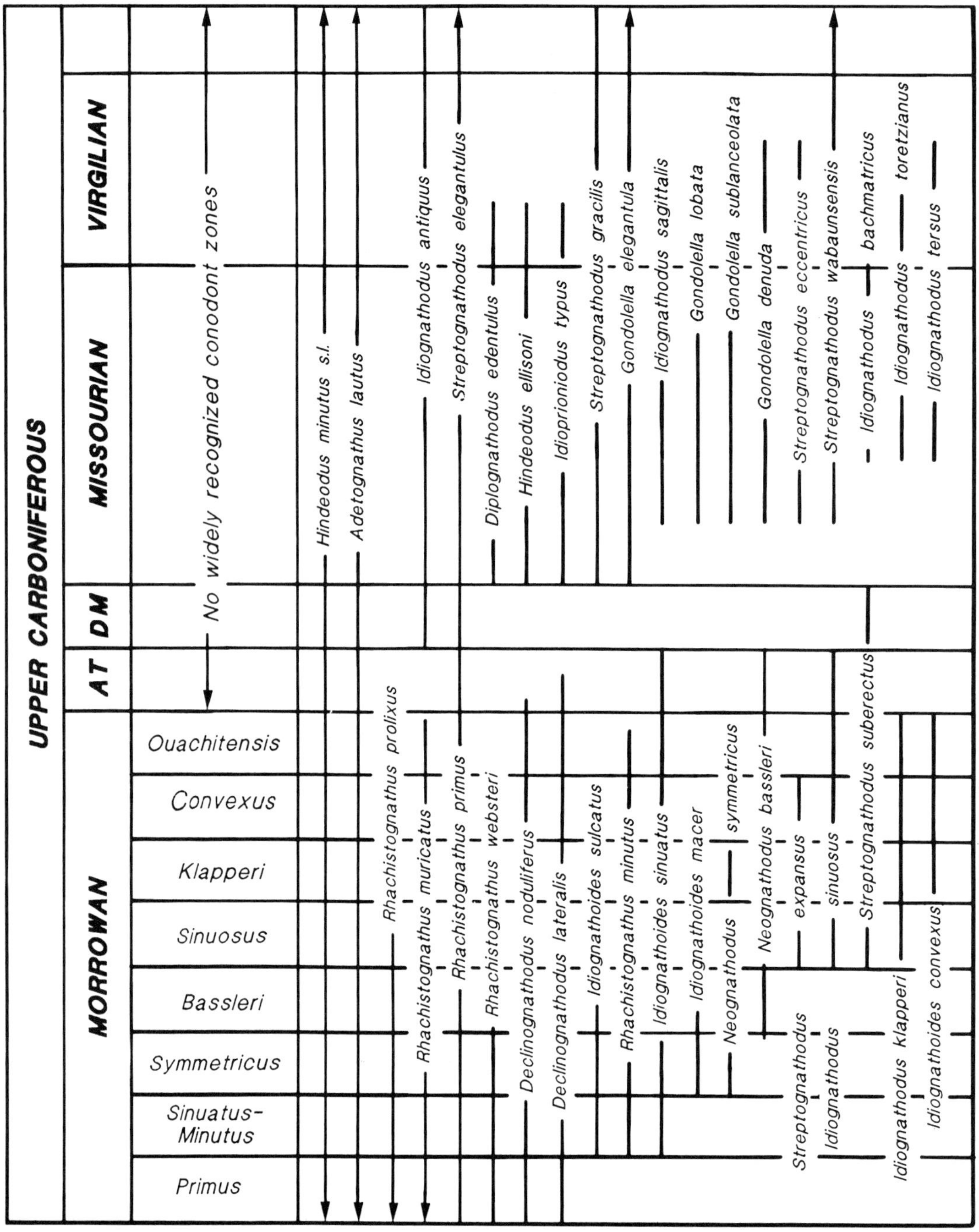

Chart 7. Ranges of selected Late Carboniferous (Pennsylvanian) conodont species. AT = Atokan; DM = Desmoinesian *(left and right pages).*

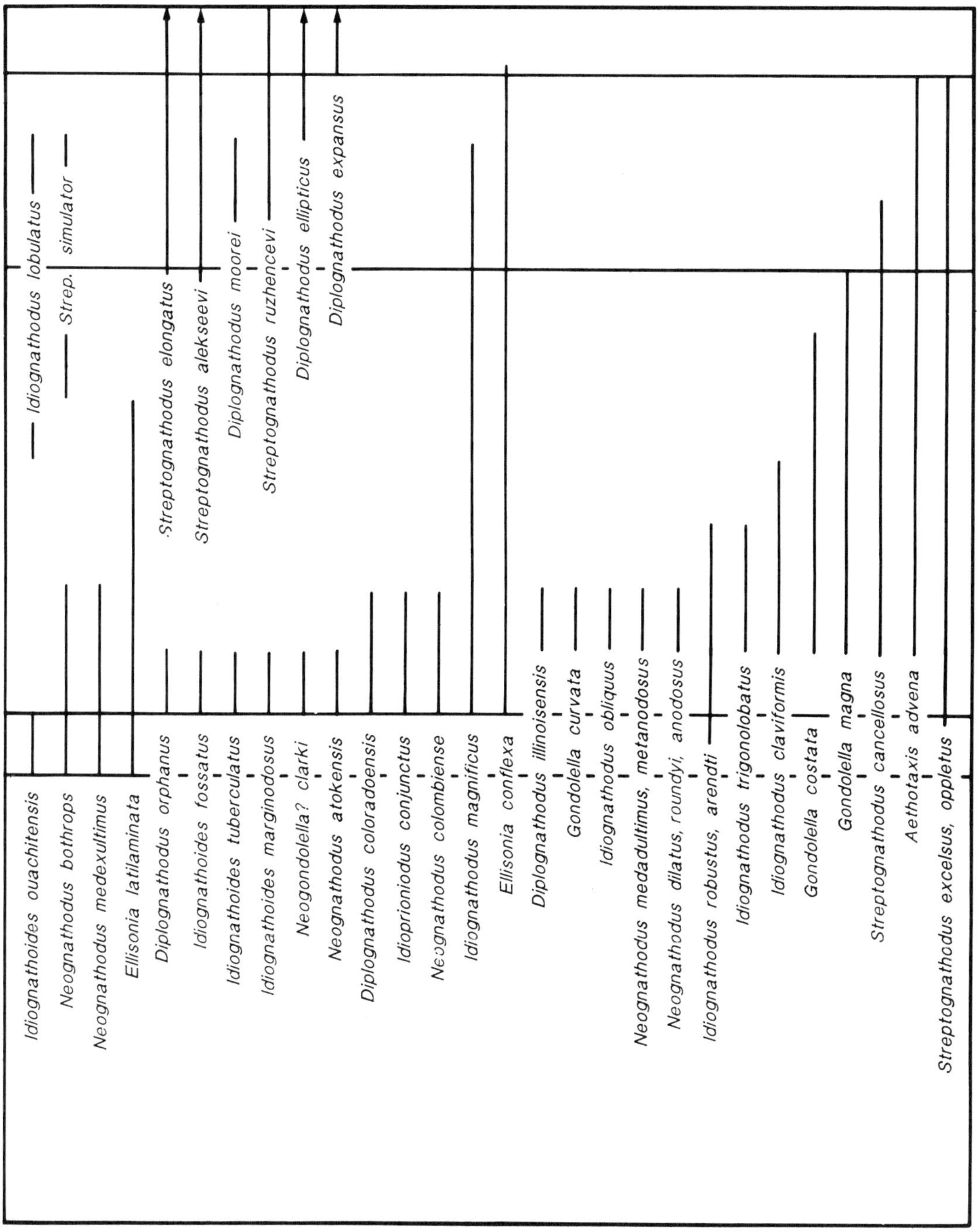
Idiognathodus lobulatus
Strep. simulator
Streptognathodus elongatus
Streptognathodus alekseevi
Diplognathodus moorei
Streptognathodus ruzhencevi
Diplognathodus ellipticus
Diplognathodus expansus
Idiognathoides ouachitensis
Neognathodus bothrops
Neognathodus medexultimus
Ellisonia latilaminata
Diplognathodus orphanus
Idiognathoides fossatus
Idiognathoides tuberculatus
Idiognathoides marginodosus
Neogondolella? clarki
Neognathodus atokensis
Diplognathodus coloradoensis
Idioprioniodus conjunctus
Neognathodus colombiense
Idiognathodus magnificus
Ellisonia conflexa
Diplognathodus illinoisensis
Gondolella curvata
Idiognathodus obliquus
Neognathodus medadultimus, metanodosus
Neognathodus dilatus, roundyi, anodosus
Idiognathodus robustus, arendti
Idiognathodus trigonolobatus
Idiognathodus claviformis
Gondolella costata
Gondolella magna
Streptognathodus cancellosus
Aethotaxis advena
Streptognathodus excelsus, oppletus

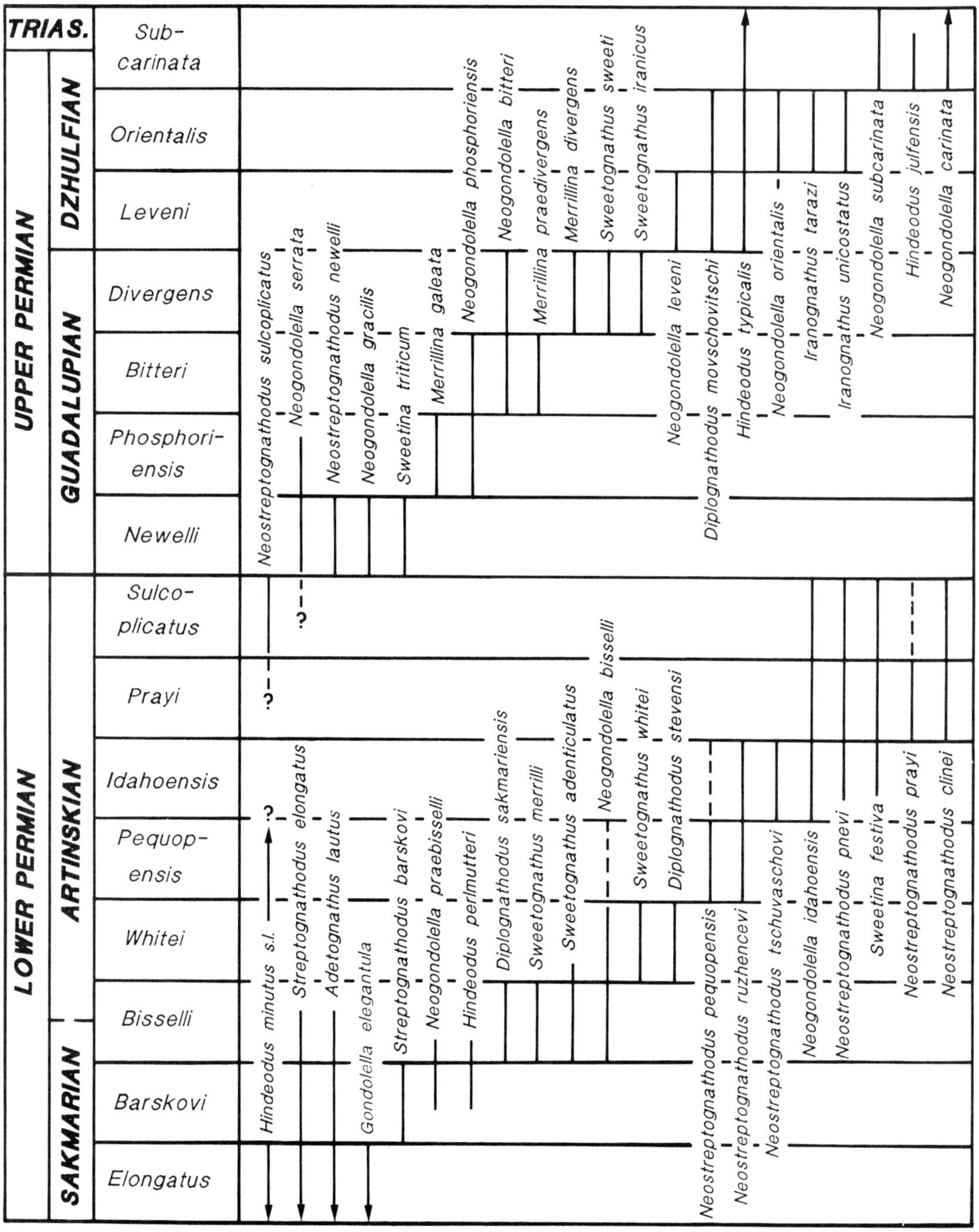

Chart 8. Ranges of selected Permian conodont species.

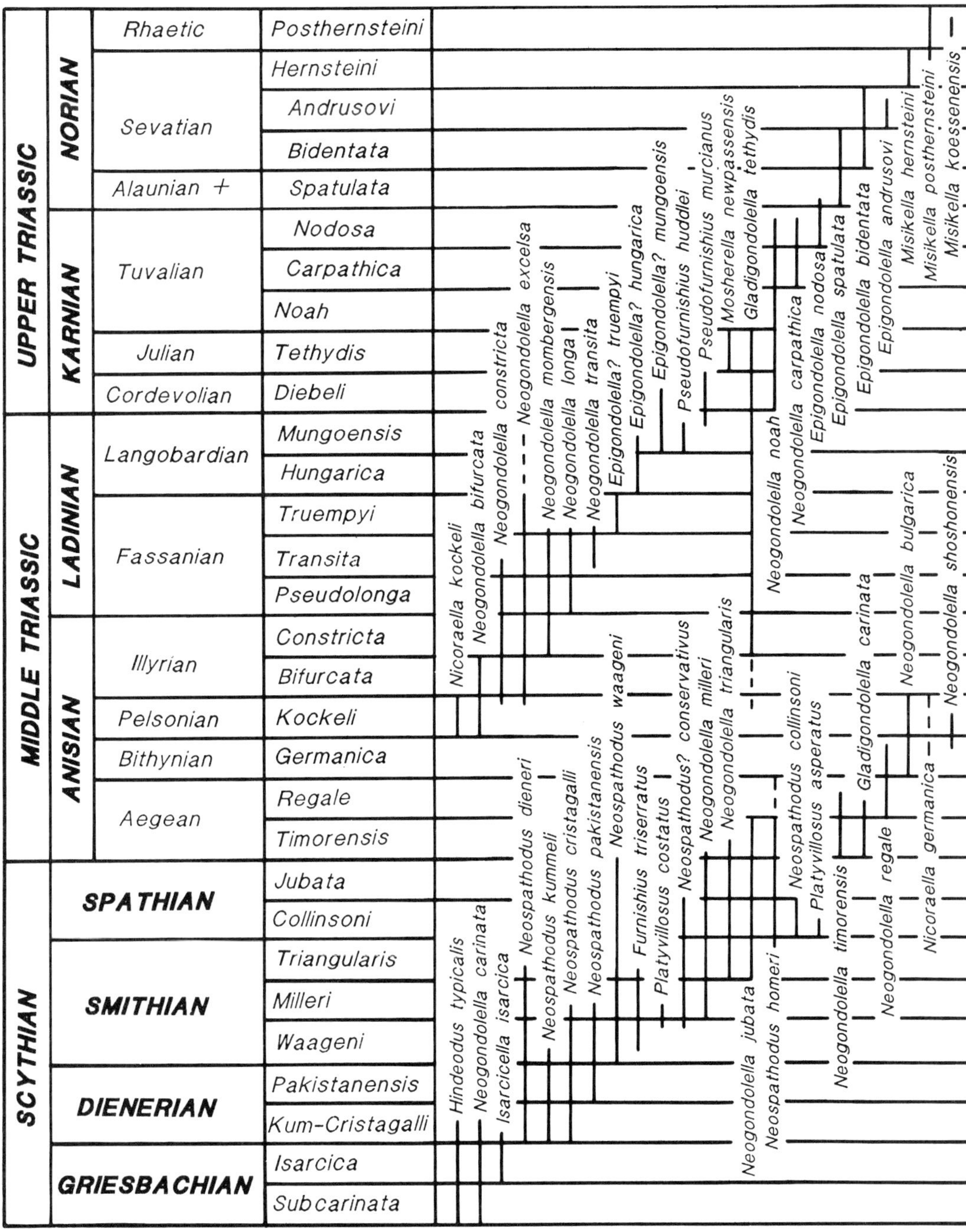

Chart 9. Ranges of selected Triassic conodont species.

Index